Zeroing Dynamics, Gradient Dynamics, and Newton Iterations

Zeroing Dynamics, Gradient Dynamics, and Newton Iterations

Yunong Zhang
Lin Xiao
Zhengli Xiao
Mingzhi Mao

CRC Press
Taylor & Francis Group
Boca Raton London New York

CRC Press is an imprint of the
Taylor & Francis Group, an **informa** business

CRC Press
Taylor & Francis Group
6000 Broken Sound Parkway NW, Suite 300
Boca Raton, FL 33487-2742

First issued in hardback 2020

© 2016 by Taylor & Francis Group, LLC
CRC Press is an imprint of Taylor & Francis Group, an Informa business

No claim to original U.S. Government works

ISBN-13: 978-1-4987-5376-0 (hbk)

Visit the Taylor & Francis Web site at
http://www.taylorandfrancis.com

and the CRC Press Web site at
http://www.crcpress.com

To our parents and
ancestors, as always

Contents

16 System of Time-Varying Nonlinear Inequalities Solving   255

VII   Application to Fractals   273

17 Fractals Yielded via Static Nonlinear Equation   275

18 Fractals Yielded via Time-Varying Nonlinear Equation   287

List of Figures

List of Tables

Preface

The online solution to mathematical problems arises in many fields of science, engineering, and business. It is usually an essential part of many solutions, e.g., matrix/vector computation, optimization, control theory, kinematics, signal processing, and pattern recognition. In recent years, due to the in-depth research on neural networks, numerous recurrent neural networks (RNNs) based on the gradient-based method have been developed and investigated. In particular, some simple neural networks were proposed to solve linear programming problems in real time and implemented on analog circuits. The neural-dynamic approach is now regarded as a powerful alternative for online computation and optimization because of its parallel distribution nature, self-adaptation ability, and potential hardware implementation.

However, the conventional gradient neural networks (GNNs) were designed intrinsically for constant (or termed static, time-invariant) coefficient matrices and vectors rather than time-varying (or termed, time-variant, nonstationary) ones. Then, time-varying problems were traditionally handled by approximating them as constant problems via a short-time invariance hypothesis. In other words, the effect of time variation is often ignored in dealing with time-varying problems in practice. Therefore, the obtained results cannot fit exactly with practical and engineering requirements. This is because GNNs do not make full use of the important time-derivative information of the problems and coefficients. There always exist lagging-behind errors between the GNN-obtained solutions and the theoretical solutions. As a result, GNNs belong to a passively tracking approach, which can only adapt to the change of problems and coefficients in a posterior passive manner.

As we know, time-varying problems are frequently encountered in scientific and engineering applications, such as aerodynamic coefficients in high-speed aircraft, circuit parameters in electronic circuits, and mechanical parameters in machinery. In order to achieve better performance (including higher precision) for solving such time-varying problems, it is necessary for us to investigate the effect of time variation. It follows from a variety of theoretical-analysis and computer-simulation results that the GNN approach cannot always solve time-varying problems effectively and efficiently, even if stringent restrictions on design parameters are imposed.

With the development of related technological societies and the introduction of new notions and innovative tools in the field of intelligent systems, research on neural networks is undergoing an enormous evolution. Since March, 12 2001, Zhang et al. have formally proposed, investigated, and developed a special class of recurrent neural network (or termed, Zhang neural network, ZNN), which has been analyzed theoretically and substantiated comparatively for solving online time-varying problems precisely and efficiently. Simply put, ZNN is a class of RNN, which originated from the research of the Hopfield neural network. It is viewed as a systematic approach to solving time-varying problems. It differs from the conventional GNN in terms of the problem to be solved, error function, design formula, dynamic equation, and the utilization of time derivatives. Following the authors' previous work on ZNN theory, zeroing dynamics (ZD) has been generalized and developed since 2008, of which the state dimension can be multiple or one. Specifically, considering that the new recurrent neural network zeros out each element of the error function in a neural-dynamic manner, we name it as zeroing dynamics, including scalar-valued situations. Accordingly, it is viewed as a systematic and methodological approach to solving various time-varying problems with the scalar situations included.

In this book, the resultant various ZD models are designed, proposed, developed, analyzed, modeled, simulated, and compared for online solution of time-varying problems, such as time-varying root finding, time-varying nonlinear equation solving, time-varying matrix inversion, time-varying matrix square root finding, time-varying quadratic optimization, and time-varying inequality solving. In view of fully utilizing the time-derivative information of problems and coefficients, the ZD method belongs to an actively predictive approach, which is more effective on the model convergence to a time-varying theoretical solution in comparison with GD (gradient dynamics) ones. For the purpose of possible hardware implementation (e.g., on digital circuits), we discretize the proposed continuous-time ZD (CTZD) models using the Euler forward-difference rule. Thus, the discrete-time models of CTZD can be obtained. In addition, by focusing on the corresponding constant problems solving, Newton iteration is found to be a special case of the discrete-time ZD (DTZD) model by utilizing the linear activation function and fixing the step-size value to be 1.

The idea for this book on neural networks and neural dynamics was conceived during classroom teaching as well as research discussion in the laboratory and at international scientific meetings. Most of the materials of this book are derived from the authors' papers published in journals and proceedings of the international conferences. In fact, since the early 1980s, the field of neural networks has undergone phases of exponential growth, generating many new theoretical concepts and tools (including the authors'). At the same time, these theoretical results have been applied successfully to the solution of many practical problems. Our first priority is thus to cover each central topic in enough detail to make the material clear and coherent; in other words, each part (and even each chapter) is written in a relatively self-contained manner.

This book contains 18 chapters that are classified into the following 7 parts:

Part I: Time-Varying Root Finding (Chapters 1 through 4)

Part II: Nonlinear Equation Solving (Chapters 5 through 7)

Part III: Matrix Inversion (Chapters 8 and 9)

Part IV: Matrix Square Root Finding (Chapters 10 and 11)

Part V: Time-Varying Quadratic Optimization (Chapters 12 through 14)

Part VI: Time-Varying Inequality Solving (Chapters 15 and 16)

Part VII: Application to Fractals (Chapters 17 and 18)

Chapter 1 — In this chapter, we propose, generalize, develop, investigate and compare the CTZD and GD models for online time-varying square root finding. In addition, the simplified CTZD (S-CTZD) model and the simplified DTZD (S-DTZD) model are generated for constant scalar-valued square root finding. In terms of constant square root finding, the Newton iteration is found to be a special case of the S-DTZD model. Simulative and numerical results via a power-sigmoid activation function further substantiate the efficacy of the ZD models for online time-varying and constant square root finding, in addition to the link and new explanation to Newton iteration.

Chapter 2 — In this chapter, we propose and investigate the CTZD model and its discrete-time models (i.e., DTZD models) with two situations (i.e., the time-derivative of the coefficient being known or unknown) for time-varying cube root finding. To find the constant cube root, a S-CTZD model and its discrete-time model (i.e., the S-DTZD model) are generated. By focusing on such a constant problem solving, the Newton iteration is found again to be a special case of the S-DTZD model by utilizing the linear activation function and fixing the step-size value to be 1.

Chapter 3 — For the purpose of online solution of the time-varying 4th root, in this chapter both of the CTZD and DTZD models are developed and investigated. In addition, the power-sigmoid activation function is exploited in the ZD models, which makes them possess the property of superior convergence and better precision (as compared with the standard situation of using the linear activation function). Computer-simulation and numerical-experiment results (i.e., simulative and numerical results) illustrate the efficacy of the ZD models for finding online the time-varying 4th root.

Chapter 4 — In this chapter, the ZD models are proposed to solve online for the time-varying 5th root. In order to get superior convergence and better precision of the ZD models, the power-sigmoid activation function is exploited again. In addition, to solve the special case of the 5th root finding (e.g., the constant 5th root finding), we develop the S-CTZD and S-DTZD models. Illustrative examples substantiate the efficacy of the ZD models for the online time-varying and constant 5th root finding.

Chapter 5 — In this chapter, we generalize the ZD design method to solve online the time-varying nonlinear equation in the form of $f(x,t) = 0 \in \mathbb{R}$. For comparative purposes, the GD model is also generalized for solving such a time-varying equation. In addition, different types of activation functions (including linear, sigmoid, power functions, or their variants, e.g., power-sigmoid function) are investigated. Computer simulation results substantiate the theoretical analysis and efficacy of the ZD models for solving online time-varying nonlinear equations.

Chapter 6 — In this chapter, we analyze, investigate and compare the characteristics of the ZD and GD models for static nonlinear equation solving. Then, the corresponding DTZD model is developed and investigated. In terms of nonlinear equation solving, the Newton iteration is found once more to be a special case of the DTZD model (by focusing on the constant problem solving, utilizing the linear activation function and fixing the step size to be 1). It is also discovered that if a nonlinear equation possesses a local minimum point, the neural state of the CTZD model, starting from some initial value close to it, may move toward the local minimum point and then stop with warning information. In contrast, the neural state of the GD model falls into the local minimum point (with no warning). Inspired by Wu's work, the improved CTZD model is proposed by defining two modified error functions and generating new neural-dynamic forms to overcome such a local-minimum problem.

Chapter 7 — In this chapter, a general CTZD model is proposed and investigated for solving the system of time-varying nonlinear equations. The solution error of such a CTZD model can exponentially converge at a large scale to zero, which is analyzed with different activation functions considered. In addition, such a CTZD model is discretized for potential digital hardware implementation. This produces the DTZDK and DTZDU models according to the criterion of whether the time-derivative information is known or not. The DTZD models are further improved by using the Broyden method, and thus DTZDK-B and DTZDU-B models are generated. The efficacy of such ZD models is verified via simulative and numerical examples.

Chapter 8 — In this chapter, we develop and investigate a DTZD model for matrix inversion, which is depicted by a system of difference equations. Comparing with the Newton iteration, we find and confirm that the DTZD model incorporates the Newton iteration as a special case. Noticing this relationship, we perform numerical comparisons on different situations by using the DTZD model and Newton iteration for matrix inversion. Different types of activation functions and different step-size values are examined for superior convergence and better stability of the DTZD model.

Chapter 9 — In this chapter, we generalize, investigate, and analyze ZD models for online time-varying full-rank matrix Moore–Penrose inversion. The computer simulation results and the application to inverse kinematic control of a redundant robot arm illustrate the feasibility and effectiveness of the ZD models for online time-varying full-rank matrix Moore–Penrose inversion.

Chapter 10 — In this chapter, we further investigate the ZD models for time-varying matrix square root finding. For the purpose of potential hardware (e.g., digital circuits) implementation, the DTZD model is generated that incorporates Newton iteration as a special case. Besides, to obtain an appropriate step-size value (in each iteration), a line-search algorithm is employed for the DTZD model. Numerical experiment results substantiate the effectiveness of the proposed ZD models aided by a line-search algorithm, in addition to the confirmed connection and explanation to the Newton iteration for matrix square root finding.

Chapter 11 — In this chapter, to pursue the superior convergence and robustness properties, a special type of activation function (i.e., hyperbolic sine activation function) is applied to the ZD model for online solution of the time-varying matrix square root. Theoretical analysis and computer simulation results further show the superior performance of the ZD model using hyperbolic sine activation functions in the context of large model implementation errors, in comparison with that using linear activation functions.

Chapter 12 — For potential digital hardware implementation, DTZD models are proposed and investigated for time-varying quadratic minimization (QM) in this chapter. Note that the DTZD models utilize the time-derivative information of the QM's time-varying coefficients. For comparison, a discrete-time GD model is also presented to solve the same time-varying QM problem. Simulative and numerical results illustrate the efficacy and superiority of the DTZD models for time-varying QM, in comparison with the discrete-time GD model.

Chapter 13 — In this chapter, we generalize and investigate the CTZD model for the online solution of time-varying convex quadratic programming (QP) subject to a time-varying linear equality constraint. For the purpose of hardware (e.g., digital circuits) implementation, DTZD models are constructed and developed by using the Euler forward-difference rule, which are also effective numerical algorithms if implemented on digital computers directly. Computer simulation and numerical-experiment results illustrate the efficacy (especially precision) of the presented CTZD and DTZD models for solving online time-varying QP problems.

Chapter 14 — In this chapter, the ZD model for online time-varying QP problem solving is developed, investigated, and applied to the redundant arm's kinematic control. Computer simulations performed on a four-link planar robot arm substantiate the superiority of the ZD model for online time-varying QP solution, as compared with the GD model. Moreover, practical physical experiments are conducted on an actual six-link planar redundant robot arm, which substantiates well the physical realizability and effectiveness of the ZD model.

Chapter 15 — In this chapter, we investigate ZD models for an online solution of scalar-valued and vector-valued time-varying linear inequalities. For the situation of the scalar-valued time-varying linear inequality, three different activation functions are exploited in the ZD models, and the corresponding DTZD models are derived and proposed for numerical experimentation or hardware implementation. For the situation of the vector-valued time-varying linear inequality, a general ZD model is presented, investigated, and analyzed with the theoretical analysis provided, and the conventional GD model is developed and exploited for comparative purposes.

Chapter 16 — To solve the system of time-varying nonlinear inequalities, this chapter investigates two ZD models. The first model is based on the conventional Zhang et al.'s neural-dynamic design method (i.e., the conventional zeroing dynamics method) and is termed a conventional ZD (CZD) model. The other one is based on a novel variant of the conventional Zhang et al.'s neural-dynamic design method, and is termed a modified ZD (MZD) model. The theoretical analysis of both CZD and MZD models is presented to show their excellent convergence performance. Compared with the CZD model for solving the system of time-varying nonlinear inequalities, it is discovered that the MZD model incorporates the CZD model as a special case (i.e., MZD model using linear activation functions (MZDL) reduces to the CZD exactly). Besides, the MZD model using power-sum activation functions (MZDP) possesses superior convergence performance to the CZD model.

Chapter 17 — In this chapter, a novel kind of new fractal is yielded by using the complex-valued DTZD (CVDTZD) model to solve a nonlinear equation in the complex domain. Comparing it with the well-known Newton fractal (i.e., the famous fractal generated by the well-known Newton iteration), we also find correspondingly that the novel fractal yielded by the CVDTZD model incorporates the Newton fractal as a special case. Note that the fractal yielded by the CVDTZD model can be completely different from Newton fractal. The CVDTZD model using different types of activation functions is thus seen as a new iterative paradigm to generate a new fractal.

Chapter 18 — In this chapter, the new fractal is yielded by using the CVDTZD model to solve a time-varying nonlinear equation in the complex domain. In addition, by comparing the area and degree of blue color in the new fractal under the same conditions, the effectiveness of the CVDTZD model using different activation functions for solving the time-varying nonlinear complex equation is illustrated.

This book is written for graduate students as well as academic and industrial researchers studying the developing fields of neural networks, neural dynamics, computer mathematics, numerical algorithms, time-varying computation and optimization, simulation and modeling, analog and digital hardware, and fractals. It provides a comprehensive view of the combined research of these fields, in addition to its accomplishments, potentials, and perspectives. We do hope that this book will generate curiosity and also enthusiasm to its readers for learning more in the fields and the research, and that it will provide new challenges to seek new theoretical tools and practical applications. At the end of the preface, it is worth pointing out that, in this book, some important figure ranges are presented in various forms to make them easier for reading and identifying.

MATLAB® is a registered trademark of The MathWorks, Inc. For product information, please contact:

The MathWorks, Inc.
3 Apple Hill Drive
Natick, MA 01760-2098 USA
Tel: 508-647-7000
Fax: 508-647-7001
E-mail: info@mathworks.com
Web: www.mathworks.com

Author Biographies

Yunong Zhang earned a BS degree from Huazhong University of Science and Technology, Wuhan, China, in 1996; an MS degree from South China University of Technology, Guangzhou, China, in 1999; and a PhD degree from the Chinese University of Hong Kong, Shatin, Hong Kong, China, in 2003. He is currently a professor at the School of Information Science and Technology, Sun Yat-sen University, Guangzhou, China. Before joining Sun Yat-sen University in 2006, Yunong had been with the National University of Singapore, University of Strathclyde, and National University of Ireland at Maynooth since 2003. In addition, he is also currently with the SYSU-CMU Shunde International Joint Research Institute, Foshan, China, for cooperative research. His main research interests include neural networks, robotics, computation, and optimization. He has been working in the research and application of neural networks/dynamics field for 16 years. He has published more than 375 scientific works of various types, including 15 monographs/books/chapters, 5 authorized patents, more than 200 first-authored works, 27 IEEE Transactions/magazine papers, 10 solely authored works, and 105 SCI-indexed papers (with 74 SCI-indexed papers published in the past 5 years and with 1213 SCI citations). Dr. Zhang was supported by the Program for New Century Excellent Talents in University in 2007, was presented the Best Paper Award of ISSCAA in 2008 and the Best Paper Award of ICAL in 2011, and was among the 2014 Highly Cited Scholars of China selected by Elsevier published in 2015. His Web page is now available at http://sist.sysu.edu.cn/~zhynong.

Lin Xiao earned a BS degree in electronic information science and technology from Hengyang Normal University, Hengyang, China, in 2009, and a PhD degree from Sun Yat-sen University, Guangzhou, China, in 2014. He is currently a lecturer at the College of Information Science and Engineering, Jishou University, Jishou, China. His current research interests include neural networks, intelligent information processing, robotics, and related areas.

Zhengli Xiao earned a BS degree in software engineering from Changchun University of Science and Technology, Changchun, China, in 2013. He is currently pursuing an MS degree at the Department of Computer Science at the School of Information Science and Technology, Sun Yat-sen University, Guangzhou, China. In addition, he is also currently with the SYSU-CMU Shunde International Joint Research Institute, Foshan, China, for cooperative research. His current research interests include neural networks, intelligent information processing, and learning machines.

Mingzhi Mao earned BS, MS, and PhD degrees from the Department of Computer Science at Sun Yat-sen University, Guangzhou, China, in 1988, 1998, and 2008, respectively. He is currently an associate professor at the School of Information Science and Technology, Sun Yat-sen University, Guangzhou, China. His main research interests include intelligence algorithms, software engineering, and management information systems.

Acknowledgments

This book comprises the results of many original research papers of the authors' research group, in which many authors of these original papers have done a great deal of detailed and creative research work. Therefore, we are much obliged to our contributing authors for their high-quality work. The continuous support of our research by the National Natural Science Foundation of China (under grant 61473323); the Foundation of Key Laboratory of Autonomous Systems and Networked Control, Ministry of Education, China (under grant 2013A07); the National Innovation Training Program for University Students (under grants 201410558065 and 201410558069); and the Science and Technology Program of Guangzhou, China (under grant 2014J4100057) is gratefully acknowledged. We also thank project editor, Robin Lloyd-Starkes and the rest of the staff at CRC, including Katy Smith, Hayley Ruggieri, Karen Schoker, and Rich O'Hanley, for the time and effort they spent preparing the book for press, as well as the constructive comments and suggestions provided. We are very grateful to CRC's person Runjun He, for his strong support during the preparation and publication of this book, and for being such a kind person. We owe our deep gratitude to all the wonderful people who have helped this book become a reality, especially now that the research projects and the book have been completed.

Part I

Time-Varying Root Finding

Chapter 1

Time-Varying Square Root Finding

Abstract

Different from the conventional gradient-based neural dynamics, a special class of neural dynamics has been proposed by Zhang *et al.* for online solution of time-varying and constant problems (e.g., nonlinear equations). The design of zeroing dynamics (ZD) is based on the elimination of an indefinite error function, instead of the elimination of a square-based nonnegative, or at least lower-bounded energy function usually associated with the design of gradient dynamics (GD). In this chapter, we propose, generalize, develop, investigate, and compare the continuous-time ZD (CTZD) and GD models for online solution of time-varying and constant square roots. In addition, a simplified continuous-time ZD (S-CTZD) model and a discrete-time ZD (DTZD) model are generated for constant scalar-valued square root finding. In terms of constant square root finding, Newton iteration is found to be a special case of the S-DTZD model (by focusing on the constant problem solving, utilizing the linear activation function, and fixing the step size to be 1). Computer-simulation and numerical-experiment results via a power-sigmoid activation function further illustrate the efficacy of the ZD models for online time-varying and constant square roots finding, in addition to the link and new explanation to Newton iteration.

1.1   Introduction

The problem of online solution of time-varying square root in the form of $x^2(t) - a(t) = 0$ or constant (or termed time-invariant) square root in the form of $x^2 - a = 0$ is considered to be a basic problem arising in science and engineering fields. Thus, many numerical algorithms are presented for such a problem solving [1,3,22,25,32,40,51,57,68]. However, it may not be efficient enough for most numerical algorithms due to their serial-processing nature performed on digital computers [115]. Suitable for analog very large scale integration (VLSI) implementation [8,119] and in view of potential

3

high-speed parallel processing, the neural-dynamic approach is now regarded as a powerful alternative to online problems solving [99–102, 104, 107, 114, 117, 118, 122, 128, 129, 132, 134]. Besides, it is worth mentioning that most reported computational schemes were theoretically/intrinsically designed for constant problems solving or related to gradient methods.

Different form the gradient-based neural-dynamic approach, a special class of neural dynamics has been formally proposed by Zhang *et al.* [99–102, 104, 107, 114, 117, 118, 122, 128, 129, 134] for time-varying or constant problems solving, such as time-varying convex quadratic program [117], matrix inversion [99, 102, 104, 118], and Sylvester equation solving [107]. In addition, the design of ZD is based on an indefinite error function, instead of a square-based nonnegative (or at least lower-bounded) energy function usually associated with the design of GD models or the Lyapunov method [78, 132]. The concept of ZD is presented as follows.

Concept. ZD has been generalized from Zhang neural network (ZNN) [132] formally since 2008 [114, 128, 134], of which the state dimension can be multiple or one. Specifically, considering that the new recurrent neural network zeros out each element of the error function in a neural-dynamic manner, we name it as ZD, including scalar-valued situations. It is viewed as a systematic and methodological approach to solving various time-varying problems, with the scalar situations included as well. It differs from conventional GD in terms of the problem to be solved, error function, design formula, dynamic equation, and the utilization of time-derivative information.

In this chapter, a novel CTZD model is introduced, generalized, developed and investigated for online time-varying square root finding [114]. For comparative purposes, the GD model is also generated and investigated for this time-varying problem solving. As we know, constant (or termed static, time-invariant) problems solving can be viewed as a special case of time-varying problems solving. Correspondingly, S-CTZD model [134] is generated. In addition, we also develop and study the simplified (S-DTZD) model for constant square root finding. Comparing with Newton iteration [57] for finding square root, we discover that the S-DTZD model contains Newton iteration as its special case. In details, Newton iteration can also be obtained (different from the traditional explanation [1, 3, 22, 25, 32, 40, 51, 57, 68]) by utilizing the linear activation function and fixing the step size to be 1 in the S-DTZD model. To the best of the authors' knowledge, most reported computational-schemes were theoretically/intrinsically designed for constant problems solving or related to gradient methods. There is almost no others' literature dealing with such a specific problem solving, i.e., online solution of time-varying square root, at present stage.

1.2   Problem Formulation and Continuous-Time (CT) Models

Let us consider the following time-varying square root problem (which is written in the form of a time-varying nonlinear equation [114] for convenience):

$$x^2(t) - a(t) = 0, \quad t \in [0, +\infty), \tag{1.1}$$

where $a(t) \in \mathbb{R}$ denotes a smoothly time-varying positive scalar, which, together with its time derivative $\dot{a}(t) \in \mathbb{R}$, is assumed to be known numerically or can be measured accurately. Our objective in this chapter is to find $x(t) \in \mathbb{R}$ in real time t, such that the above smoothly time-varying nonlinear Equation (1.1) holds true. For presentation convenience, let $x^*(t) \in \mathbb{R}$ denote a theoretical time-varying square root of $a(t)$.

1.2.1 CTZD Model

By following Zhang *et al.*'s neural-dynamic design method (i.e., zeroing dynamics design method) [102, 104, 107, 117, 118, 122, 129, 134], to monitor and control the process of time-varying square root finding, we first define an indefinite error function as follows (of which the word "indefinite" means that such an error function can be positive, zero, negative, bounded, unbounded, or even lower-unbounded):

$$e(t) = x^2(t) - a(t) \in \mathbb{R}.$$

Secondly, the time-derivative $\dot{e}(t)$ of error function $e(t)$ can be chosen and forced, such that $e(t)$ mathematically converges to zero. Specifically, we can choose $e(t)$ via the following ZD design formula [102, 104, 107, 117, 118, 122, 129, 134]:

$$\frac{\mathrm{d}e(t)}{\mathrm{d}t} = \dot{e}(t) = -\gamma\phi\big(e(t)\big),$$

where design parameter $\gamma > 0$ is used to scale the convergence rate of the CTZD model, and $\phi(\cdot)$: $\mathbb{R} \to \mathbb{R}$ denotes a general activation function mapping. Expanding the above design formula, we thus obtain

$$2x(t)\dot{x}(t) - \dot{a}(t) = -\gamma\phi\big(x^2(t) - a(t)\big),$$

which leads to the following CTZD model for time-varying square root finding:

$$\dot{x}(t) = \frac{\dot{a}(t) - \gamma\phi\big(x^2(t) - a(t)\big)}{2x(t)}, \tag{1.2}$$

where $x(t)$, starting from randomly generated initial state (or termed initial condition) $x(0) = x_0 \in \mathbb{R}$, is the neural state corresponding to theoretical solution $x^*(t)$ of (1.1). In view of CTZD model (1.2), different choices of activation function $\phi(\cdot)$ and design parameter γ may lead to different performance of the neural dynamics. Generally speaking, any monotonically increasing odd activation function $\phi(\cdot)$ can be used for the construction of the neural dynamics. Since March 2001 [107], we have introduced and used multiple types of activation functions (e.g., linear activation function, power activation function, power-sum activation function, bipolar sigmoid activation function and power-sigmoid activation function) for the research. For more details, see [99, 102, 104, 117, 118, 122, 129, 134]. Similar to usual neural-dynamic approaches, design parameter γ in CTZD model (1.2), being the reciprocal of a capacitance parameter in the hardware implementation, should be set as large as the hardware would permit (e.g., in analog circuits or VLSI [8, 119]) or selected appropriately (e.g., between 10^3 and 10^8) for simulative or experimental purposes.

As for the convergence property of CTZD model (1.2), we have the following proposition [114, 128].

Proposition 1 *Consider a smoothly time-varying positive scalar $a(t) \in \mathbb{R}$ in Equation (1.1). If a monotonically increasing odd activation function $\phi(\cdot)$ is used, then*

    1) neural state $x(t) \in \mathbb{R}$ of CTZD model (1.2), starting from randomly generated positive initial state $x_0 \in \mathbb{R}^+$, converges to theoretical positive time-varying square root $x^*(t)$ [or denoted by $a^{1/2}(t)$] of $a(t)$; and,

    2) neural state $x(t) \in \mathbb{R}$ of CTZD model (1.2), starting from randomly generated negative initial state $x_0 \in \mathbb{R}^-$, converges to theoretical negative time-varying square root $x^*(t)$ [or denoted by $-a^{1/2}(t)$] of $a(t)$.

1.2.2   GD Model

For comparative purposes, the GD model is also developed and exploited for online solution of the above-presented time-varying square root problem (1.1). To solve Equation (1.1), following the conventional gradient-based design method, we first define a scalar-valued square-based energy function, such as $\mathscr{E} = (x^2(t) - a(t))^2/2$. Then, a typical continuous-time learning rule based on the negative-gradient information leads to the following differential equation (which is the linear-activation form of our presented GD model [99, 100, 117, 122, 129, 134]):

$$\frac{\mathrm{d}x(t)}{\mathrm{d}t} = \dot{x}(t) = -\gamma\frac{\partial\mathscr{E}}{\partial x} = -2\gamma x(t)\big(x^2(t) - a(t)\big),$$

where $x(t)$, starting from randomly generated initial condition $x(0) = x_0 \in \mathbb{R}$, is the neural state of the GD model. Under the inspiration of [104], from the above we can obtain the nonlinear-activation form of a general GD model (to say, nonlinear-activation GD model) by imposing a nonlinear activation function $\phi(\cdot)$ as follows:

$$\dot{x}(t) = -2\gamma x(t)\phi\big(x^2(t) - a(t)\big). \tag{1.3}$$

It is worth comparing the two design methods of CTZD model (1.2) and GD model (1.3), both of which are exploited for finding the time-varying square root of (1.1). The differences lie in the following facts.

1) The design of CTZD model (1.2) is based on the elimination of an indefinite error function $e(t) = x^2(t) - a(t) \in \mathbb{R}$, which can be negative and lower-unbounded. In contrast, the design of GD model (1.3) is based on the elimination of the square-based nonnegative energy function $\mathscr{E} = (x^2(t) - a(t))^2/2 \geqslant 0$.

2) The design of CTZD model (1.2) is based on a new method intrinsically for time-varying problems solving. Thus, it can converge to an exact/theoretical time-varying square root of (1.1). In contrast, the design of GD model (1.3) is based on a method intrinsically for constant problems solving. It is thus only able to approximately approach the theoretical time-varying square root of (1.1).

3) CTZD model (1.2) exploits the time-derivative information of $a(t)$ [i.e., $\dot{a}(t)$] in a methodological and systematical manner during its real-time finding process. This is the main reason why CTZD model (1.2) can converge to an exact/theoretical time-varying square root of (1.1). In contrast, GD model (1.3) does not exploit the important time-derivative information of $a(t)$, and thus is not effective enough on solving the time-varying square root problem.

Furthermore, let us consider the following constant square root problem formulated in most textbooks, e.g., in [25, 51, 57], which is a special case of time-varying nonlinear Equation (1.1):

$$x^2(t) - a = 0, \tag{1.4}$$

where $a \in \mathbb{R}$ is a positive constant. In view of a being constant in (1.4), i.e., $\dot{a}(t) = 0$, CTZD model (1.2) reduces to

$$\dot{x}(t) = -\gamma\frac{\phi\big(x^2(t) - a\big)}{2x(t)}, \tag{1.5}$$

which is termed as S-CTZD model used to find the constant square root. According to Proposition 1, we have the following proposition on the convergence property of S-CTZD model (1.5).

Proposition 2 *Consider a constant positive scalar $a \in \mathbb{R}$ in Equation (1.4). If a monotonically increasing odd activation function $\phi(\cdot)$ is used, then*

1) *neural state $x(t) \in \mathbb{R}$ of S-CTZD model (1.5), starting from randomly generated positive initial state $x_0 \in \mathbb{R}^+$, converges to theoretical positive square root x^* (or denoted by $a^{1/2}$) of a; and*

2) *neural state $x(t) \in \mathbb{R}$ of S-CTZD model (1.5), starting from randomly generated negative initial state $x_0 \in \mathbb{R}^-$, converges to theoretical negative square root x^* (or denoted by $-a^{1/2}$) of a.*

1.3 S-DTZD Model and Newton Iteration

For potential hardware implementation (e.g., on digital circuits) of S-CTZD used for constant square root finding, we can discretize the S-CTZD model (1.5) by using Euler forward-difference rule [1,3,22,25,32,40,51,57,68]:

$$\dot{x}(t = k\tau) \approx (x((k+1)\tau) - x(k\tau))/\tau,$$

where $\tau > 0$ denotes the sampling gap, and $k = 0, 1, 2, \cdots$ denotes the update index (or termed updating index, or iteration index, here). In general, we denote $x_k = x(t = k\tau)$ for presentation convenience. Then, the (DTZD) model for the constant square root finding can be derived from S-CTZD model (1.5) as follows:

$$(x_{k+1} - x_k)/\tau = -\gamma \frac{\phi(x_k^2 - a)}{2x_k},$$

which can be further formulated as

$$x_{k+1} = x_k - h \frac{\phi(x_k^2 - a)}{2x_k}, \tag{1.6}$$

where parameter $h = \tau\gamma$ denotes the step size. Generally speaking, different choices of step size h and activation function $\phi(\cdot)$ may lead to different convergence performance of DTZD model (1.6). From an important theorem in [128], the following proposition on the convergence property of DTZD model (1.6) can be presented.

Proposition 3 *Consider the constant square root problem (1.4). When step size h of DTZD model (1.6) satisfies $0 < h < 2/\phi'(0)$ [with $\phi'(0)$ being the derivative of $\phi(\cdot)$ with respect to its input argument and taken at the zero point], there generally exists a positive number δ, such that the sequence $\{x_k\}_{k=0}^{+\infty}$ generated by DTZD model (1.6) can converge to a theoretical root x^* of (1.4) for any initial state $x_0 \in [x^* - \delta, x^* + \delta]$.*

By reviewing DTZD model (1.6), if we utilize linear activation function $\phi(e) = e$, DTZD model (1.6) reduces to

$$x_{k+1} = x_k - h \frac{x_k^2 - a}{2x_k}. \tag{1.7}$$

In addition, if $h = 1$, DTZD model (1.7) further reduces to

$$x_{k+1} = x_k - \frac{x_k^2 - a}{2x_k}, \text{ i.e., } x_{k+1} = \frac{x_k + a/x_k}{2}, \tag{1.8}$$

which is exactly Newton iteration for constant square root finding widely presented in textbooks and literature [1,3,22,25,32,40,51,57,68]. In other words, we discover that a general form of Newton

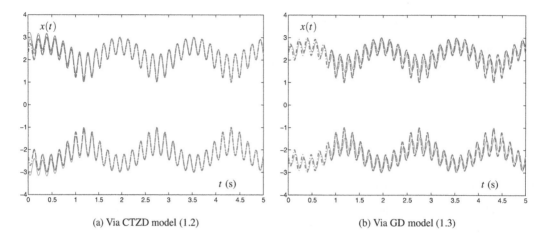

(a) Via CTZD model (1.2)   (b) Via GD model (1.3)

FIGURE 1.1 Time-varying square root finding of (1.9) via CTZD model (1.2) and GD model (1.3) with $\gamma = 1$.

iteration can be given by discretizing the S-CTZD model (1.5). Specifically, Newton iteration for constant square root finding can be derived by focusing on the constant problem solving (1.4), discretizing (1.5) via Euler forward difference, utilizing a linear activation function, and fixing the step size to be 1. Evidently, this chapter shows a new explanation to Newton iteration for constant square root finding, which is different from the traditional (or to say, standard) explanation appearing in almost all literature and textbooks, i.e., via Taylor series expansion [1,3,22,25,32,40,51,57,68].

1.4   Illustrative Examples

The previous two sections have proposed the CTZD and DTZD models and their links to Newton iteration for online solution of scalar square root. In this section, computer-simulation and numerical-experiment results and observations are provided for the substantiation of the proposed link (to Newton method), as well as the efficacy of ZD models with the aid of suitable initial state x_0, a power-sigmoid activation function (with design parameters $\xi = 4$ and $p = 3$), and step size h.

1.4.1   Time-Varying Square Root Finding

For illustration and comparison, let us consider the following time-varying square root problem (written in the form of a nonlinear equation):

$$x^2(t) - 4\sin(20t)\cos(16t) - 5 = 0. \tag{1.9}$$

Both dynamic models [i.e., CTZD model (1.2) and GD model (1.3)] are exploited for solving online the time-varying square root of (1.9) with design parameter $\gamma = 1$. The computer-simulation results are shown in Figure 1.1. Note that, to check the two dynamic-models' solution effectiveness, the theoretical roots $x_1^*(t)$ and $x_2^*(t)$ [of which $x_1^*(t) = -x_2^*(t)$] of (1.9) are shown and denoted via dash-dotted curves in the figure. Specifically, as seen from Figure 1.1(a), starting from three initial states randomly generated in $[-4,4]$, the neural states of CTZD model (1.2) denoted by $x(t)$ fit well with

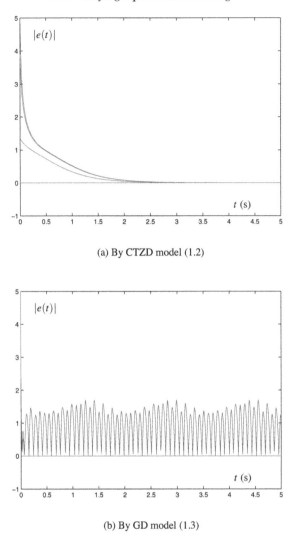

(a) By CTZD model (1.2)

(b) By GD model (1.3)

FIGURE 1.2 Residual errors $|x^2(t) - 4\sin(20t)\cos(16t) - 5|$ synthesized by CTZD model (1.2) and GD model (1.3) with $\gamma = 1$.

the theoretical square root $x_1^*(t)$ or $x_2^*(t)$ after a short time (e.g., 3 seconds or so, i.e., 3 s for short). In contrast, the neural states $x(t)$ of GD model (1.3) depicted in Figure 1.1(b), also starting from three randomly generated initial states in $[-4, 4]$, do not fit well with any one of the theoretical square roots [i.e., $x_1^*(t)$ and $x_2^*(t)$] even after a long period of time. Simply put, the steady states $x(t)$ of GD model (1.3) lag behind the theoretical square root $x_1^*(t)$ or $x_2^*(t)$ of (1.9).

Besides, in order to further investigate the convergence performance, we monitor the residual errors $|e(t)| = |x^2(t) - 4\sin(20t)\cos(16t) - 5|$ during the problem solving processes of both dynamic-models [i.e., CTZD model (1.2) and GD model (1.3)]. Note that symbol $|\cdot|$ denotes the absolute value of a scalar input argument. Specifically, it is seen from Figure 1.2(a) that, by applying CTZD model (1.2) to time-varying square root finding of (1.9), the residual errors $|e(t)|$ converge to zero within around 3 s. In contrast, it follows from Figure 1.2(b) that, by applying GD model (1.3) to online solution of (1.9) under the same simulation conditions (e.g., with design parameter $\gamma = 1$ for both dynamic-models), its residual errors $|x^2(t) - 4\sin(20t)\cos(16t) - 5|$ are still quite large even

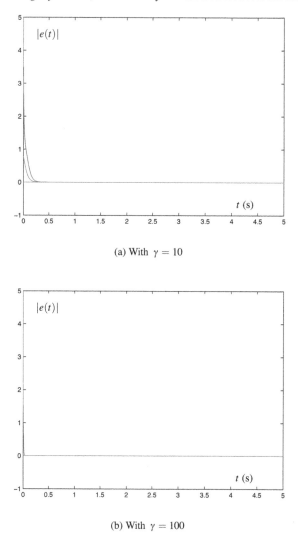

(a) With $\gamma = 10$

(b) With $\gamma = 100$

FIGURE 1.3 Residual errors $|x^2(t) - 4\sin(20t)\cos(16t) - 5|$ synthesized by CTZD model (1.2) with different γ.

after a long period of time. In the authors' opinion, the reason why the CTZD model does well in online time-varying square root finding (compared with the GD model) is that the time-derivative information of the time-varying coefficient in (1.9) is fully considered and utilized in the ZD design method and model. In addition, it is worth pointing out that, as shown in Figures 1.2(a) and 1.3, the convergence time of CTZD model (1.2) can be expedited from 3 s to 0.5 s and even to 0.05 s, as design parameter γ is increased from 1 to 10 and to 100, respectively. Furthermore, if $\gamma = 10^4$, the convergence time is only 0.5 ms (millisecond). This may show that CTZD model (1.2) actually has an exponential-convergence property, which can be expedited effectively by increasing the value of γ.

Thus, we summarize that CTZD model (1.2) is more efficient and effective for online time-varying square root finding, as compared with GD model (1.3).

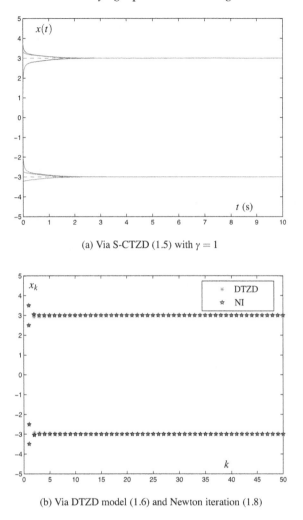

(a) Via S-CTZD (1.5) with $\gamma = 1$

(b) Via DTZD model (1.6) and Newton iteration (1.8)

FIGURE 1.4 Constant square root finding of (1.10) via S-CTZD model (1.5), DTZD model (1.6), and Newton iteration (NI) (1.8).

1.4.2 Constant Square Root Finding

Now, let us consider the following constant square root problem:

$$x^2 - 9 = 0. \tag{1.10}$$

Evidently, the theoretical square roots of problem (1.10) are $x_1^* = 3$ and $x_2^* = -3$. Correspondingly, as seen from Figure 1.4(a), starting from randomly-generated initial states in $[-4, 4]$, the neural states $x(t)$ of S-CTZD model (1.5) converge to a theoretical square root $x_1^* = 3$ or $x_2^* = -3$ rapidly and accurately. In addition, by employing DTZD model (1.6) and Newton iteration (1.8) to find the constant square root of (1.10), the numerical-experiment results are shown in Figure 1.4(b), where four initial states $x_0 = \pm 3.5$ and ± 2.5 are chosen for DTZD model (1.6) and Newton iteration (1.8) to start. As shown in Figure 1.4(b), setting step size $h = 0.1$ [128] for DTZD model (1.6), we find that the DTZD model's neural states (denoted by symbol "∗") converge to a theoretical square root (i.e., $x_1^* = 3$ or $x_2^* = -3$) within a few iterations. Besides, denoted by symbol "⋆" in Figure 1.4(b), the solutions of Newton iteration (1.8) converge to the theoretical square root of problem (1.10) in a rapid and accurate manner similar to the DTZD model. This point confirms the new explanation to Newton iteration. In other words, Newton iteration can be viewed as a special case of DTZD model (1.6) when the linear activation function and step size $h = 1$ are used.

1.5   Summary

In this chapter, a special class of neural dynamics (termed zeroing dynamics, ZD) has been developed and investigated for time-varying square root finding in real time. Such neural dynamics have been elegantly introduced by defining an indefinite error-monitoring function, instead of the traditional nonnegative (or at least low-bounded) energy function usually associated with GD. In addition, we have generated the DTZD model for constant square root finding, and discovered that such a DTZD model contains Newton iteration as a special case in some sense, presenting a new reasonable interpretation about Newton iteration. Computer-simulation and numerical-experiment results via two illustrative examples have further substantiated the efficacy of the ZD models on the scalar-valued square root finding, as well as the DTZD model's link and new explanation to Newton iteration.

Chapter 2

Time-Varying Cube Root Finding

Abstract

In this chapter, we generalize, propose and investigate the continuous-time zeroing dynamics (CTZD) model and its discrete-time ZD (DTZD) models with two situations considered (i.e., the time-derivative of the coefficient being known or unknown) for time-varying cube root finding. To find the constant cube root, a simplified CTZD (S-CTZD) model and its discrete-time model (S-DTZD) are generated. By focusing on such a constant problem solving, Newton iteration is found again to be a special case of the S-DTZD model by utilizing the linear activation function and fixing the step-size value to be 1. Computer-simulation and numerical-experiment results via the power-sigmoid activation function further substantiate the efficacy of the ZD models for the time-varying and constant cube roots finding, as well as the link and new explanation to Newton iteration.

2.1 Introduction

Online solution of time-varying cube root in the form of $x^3(t) - a(t) = 0$ or constant cube root in the form of $x^3 - a = 0$ is considered to be a basic problem arising in science and engineering fields, e.g., computer graphics [67, 74, 82], scientific computing [67], and field-programmable gate arrays (FPGA) implementations [24]. It is usually a fundamental part of many solutions. Thus, many numerical algorithms have been presented for the constant cube root problem solving [24, 51, 66, 67, 74, 82].

In this chapter, the continuous-time ZD (CTZD) model is generalized, developed, and investigated for online solution of time-varying cube root [111]. As mentioned previously, the constant problem solving can be viewed as a special case of the time-varying problem solving. Correspondingly, the S-CTZD model is generated and investigated to find the constant cube root. For the purpose of potential hardware (e.g., digital circuits or digital computers) implementation, the DTZD models, including the S-DTZD model, are developed and studied for time-varying and constant cube roots finding. Comparing with Newton iteration for constant cube root finding, we discover for the second time in this book that the S-DTZD model incorporates Newton iteration as a special case.

2.2   ZD Models for Time-Varying Case

In this section, the CTZD model and its discrete-time models are proposed for time-varying cube root finding. Let us consider the following time-varying cube root problem (which is written in the form of a time-varying nonlinear equation [129] for general purposes):

$$x^3(t) - a(t) = 0, \ \ t \in [0, +\infty), \tag{2.1}$$

where $a(t) \in \mathbb{R}$ denotes a smoothly time-varying scalar, which, together with its time-derivative $\dot{a}(t) \in \mathbb{R}$, is assumed to be known or can be measured numerically. Our objective is to find $x(t) \in \mathbb{R}$ in real time t such that the above smoothly time-varying nonlinear Equation (2.1) holds true. For presentation convenience, let $x^*(t) \in \mathbb{R}$ or $a^{1/3}(t) \in \mathbb{R}$ denote a theoretical time-varying cube root of $a(t)$.

2.2.1   CTZD Model

By following Zhang *et al.*'s neural-dynamic design method (i.e., zeroing dynamics design method) [99–102, 104, 107, 114, 117, 118, 122, 128, 129, 134] to monitor and control the time-varying cube root finding process, we first define an indefinite error function:

$$e(t) = x^3(t) - a(t) \in \mathbb{R}.$$

Secondly, the time-derivative $\dot{e}(t)$ of error function $e(t)$ can be chosen and forced, such that $e(t)$ exponentially converges to zero. Specifically, the time derivative $\dot{e}(t)$ is chosen via the following ZD design formula [99–102, 104, 107, 114, 117, 118, 122, 128, 129, 134]:

$$\frac{de(t)}{dt} = -\gamma\phi\big(e(t)\big),$$

where design parameter $\gamma > 0$, and $\phi(\cdot) : \mathbb{R} \to \mathbb{R}$ denotes the activation function, being the same as those in Chapter 1. Note that different choices of design parameter γ and activation function $\phi(\cdot)$ lead to different performance of the neural dynamics.

Thirdly, expanding the above ZD design formula, we have the following differential equation of the CTZD model for time-varying cube root finding:

$$3x^2(t)\dot{x}(t) - \dot{a}(t) = -\gamma\phi\big(x^3(t) - a(t)\big),$$

or equivalently, if $x(t) \neq 0$,

$$\dot{x}(t) = \frac{-\gamma\phi\big(x^3(t) - a(t)\big) + \dot{a}(t)}{3x^2(t)}, \tag{2.2}$$

where $x(t)$, starting from randomly generated nonzero initial condition $x(0) = x_0$, is the neural state corresponding to the time-varying theoretical cube root $a^{1/3}(t)$ of (2.1).

As for the convergence property of CTZD model (2.2), the following theorem and its proof can be established, summarized, and presented [128, 129].

Theorem 1 *Consider a smoothly time-varying positive (or negative) scalar $a(t)$ in (2.1). If a monotonically increasing odd activation function $\phi(\cdot)$ is used, then neural state $x(t) \in \mathbb{R}$ of CTZD model (2.2), starting from a randomly generated positive (or negative) initial state $x_0 \in \mathbb{R}$, can converge to theoretical time-varying cube root $x^*(t)$ [or to say, $a^{1/3}(t)$] of $a(t)$.*

Proof For CTZD model (2.2), defining a Lyapunov function candidate $v(x(t),t) = (x^3(t) - a(t))^2/2 \geqslant 0$ with respect to $e(t) = x^3(t) - a(t)$, we have its time derivative

$$\dot{v}(x(t),t) = \frac{\mathrm{d}v(x(t),t)}{\mathrm{d}t} = \left(x^3(t) - a(t)\right)\left(3x^2(t)\dot{x}(t) - \dot{a}(t)\right)$$
$$= -\gamma\left(x^3(t) - a(t)\right)\phi\left(x^3(t) - a(t)\right).$$

Because a monotonically increasing odd activation function is used, we have $\phi(-x^3(t) + a(t)) = -\phi(x^3(t) - a(t))$ and

$$\phi\left(x^3(t) - a(t)\right) \begin{cases} > 0, & \text{if } x^3(t) - a(t) > 0, \\ = 0, & \text{if } x^3(t) - a(t) = 0, \\ < 0, & \text{if } x^3(t) - a(t) < 0. \end{cases}$$

Therefore, the following results can be obtained:

$$\left(x^3(t) - a(t)\right)\phi\left(x^3(t) - a(t)\right) \begin{cases} > 0, & \text{if } x^3(t) - a(t) \neq 0, \\ = 0, & \text{if } x^3(t) - a(t) = 0, \end{cases}$$

which can guarantee the final negative-definiteness of $\dot{v}(x(t),t)$. In other words, $\dot{v}(x(t),t) < 0$ for $x^3(t) - a(t) \neq 0$ [equivalently, $x(t) \neq x^*(t)$], and $\dot{v}(x(t),t) = 0$ for $x^3(t) - a(t) = 0$ [equivalently, $x(t) = x^*(t)$]. By Lyapunov theory [78, 100, 132], residual error $e(t) = x^3(t) - a(t)$ can globally converge to zero; or equivalently, with $a(t)$ and $x(0)$ being nonzero, starting from randomly generated initial state $x(0)$, neural state $x(t)$ of CTZD model (2.2) can converge to the time-varying theoretical cube root $a^{1/3}(t)$ of $a(t)$. The proof is thus complete.

2.2.2 DTZD Models

For the purpose of potential hardware implementation (e.g., on digital circuits) of CTZD model (2.2) used for time-varying cube root finding, the DTZD models are generalized and developed for solving the time-varying cube root problem (2.1). Note that the DTZD models are divided into two formulations according to the criterion, whether the time-derivative of coefficient $a(t)$ [i.e., $\dot{a}(t)$] is known or not.

Situation 1: DTZD model assuming $\dot{a}(t)$ known

To discretize CTZD model (2.2) for solving (2.1), we refer to Euler forward-difference rule [1, 3, 25, 32, 51, 57, 118]:

$$\dot{x}(t = k\tau) \approx (x((k+1)\tau) - x(k\tau))/\tau,$$

where, again, $\tau > 0$ denotes the sampling gap, and $k = 0,1,2,\cdots$ denotes the update index. In general, we denote $x_k = x(t = k\tau)$ for presentation convenience. Besides, $a(t)$ and $\dot{a}(t)$ (assumed to be known in this situation) are discretized by the standard sampling method, of which the sampling gap is the same as and denoted by $\tau = t_{k+1} - t_k$, and $a(t_k) = a(t = k\tau)$ and $\dot{a}(t_k) = \dot{a}(t = k\tau)$. For convenience and also for consistency with x_k, we use a_k standing for $a(t_k)$ and $\dot{a}_k$ standing for $\dot{a}(t_k)$.

Thus, the DTZD model with $\dot{a}(t)$ known (i.e., the DTZDK model) for cube root finding is derived from the CTZD model (2.2) as

$$(x_{k+1} - x_k)/\tau = \frac{-\gamma\phi\left(x_k^3 - a_k\right) + \dot{a}_k}{3x_k^2},$$

which can be further formulated as

$$x_{k+1} = x_k - \frac{h\phi\left(x_k^3 - a_k\right) - \tau\dot{a}_k}{3x_k^2}, \tag{2.3}$$

where x_k corresponds to the kth update (or to say, sampling, iteration) of $x(t = k\tau)$; and, again, $h = \tau\gamma > 0$ denotes the step size, and $\phi(\cdot)$ denotes the activation function. In addition, $\tau > 0$ should be set appropriately small for convergence purposes.

Situation 2: DTZD model assuming $\dot{a}(t)$ unknown

Generally speaking, in many science and engineering fields, $\dot{a}(t)$ may be difficult to know its analytical form or value. Thus, we investigate the discrete-time one of the CTZD model when $\dot{a}(t)$ is unknown. In this situation, $\dot{a}(t)$ might generally be estimated from $a(t)$ by employing Euler backward-difference rule [1, 3, 25, 32, 51, 57, 118]:

$$\dot{a}(t = k\tau) \approx (a(k\tau) - a((k-1)\tau))/\tau,$$

where $\tau > 0$ is defined the same as before. Similarly, we denote $a_k = a(t = k\tau)$ and $\dot{a}_k = \dot{a}(t = k\tau)$ for presentation convenience. Thus, from CTZD model (2.2), we derive the DTZD model with $\dot{a}(t)$ unknown (i.e., the DTZDU model) for solving problem (2.1) as

$$(x_{k+1} - x_k)/\tau = \frac{-\gamma\phi\left(x_k^3 - a_k\right)}{3x_k^2} + \frac{a_k - a_{k-1}}{3\tau x_k^2},$$

which can be further formulated as

$$x_{k+1} = x_k - \frac{h\phi\left(x_k^3 - a_k\right)}{3x_k^2} + \frac{a_k - a_{k-1}}{3x_k^2}, \tag{2.4}$$

with $k = 1, 2, 3, \cdots$, where $h > 0$ denotes a suitable step size. Note that, from the above Euler backward-difference rule, we cannot obtain $\dot{a}(t = 0)$, since t is defined within $[0, +\infty)$ and a_{-1} is undefined. Thus, in this situation, we can choose $\dot{a}(t = 0) = 0$ or any other value to start the DTZDU model (2.4).

2.3   Simplified ZD Models for Constant Case and Newton Iteration

In this section, the simplified CTZD model and its discrete-time model are generalized for constant cube root finding.

2.3.1   S-CTZD Model

Being a special case of time-varying nonlinear Equation (2.1), let us consider the following static nonlinear equation (which corresponds to the constant cube root problem formulated in most textbooks and literature, e.g., in [25, 51, 57]):

$$x^3(t) - a = 0, \tag{2.5}$$

where $a \in \mathbb{R}$ is a positive (or negative) constant. By considering the constant a in (2.5) and its time derivative $\dot{a}(t) = 0$, CTZD model (2.2) can be simplified as

$$\dot{x}(t) = -\gamma \frac{\phi\left(x^3(t) - a\right)}{3x^2(t)}, \tag{2.6}$$

which is termed as an S-CTZD model used to find the constant cube root. Note that $x(t)$, γ and $\phi(\cdot)$ are defined the same as before. Based on Theorem 1, we have the following corollary on the convergence property of S-CTZD model (2.6) with its proof omitted due to the similarity to the one of Theorem 1.

Corollary 1 *Consider a constant positive (or negative) scalar-valued constant $a \in \mathbb{R}$ in (2.5). If a monotonically increasing odd activation function $\phi(\cdot)$ is used, then neural state $x(t) \in \mathbb{R}$ of S-CTZD model (2.6), starting from a randomly generated positive (or negative) initial state $x_0 \in \mathbb{R}$, can converge to the theoretical cube root x^* (or to say, $a^{1/3}$) of constant a.*

2.3.2 S-DTZD Model

For the purpose of potential hardware implementation (e.g., based on digital circuits) for constant cube root finding, we also discretize S-CTZD model (2.6) by using the aforementioned Euler forward-difference rule [1,3,25,32,51,57,118]. Thus, an S-DTZD model for the constant cube root finding is derived from S-CTZD model (2.6) as

$$(x_{k+1} - x_k)/\tau = -\gamma \frac{\phi\left(x_k^3 - a\right)}{3x_k^2},$$

which can be further formulated as

$$x_{k+1} = x_k - h \frac{\phi\left(x_k^3 - a\right)}{3x_k^2}. \tag{2.7}$$

From an important theorem in [128], the following proposition on the convergence property of S-DTZD model (2.7) [111] is achieved and presented.

Proposition 4 *Consider the constant cube root problem (2.5). When step size h of S-DTZD model (2.7) satisfies $0 < h < 2/\phi'(0)$, there generally exists a positive number δ, such that the sequence $\{x_k\}_{k=0}^{+\infty}$ generated by S-DTZD model (2.7) can converge to a theoretical root x^* of (2.5) for any initial state $x_0 \in [x^* - \delta, x^* + \delta]$.*

As for specific types of activation function $\phi(\cdot)$, we thus have the following ranges of step size h according to the formula $0 < h < 2/\phi'(0)$.

1) If the linear activation function $\phi(e) = e$ is used, we get $\phi'(0) = 1$. Then the step size h should satisfy the range $0 < h < 2$.

2) If the bipolar sigmoid activation function

$$\phi(e) = \frac{1 - \exp(-\xi e)}{1 + \exp(-\xi e)}$$

is used with $\xi = 4$, we get $\phi'(0) = \xi/2 = 2$. Then the step size h should satisfy the range $0 < h < 1$.

3) If the power-sigmoid activation function

$$\phi(e) = \begin{cases} e^p, & \text{if } |e| \geqslant 1 \\ \frac{1+\exp(-\xi)}{1-\exp(-\xi)} \frac{1-\exp(-\xi e)}{1+\exp(-\xi e)}, & \text{otherwise} \end{cases}$$

is used with $\xi = 4$ and $p = 3$, we get $\phi'(0) = 2.0746$. Then the step size h should satisfy the range $0 < h < 0.9640$.

2.3.3 Link and New Explanation to Newton Iteration

Look at S-DTZD model (2.7). When we utilize the linear activation function $\phi(e) = e$, S-DTZD model (2.7) is written as

$$x_{k+1} = x_k - h\frac{x_k^3 - a}{3x_k^2}. \tag{2.8}$$

In addition, for $h = 1$, the linearly activated S-DTZD model (2.8) further reduces to

$$x_{k+1} = \frac{2x_k + a/x_k^2}{3}, \tag{2.9}$$

which is exactly Newton iteration for constant cube root finding widely presented in textbooks and literature [66, 82]. In other words, in addition to Section 1.3 of Chapter 1, we discover that a general form of Newton iteration is given by discretizing S-CTZD model (2.6). Specifically, Newton iteration for constant cube root finding is derived by discretizing S-CTZD model (2.6) via Euler forward difference, utilizing the linear activation function, and fixing the step size to be 1. Evidently, this chapter shows a new explanation to Newton iteration for constant cube root finding, which is evidently different from the traditional (or to say, standard) explanations appearing in almost all literature and textbooks, i.e., via Taylor series expansion [1, 3, 25, 32, 51, 57, 118].

2.4 Illustrative Examples

The previous two sections have presented the CTZD, S-CTZD, DTZD, and S-DTZD models and the link to Newton iteration for online solution of cube root (including time-varying and constant cases). In this section, computer-simulation and numerical-experiment results are provided for verifying the efficacy of the ZD models with the aid of suitable initial state $x(0)$, step size h, and sampling gap τ, where a power-sigmoid activation function is exploited with design parameters $\xi = 4$ and $p = 3$ [104, 107, 117, 118, 129].

2.4.1 Time-Varying Cube Root Finding

In this subsection, the CTZD model and its discrete-time models [with time-derivative $\dot{a}(t)$ known or unknown] are investigated for online time-varying cube root finding.

Example 2.1 By CTZD model (2.2) and DTZDK model (2.3)

For illustration and comparison, let us consider the following time-varying cube root problem (written in the general form of a nonlinear equation):

$$x^3(t) - 2\sin(3t) - 5 = 0. \tag{2.10}$$

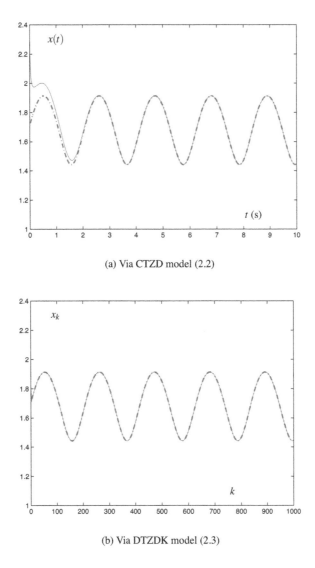

(a) Via CTZD model (2.2)

(b) Via DTZDK model (2.3)

FIGURE 2.1 Time-varying cube root finding of (2.10) via CTZD model (2.2) with $\gamma = 1$ and DTZDK model (2.3) with $h = 0.8$.

First, CTZD model (2.2) is exploited for online solution of (2.10) with the simulation results shown in Figure 2.1(a). Note that, to check the solution correctness (or to say, effectiveness) of CTZD model (2.2), the theoretical root $x^*(t)$ of (2.10) is shown and denoted by dash-dotted curves in the figure. As seen from Figure 2.1(a), starting from a randomly generated initial state in $[0, 10]$, the neural state $x(t)$ of CTZD model (2.2) fits well with the theoretical cube root $x^*(t)$ [i.e., $a^{1/3}(t)$] after a short time (e.g., 3 s or so). This illustrates the efficacy of CTZD model (2.2) for the time-varying cube root problem solving.

Secondly, DTZDK model (2.3) is exploited for online solution of (2.10) with the experiment result shown in Figure 2.1(b). Note that the time derivative of the coefficient $a(t) = 2\sin(3t) + 5$ can be obtained via a simple differentiation operation. In other words, $\dot{a}(t)$ is known in this example

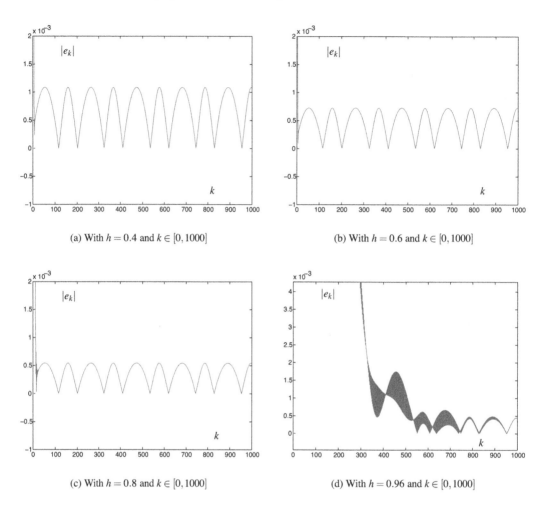

(a) With $h = 0.4$ and $k \in [0, 1000]$

(b) With $h = 0.6$ and $k \in [0, 1000]$

(c) With $h = 0.8$ and $k \in [0, 1000]$

(d) With $h = 0.96$ and $k \in [0, 1000]$

FIGURE 2.2 Residual errors $|e_k| = |x_k^3 - 2\sin(3t_k) - 5|$ synthesized by DTZDK model (2.3) with different h, where labeled ordinate ranges of subfigures are $[-1, 2] \cdot 10^{-3}$, $[-1, 2] \cdot 10^{-3}$, $[-1, 2] \cdot 10^{-3}$, and $[0, 4] \cdot 10^{-3}$.

(that is the reason why we choose such a discrete-time model as a root finder). From Figure 2.1(b), we see that the neural state x_k of DTZDK model (2.3) with step size $h = 0.8$, starting from a suitable initial state generated in [1,2], converges well to the theoretical time-varying cube root within a small number of updates (or to say similarly, iterations). Note that, in such an experiment, update index $k = t/\tau$, with $t \in [0, 10]$ s and sampling gap $\tau = 0.01$ s.

In order to further investigate the convergence performance of DTZDK model (2.3), we also monitor the residual error $|e_k| = |x_k^3 - 2\sin(3t_k) - 5|$ during the problem solving process of DTZDK model (2.3) with respect to different values of step size h. By applying DTZDK model (2.3) to time-varying cube root finding of (2.10), the maximal steady-state residual errors for different step-size values (i.e., $h = 0.4$, 0.6, 0.8, 0.95 and 0.96) are shown in Figure 2.2 and Table 2.1. These show that DTZDK model (2.3) has good convergence property (with acceptably small residual errors).

TABLE 2.1   MSSREs of $|e_k|$ of DTZDK Model (2.3) Solving (2.10) with Different h Values

	h	τ	Maximal steady-state residual error (MSSRE, 10^{-3})
	0.4	0.01	1.08
	0.6	0.01	0.72
$x^3(t) - 2\sin(3t) - 5$	0.8	0.01	0.54
	0.95	0.01	0.45 (best)
	0.96	0.01	0.49

TABLE 2.2   MSSREs of $|e_k|$ of DTZDK Model (2.3) Solving (2.11) with Different h Values

	h	τ	Maximal steady-state residual error (10^{-3})
	0.4	0.01	1.5
	0.6	0.01	0.99
$x^3(t) - \sin(5t) - 4$	0.8	0.01	0.75
	0.95	0.01	0.63 (best)
	0.96	0.01	0.71

Example 2.2   Choice of step size h

From the above example, we know that different step-size values lead to different convergence performance of DTZDK model (2.3). For further investigation and verification, let us consider the following time-varying cube root problem with a higher frequency (written in the general form of a nonlinear equation):

$$x^3(t) - \sin(5t) - 4 = 0. \tag{2.11}$$

Table 2.2 illustrates the experiment results, which are synthesized by DTZDK model (2.3) with different values of step size h used (i.e., $h = 0.4, 0.6, 0.8, 0.95,$ and 0.96). In addition to Table 2.1, from Table 2.2 we see that good convergence is achieved with step size $h \in [0.4, 0.96]$; the maximal steady-state residual error $\max_{k \to \infty} |e_k|$ decreases when h increases within range $[0.4, 0.96)$; and, the best value of step size h appears to be 0.95, which corresponds to the smallest $\max_{k \to \infty} |e_k|$.

Example 2.3   Choice of sampling gap τ

As mentioned previously, the sampling gap τ [shown in Equation (2.3)] has an effect on the convergence performance of the DTZDK model. In this example, to investigate the effect of τ, let us consider the following time-varying cube root problem with an even higher frequency:

$$x^3(t) - 2\sin(7t) - 5 = 0. \tag{2.12}$$

Based on the previous numerical-experiment results and observations, we choose DTZDK model (2.3) with the best step size $h = 0.95$ to solve the nonlinear Equation (2.12). Table 2.3 and Figure 2.3 show the results synthesized by DTZDK model (2.3) with different values of τ. As seen from the table and the figure, the maximal steady-state residual error of $\tau = 0.01$ is 2.5×10^{-3}, while that of $\tau = 0.001$ is about 2.5×10^{-5}. Furthermore, the maximal steady-state residual error of $\tau = 0.0001$ is about 2.5×10^{-7}, and that of $\tau = 0.00001$ is about 2.5×10^{-9}. The results show the relationship between the maximal steady-state residual error and the sampling gap τ in an $O(\tau^2)$ manner. That is, the maximal steady-state residual error reduces by 100 times when the sampling gap τ decreases by 10 times, which implies that τ can be selected appropriately small to satisfy the usual precision we need in practice. Thus, we can have the important conclusion that the maximal steady-state residual error of DTZDK model (2.3) is of order $O(\tau^2)$.

TABLE 2.3   MSSREs of $|e_k|$ of DTZDK Model (2.3) Solving (2.12) with Different τ Values

	h	τ	**Maximal steady-state residual error**
	0.95	0.01	2.50×10^{-3}
$x^3(t) - 2\sin(7t) - 5$	0.95	0.001	2.48×10^{-5}
	0.95	0.0001	2.48×10^{-7}
	0.95	0.00001	2.48×10^{-9}

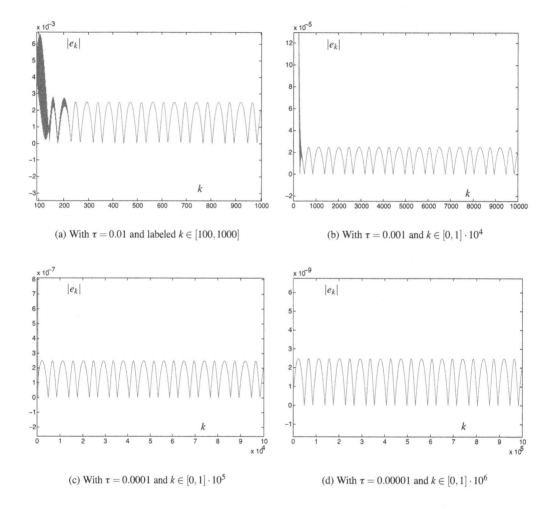

(a) With $\tau = 0.01$ and labeled $k \in [100, 1000]$

(b) With $\tau = 0.001$ and $k \in [0,1] \cdot 10^4$

(c) With $\tau = 0.0001$ and $k \in [0,1] \cdot 10^5$

(d) With $\tau = 0.00001$ and $k \in [0,1] \cdot 10^6$

FIGURE 2.3 Residual errors $|e_k| = |x_k^3 - 2\sin(7t_k) - 5|$ synthesized by DTZDK model (2.3) with $h = 0.95$ and different τ, where labeled ordinate ranges of subfigures are $[-3, 6] \cdot 10^{-3}$, $[-2, 12] \cdot 10^{-5}$, $[-2, 8] \cdot 10^{-7}$, and $[-1, 6] \cdot 10^{-9}$.

Example 2.4   By DTZDK model (2.3) and DTZDU model (2.4)

In this example, to investigate the effectiveness of DTZDU model (2.4), let us consider the following time-varying cube root problem with a much higher frequency:

$$x^3(t) - \sin(20t) - 4 = 0. \tag{2.13}$$

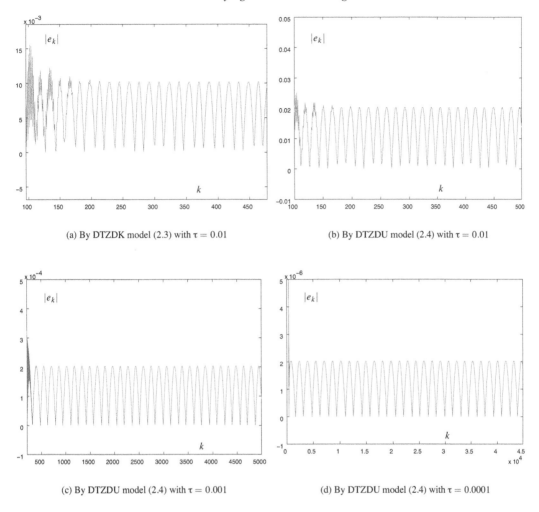

(a) By DTZDK model (2.3) with $\tau = 0.01$

(b) By DTZDU model (2.4) with $\tau = 0.01$

(c) By DTZDU model (2.4) with $\tau = 0.001$

(d) By DTZDU model (2.4) with $\tau = 0.0001$

FIGURE 2.4 Residual errors $|e_k| = |x_k^3 - \sin(20t_k) - 4|$ synthesized by DTZDK model (2.3) and DTZDU model (2.4) with $h = 0.95$ and different τ, where labeled ranges of subfigures are $10^2 \cdot [1, 4.5] \times [-5, 15] \cdot 10^{-3}$, $10^2 \cdot [1, 5] \times [-1, 5] \cdot 10^{-2}$, $10^3 \cdot [0.5, 5] \times [-1, 5] \cdot 10^{-4}$, and $10^4 \cdot [0, 4.5] \times [-1, 5] \cdot 10^{-6}$.

For comparative purposes, both DTZDK model (2.3) and DTZDU model (2.4) are exploited for the time-varying cube root finding of (2.13). Note that the time derivative of $a(t) = \sin(20t) + 4$ for DTZDK model (2.3) is obtained by the simple differentiation operation, while that of DTZDU model (2.4) is generated by using the backward difference rule. With $\tau = 0.01$, the comparative numerical-experiment results are illustrated in Figure 2.4(a) and (b), in which the maximal steady-state residual error of DTZDK model (2.3) is 1.02×10^{-2}, and that of DTZDU model (2.4) is 2.07×10^{-2}. In addition, from Figure 2.4(b), (c) and (d), we compare and see that the maximal steady-state residual error of DTZDU model (2.4) reduces when the sampling gap τ decreases (which is similar to the observation in Example 2.3). That is, when the sampling gap τ changes from 0.01 to 0.0001, the maximal steady-state residual error reduces from 2.07×10^{-2} to 2.07×10^{-6}. Based on the numerical results, we have the important conclusion that the maximal steady-state residual error of DTZDU model (2.4) is of order $O(\tau^2)$. These substantiate again the effectiveness and precision of DTZDU model (2.4) for time-varying cube root finding.

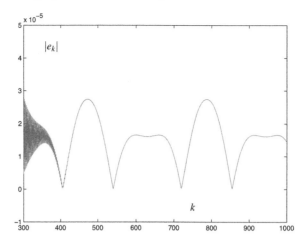

(a) By DTZDU model (2.4) with $\tau = 0.01$

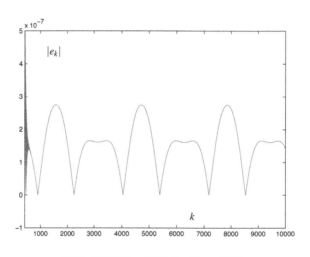

(b) By DTZDU model (2.4) with $\tau = 0.001$

FIGURE 2.5 Residual errors synthesized by DTZDU model (2.4) with $h = 0.95$ and different τ, where ordinate ranges of subfigures are $[-1, 5] \cdot 10^{-5}$ and $[-1, 5] \cdot 10^{-7}$.

Example 2.5 With time derivative $\dot{a}(t)$ assumed unknown

For further investigation of DTZDU model (2.4), let us consider the following time-varying cube root problem with $\dot{a}(t)$ unknown (or more difficult to obtain):

$$x^3(t) - \sin(\cos(\sin(\cos(t)))) - 4 = 0. \tag{2.14}$$

Evidently, DTZDK model (2.3) is quite complex for such a time-varying nonlinear equation solving, since it is much more difficult to obtain $\dot{a}(t)$ analytically in this example. Thus, it is preferable to exploit DTZDU model (2.4) for online solution of (2.14). Figure 2.5 shows the numerical-experiment results synthesized by DTZDU model (2.4). As seen from the figure, DTZDU model (2.4) is effective and precise in solving such a time-varying cube root problem, in the sense that the maximal steady-state residual error is quite small. Besides, Figure 2.5 and related numerical results (e.g.,

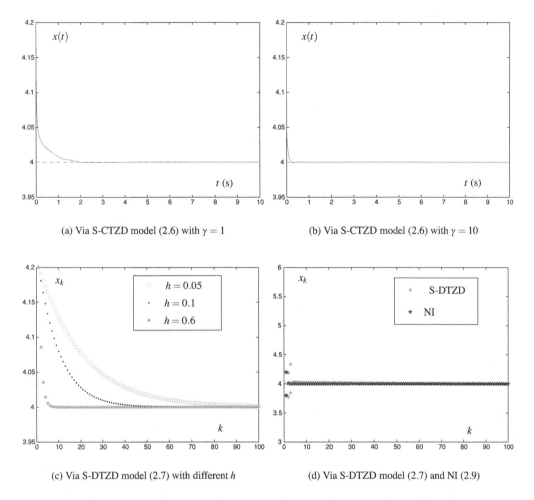

(a) Via S-CTZD model (2.6) with $\gamma = 1$

(b) Via S-CTZD model (2.6) with $\gamma = 10$

(c) Via S-DTZD model (2.7) with different h

(d) Via S-DTZD model (2.7) and NI (2.9)

FIGURE 2.6 Constant cube root finding of (2.15) via S-CTZD model (2.6), S-DTZD model (2.7), and Newton iteration (NI) (2.9), where subfigure ranges are $[0,10] \times [3.95, 4.2]$, $[0,10] \times [3.95, 4.2]$, $[0,100] \times [3.95, 4.2]$, and $[0,100] \times [3,6]$.

$\max_{k \to \infty} |e_k| = 2.74 \times 10^{-9}$ with $\tau = 0.0001$) illustrate that the maximal steady-state residual error decreases in an $O(\tau^2)$ manner when the sampling gap τ decreases.

In summary, the above five examples have substantiated that CTZD model (2.2) and its discrete-time models [i.e., DTZDK model (2.3) and DTZDU model (2.4)] are effective, efficient, and precise in time-varying cube root finding.

2.4.2 Constant Cube Root Finding

In this subsection, S-CTZD model (2.6) and S-DTZD model (2.7) are applied to constant cube root finding. Now, let us consider the following constant cube root problem:

$$x^3 - 64 = 0, \tag{2.15}$$

of which the theoretical cube root is evidently $x^* = 4$, given for comparison and verification purposes. When S-CTZD model (2.6) is used to solve such a constant problem, the simulation results (with design parameter $\gamma = 1$ and 10) are shown in Figure 2.6(a) and (b). As seen from these two subfigures, neural states $x(t)$ of S-CTZD model (2.6) converge to theoretical cube root $x^* = 4$ rapidly and accurately, which illustrates the effectiveness and efficiency of S-CTZD model (2.6) for online solution of constant cube root problem (2.15).

In addition, S-DTZD model (2.7) and Newton iteration (2.9) are employed to solve (2.15). Specifically, Figure 2.6(c) and (d) illustrate the numerical results synthesized by S-DTZD model (2.7) using the linear activation function and the power-sigmoid activation function as well as Newton iteration (2.9). Using the linear activation function with different values of step size h, it follows from Figure 2.6(c) that neural states x_k of S-DTZD model (2.7) converge to theoretical cube root $x^* = 4$ within a small number of iterations, which shows the efficacy of S-DTZD model (2.7). Besides, as shown in Figure 2.6(d), by employing the power-sigmoid activation function with $\xi = 4$ and $p = 3$, and setting step size $h = 0.022$, we find that neural states x_k computed by S-DTZD model (2.7) and Newton iteration (2.9) converge to theoretical cube root $x^* = 4$ within just a few iterations. These substantiate again the effectiveness of S-DTZD model (2.7) for constant cube root finding, in addition to the important link to Newton iteration [i.e., Newton iteration (2.9) can be viewed as a special case of S-DTZD model (2.7) when the linear activation function and step-size value $h = 1$ are used].

2.5   Summary

In this chapter, a special class of neural dynamics (termed zeroing dynamics, ZD) has been developed and investigated for time-varying cube root finding in real time or online. In addition, the S-CTZD model has been generated to find the constant cube root (which can be viewed as a special case of the time-varying problem solving). For the purpose of potential hardware (e.g., digital circuits) implementation, the discrete-time ZD models [i.e., DTZDK model (2.3), DTZDU model (2.4), and S-DTZD model (2.7)] have been developed and studied for time-varying and constant cube roots finding. Moreover, focusing on the constant problem solving, we have discovered again that the S-DTZD model contains Newton iteration as a special case, which provides a new reasonable interpretation to Newton iteration. Simulative and numerical results have finally illustrated the efficacy of the ZD models on time-varying and constant cube roots finding.

Chapter 3

Time-Varying 4th Root Finding

Abstract

In this chapter, for the purpose of online solution of the time-varying 4th root, both of the continuous-time zeroing dynamics (CTZD) and discrete-time ZD (DTZD) models are developed and investigated. In addition, the power-sigmoid activation function is exploited, which makes ZD models possess the property of superior convergence and better precision (as compared with the standard situation of using the linear activation function). Computer-simulation and numerical-experiment results further illustrate the efficacy of the ZD models for finding online the time-varying 4th root.

3.1   Introduction

Online solution of the 4th root in the form of $x^4 - a = 0$, where a is a positive scalar, is considered to be an important special case of the nonlinear equation problem arising in science and engineering fields [1,38,41,57,126]. For example, in [38], one image can be described in Torelli group by finding its fourth root of unity. In [41], Harris corner strength image may be refined from the fourth root of the original image. Thus, many related numerical algorithms have been presented for the problem solving [1,57]. It is worth mentioning here that many reported popular computational schemes are related to (or originate from) gradient-based methods or designed theoretically for solving constant problems that assume that the coefficient a does not vary as time evolves. However, when the time-varying situation is further considered, i.e., $x^4(t) - a(t) = 0$ [in which $a(t)$ denotes a time-varying positive scalar], it may not be efficient enough for gradient dynamics (GD) because of a relatively large lagging error generated.

Different from the gradient-based neural-dynamic approach, a special class of ZD has been formally proposed by Zhang et al. [99,100,117,122,126,129,134] for time-varying and/or constant problems solving. Considering the superiorness of the proposed models in Chapters 1 and 2 based on Zhang et al.'s neural-dynamic method (i.e., ZD method), in this chapter, we propose a series of ZD models for the time-varying 4th root finding. In addition, the ZD models utilize the time-derivative information of the time-varying coefficient of the 4th root nonlinear equation, such that the exact solution can be obtained in an error-free manner; whereas, without considering it, the GD model generates a rather large lagging error.

Specifically, the CTZD model is first introduced, generalized, developed, and investigated for the online time-varying 4th root finding [126]. The neural state of the CTZD model is proved to converge to the theoretical solution of the time-varying 4th root problem. Secondly, using the Euler difference rule, we generate its discrete-time models with two situations considered (i.e., according to the criterion whether the time-derivative of the related coefficient is known or not). Finally, computer-simulation and numerical-experiment results of the ZD models are provided, which illustrate the efficacy and precision of the ZD models for the real-time or online time-varying 4th root finding.

3.2 Problem Formulation and ZD Models

To lay a basis for further discussion, the ZD models are presented for finding the time-varying 4th root in this section.

3.2.1 Problem Formulation

Now consider the following mathematical problem of the 4th root [126]:

$$x^4(t) - a(t) = 0, \ t \in [0, +\infty), \tag{3.1}$$

where $a(t) \in \mathbb{R}^+$ denotes a smoothly time-varying positive scalar, which, together with its time derivative $\dot{a}(t) \in \mathbb{R}$, is assumed to be known numerically or can be measured accurately. Our first objective is to find $x(t) \in \mathbb{R}$ in real time t, such that the above smoothly time-varying nonlinear Equation (3.1) always holds true. For presentation convenience, let $x^*(t) \in \mathbb{R}$ [or to say, $\pm a^{1/4}(t)$] denote the theoretical time-varying 4th root of $a(t)$.

3.2.2 CTZD Model

Following Zhang et al.'s neural-dynamic design method [102, 104, 117, 118, 122, 129, 134], to monitor and control the process of the time-varying 4th root finding, we first define an indefinite error function as follows:

$$e(t) = x^4(t) - a(t) \in \mathbb{R}.$$

Secondly, the time-derivative $\dot{e}(t)$ of error function $e(t)$ can be chosen and forced such that $e(t)$ mathematically converges to zero. Specifically, we choose the time derivative $\dot{e}(t)$ via the following ZD design formula [102, 104, 117, 118, 122, 129, 134]:

$$\frac{de(t)}{dt} = -\gamma \phi(e(t)), \tag{3.2}$$

Thirdly, expanding the above design formula, we thus obtain

$$4x^3(t)\dot{x}(t) - \dot{a}(t) = -\gamma \phi(x^4(t) - a(t)),$$

which leads to the following CTZD model for the time-varying 4th root finding:

$$\dot{x}(t) = \frac{\dot{a}(t) - \gamma \phi(x^4(t) - a(t))}{4x^3(t)}, \tag{3.3}$$

where $x(t)$, starting from a nonzero randomly generated initial state $x(0) \in \mathbb{R}$, denotes the neural state corresponding to theoretical solution $x^*(t)$ of (3.1). In this chapter, the following two types of activation functions are used:

1) Linear activation function $\phi(e) = e$

2) Power-sigmoid activation function

$$\phi(e) = \begin{cases} \frac{(1+\exp(-\xi))}{(1-\exp(-\xi))} \frac{1-\exp(-\xi e)}{1+\exp(-\xi e)}, & \text{if } |e| \leqslant 1 \\ e^p, & \text{if } |e| \geqslant 1 \end{cases}$$

with design parameters $\xi \geqslant 2$ and odd integer $p \geqslant 3$

After presenting CTZD model (3.3) for solving the 4th root problem, detailed design consideration, and theoretical analysis are given in the following theorems.

Theorem 2 *Consider a smoothly time-varying positive scalar $a(t) \in \mathbb{R}^+$ in (3.1). If a monotonically increasing odd activation function $\phi(\cdot)$ is used, then neural state $x(t) \in \mathbb{R}$ of CTZD model (3.3), starting from a nonzero randomly generated initial state $x(0) \in \mathbb{R}$, can converge to the theoretical time-varying 4th root $x^*(t)$.*

Proof Defining a Lyapunov function candidate $v(x(t),t) = (x^4(t) - a(t))^2/2 \geqslant 0$ for CTZD model (3.3), we have its time derivative as

$$\begin{aligned} \dot{v}(x(t),t) &= \frac{\mathrm{d}v(x(t),t)}{\mathrm{d}t} \\ &= (x^4(t) - a(t))(4x^3(t)\dot{x}(t) - \dot{a}(t)) \\ &= -\gamma(x^4(t) - a(t))\phi(x^4(t) - a(t)). \end{aligned} \tag{3.4}$$

Because a monotonically increasing odd activation function is used, we readily have $\phi(-x^4(t) + a(t)) = -\phi(x^4(t) - a(t))$ and

$$\phi(x^4(t) - a(t)) \begin{cases} > 0, & \text{if } x^4(t) - a(t) > 0, \\ = 0, & \text{if } x^4(t) - a(t) = 0, \\ < 0, & \text{if } x^4(t) - a(t) < 0. \end{cases}$$

Therefore, the following results are obtained; i.e.,

$$(x^4(t) - a(t))\phi(x^4(t) - a(t)) \begin{cases} > 0, & \text{if } x^4(t) - a(t) \neq 0, \\ = 0, & \text{if } x^4(t) - a(t) = 0, \end{cases}$$

which guarantee the final negative-definiteness of $\dot{v}(x(t),t)$. That is, $\dot{v}(x(t),t) < 0$ for $x^4(t) - a(t) \neq 0$ [equivalently, $x(t) \neq x^*(t)$], and $\dot{v}(x(t),t) = 0$ for $x^4(t) - a(t) = 0$ [equivalently, $x(t) = x^*(t)$]. By the Lyapunov theory, starting from nonzero randomly generated initial state $x(0)$, neural state $x(t)$ of CTZD model (3.3) can converge to the theoretical 4th root $x^*(t)$. The proof is thus complete.

Theorem 3 *In addition to Theorem 2, when neural state $x(t)$ of CTZD model (3.3) converges to $x^*(t)$, CTZD model (3.3) possesses the following properties.*

1) If the linear activation function is used, the exponential convergence with rate γ [in terms of residual error $e(t) = x^4 - a(t) \to 0$] is achieved for CTZD model (3.3).

2) *If the power-sigmoid activation function is used, superior convergence is achieved for the whole error range $e(t) = x^4 - a(t) \in (-\infty, +\infty)$, as compared with the situation of using the linear activation function.*

Proof. Now we come to prove the additional convergence properties of CTZD model (3.3) by considering the mentioned two types of activation function $\phi(\cdot)$.

1) For the situation of using the linear activation function, it follows from (3.2) that $d(x^4(t) - a(t))/dt = -\gamma(x^4(t) - a(t))$, which yields $x^4(t) - a(t) = \exp(-\gamma t)(x^4(0) - a(0))$. This proves the exponential convergence of CTZD model (3.3) with rate γ in the sense of residual error $e(t) = x^4(t) - a(t) \to 0$ asymptotically and exponentially. Moreover, we show that neural state $x(t)$ of CTZD model (3.3) converges to the theoretical time-varying solution $x^*(t)$ of nonlinear equation $x^4(t) - a(t) = 0$ by the following procedure. From the theorem of Taylor series expansion [57], we have

$$
\begin{aligned}
x^4(t) - a(t) &= \left(x^{*4}(t) - a(t)\right) \\
&+ \left(x(t) - x^*(t)\right)\left(4x^{*3}(t)\right) + \frac{\left(x(t) - x^*(t)\right)^2}{2!}\left(12x^{*2}(t)\right) \\
&+ \frac{\left(x(t) - x^*(t)\right)^3}{3!}\left(24x^*(t)\right) + 24\frac{\left(x(t) - x^*(t)\right)^4}{4!} + \cdots
\end{aligned}
$$

Thus, in view of $x^4(t) - a(t) = \exp(-\gamma t)(x^4(0) - a(0))$ and $x^{*4} - a(t) = 0$, omitting the higher-order terms, we have

$$
\left(x(t) - x^*(t)\right)\left(4x^{*3}(t)\right) \approx \exp(-\gamma t)\left(x^4(0) - a(0)\right),
$$

which, in view of $a(t) > 0$, $4x^{*3}(t) \neq 0$ and $|x^{*3}(t)| \geq \delta > 0$, yields

$$
|x(t) - x^*(t)| \approx \exp(-\gamma t)\left|\frac{\left(x^4(0) - a(0)\right)}{4x^{*3}(t)}\right| \leq \exp(-\gamma t)\left|\frac{\left(x^4(0) - a(0)\right)}{4\delta}\right|.
$$

This implies that neural state $x(t)$ of CTZD model (3.3) itself converges to the theoretical solution $x^*(t)$ also in an exponential manner.

2) For the power-sigmoid activation function, the theoretical analysis has two parts, i.e., $|e(t)| \leq 1$ and $|e(t)| > 1$.

i) For small error $|e(t)| \leq 1$, the bipolar-sigmoid part of $\phi(\cdot)$ is activated with $\phi(|e(t)|) \geq |e(t)|$. So, by reviewing Equation (3.4), $\dot{v}(x(t), t)$ becomes more negative, which implies that the superior convergence is achieved by CTZD model (3.3) using the bipolar sigmoid activation function for such an error range, as compared with the situation of using the linear activation function [132].

ii) For large error $|e(t)| > 1$, the power part of $\phi(\cdot)$ is activated with $\phi(|e(t)|) = |e^p(t)| \geq |e(t)|$, the solution to differential Equation (3.2) becomes

$$
e(t) = e(0)\left((p-1)e^{p-1}(0)\gamma t + 1\right)^{-\frac{1}{p-1}}.
$$

Besides, for $p = 3$, residual error $e(t) = e(0)/\sqrt{2e^2(0)\gamma t + 1}$. Evidently, as $t \to \infty$, $e(t) = x^4(t) - a(t) \to 0$. By reviewing the proof of Theorem 2, especially the Lyapunov function candidate $v(x(t), t) = (x^4(t) - a(t))^2/2 \geq 0$ and its time derivative $\dot{v}(x(t), t) = -\gamma(x^4(t) - a(t))\phi(x^4(t) - a(t))$, in the error range $|x^4(t) - a(t)| > 1$, we have $(x^4(t) - a(t))^{p+1} > (x^4(t) - a(t))^2 > |x^4(t) - a(t)|$. In addition, in the error range $|x^4(t) - a(t)| \gg 1$, we have $(x^4(t) - a(t))^{p+1} \gg (x^4(t) - a(t))^2 \gg |x^4(t) - a(t)|$. In other words, the deceleration magnitude of the power activation function is much greater than that of the linear activation function. This implies that, when using the power activation function, a much faster convergence can be achieved by CTZD model (3.3) for such an error range in comparison with the situation of using the linear activation function.

It follows from the above analysis that, to achieve superior convergence for the whole error range $(-\infty, +\infty)$, CTZD model (3.3) can be activated by the power-sigmoid activation function with suitable design parameters ξ and p, as compared with the situation of using the linear activation function. The proof is now complete.

3.2.3 DTZD Models

For potential hardware implementation on digital circuits of CTZD model (3.3) exploited for the time-varying 4th root finding, the DTZD models are derived and developed for solving online the time-varying 4th root problem (3.1). Note that the formulation of the DTZD model is presented in two situations sequently in this subsection, according the criterion whether the time derivative of coefficient $a(t)$ [i.e., $\dot{a}(t)$] is known or not.

Situation 1 DTZD model assuming $\dot{a}(t)$ known

To transform the CTZD model (3.3) to its discrete-time one for solving (3.1), we refer to the well-known Euler forward-difference rule [57, 118]:

$$\dot{x}(t = k\tau) \approx (x((k+1)\tau) - x(k\tau))/\tau.$$

Therefore, the DTZD model with $\dot{a}(t)$ known (i.e., the DTZDK model) for solving the time-varying 4th root problem (3.1) is derived from CTZD model (3.3) as

$$(x_{k+1} - x_k)/\tau = \frac{-\gamma\phi\left(x_k^4 - a_k\right) + \dot{a}_k}{4x_k^3},$$

which is further written as

$$x_{k+1} = x_k - \frac{h\phi\left(x_k^4 - a_k\right) - \tau\dot{a}_k}{4x_k^3}. \tag{3.5}$$

Note again that different choices of step size h and activation function $\phi(\cdot)$ lead to different convergence performance of (3.5).

Situation 2. DTZD model assuming $\dot{a}(t)$ unknown

Generally speaking, in real-world applications, the analytical form of $\dot{a}(t)$ may be difficult to know, derive, or estimate. In view of this point, we now investigate the DTZD model with $\dot{a}(t)$ unknown. So, as a simplest way, $\dot{a}(t)$ can be obtained by employing Euler backward-difference rule [57, 118]:

$$\dot{a}(t = k\tau) \approx (a(k\tau) - a((k-1)\tau))/\tau.$$

Therefore, we develop the DTZD model with $\dot{a}(t)$ unknown (i.e., the DTZDU model) from CTZD model (3.3) as follows:

$$(x_{k+1} - x_k)/\tau = \frac{-\gamma\phi\left(x_k^4 - a_k\right)}{4x_k^3} + \frac{a_k - a_{k-1}}{4\tau x_k^3},$$

which is further formulated as

$$x_{k+1} = x_k - \frac{h\phi\left(x_k^4 - a_k\right)}{4x_k^3} + \frac{a_k - a_{k-1}}{4x_k^3}. \tag{3.6}$$

Note again that, from the above well-known Euler backward-difference rule [57], we can not obtain $\dot{a}(t=0)$ since $t \in [0, +\infty)$ in this situation. Therefore, in the situation, we can choose $\dot{a}(t=0) = 0$ or any other value to start DTZDU model (3.6).

3.3   GD Model

For comparative purposes, the GD model is developed and exploited for online solution of the above-presented time-varying 4th root problem (3.1). To solve (3.1), by following the gradient-based design method, we first define a scalar-valued square-based energy function, such as $\mathscr{E} = (x^4(t) - a(t))^2/2$. Then, a typical continuous-time adaptation rule based on the negative-gradient information leads to the following differential equation, which is the linear-activation form of the GD model [99]:

$$\frac{\mathrm{d}x(t)}{\mathrm{d}t} = -\gamma \frac{\partial \mathscr{E}}{\partial x} = -4\gamma x^3(t)\left(x^4(t) - a(t)\right).$$

Under the inspiration of [99], from the above we obtain the nonlinear-activation (or to say, nonlinearly activated) form of a general GD model by imposing nonlinear activation function $\phi(\cdot)$ as follows:

$$\dot{x}(t) = -4\gamma x^3(t)\phi\left(x^4(t) - a(t)\right), \tag{3.7}$$

where $x(t) \in \mathbb{R}$, starting from randomly generated initial condition (or to say, initial state, initial value) $x(0) \in \mathbb{R}$, denotes the neural state of GD model (3.7).

Now, it is worth comparing the two design methods and results of CTZD model (3.3) and GD model (3.7) here, both of which are exploited for finding the time-varying 4th root of (3.1). The differences lie in the following facts.

1) The design of CTZD model (3.3) is based on the elimination of an indefinite error function $e(t) = x^4(t) - a(t) \in \mathbb{R}$, which can be negative and lower-unbounded. In contrast, the design of GD model (3.7) is based on the elimination of the square-based nonnegative energy function $\mathscr{E} = (x^4(t) - a(t))^2/2$.

2) CTZD model (3.3) is based on a new method intrinsically for time-varying problems solving. Thus, it can converge to the exact theoretical time-varying 4th root of (3.1). In contrast, GD (3.7) is based on a method intrinsically for constant problems solving. It is thus only able to approximately approach the theoretical time-varying 4th root of (3.1).

3) In a methodological and systematical manner, CTZD model (3.3) exploits the time-derivative information of $a(t)$ [i.e., $\dot{a}(t)$] during its real-time solving process. This is the main reason why CTZD model (3.3) can converge to the exact theoretical time-varying 4th root of (3.1). In contrast, GD model (3.7) does not exploit the time-derivative information of $a(t)$, and thus is neither effective nor precise enough for solving time-varying 4th root problem.

3.4   Illustrative Examples

In Section 3.2, the CTZD and DTZD models are discussed for online solution of the time-varying 4th root. In this section, computer-simulation and numerical-experiment results are provided for

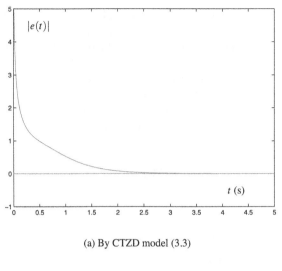

(a) By CTZD model (3.3)

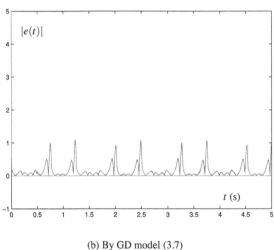

(b) By GD model (3.7)

FIGURE 3.1 Residual errors $|x^4(t) - 6\sin(20t)\cos(5t) - 7|$ synthesized by CTZD model (3.3) and GD (3.7) with $\gamma = 1$.

substantiating the efficacy of the aforementioned ZD models with suitable initial-state value $x(0)$, the power-sigmoid activation function (with design parameters $\xi = 4$ and $p = 3$), and different values of step size h.

Example 3.1 By CTZD model (3.3) and GD model (3.7)

For illustration and comparison, let us consider the following time-varying 4th root problem (written in the general form of a nonlinear equation):

$$x^4(t) - 6\sin(20t)\cos(5t) - 7 = 0. \tag{3.8}$$

With design parameter $\gamma = 1$, both CTZD model (3.3) and GD model (3.7) are exploited for finding online the time-varying 4th root of (3.8). Specifically, Figure 3.1 shows residual errors $|e(t)| = |x^4(t) - 6\sin(20)\cos(5t) - 7|$ during the solving processes of problem (3.8). It is seen from Figure 3.1(a) that, by applying CTZD model (3.3) to the time-varying 4th root finding of (3.8), residual error $|e(t)|$ diminishes to zero within around 3 s. In contrast, it follows from Figure 3.1(b) that, using GD model (3.7) for online solution of (3.8) under the same conditions (e.g., with design parameter $\gamma = 1$), residual error $|e(t)|$ is rather large all the time, which means that the solution of GD model (3.7) does not fit well with the theoretical solution of (3.8). Once more, in the authors' opinion, the main reason why CTZD model (3.3) has a better performance for the online time-varying 4th root finding [compared with GD model (3.7)] lies in the fact that the time-derivative information of the time-varying coefficient in (3.8) is fully utilized in CTZD model (3.3). Thus, CTZD model (3.3) is more effective, efficient, and precise for the time-varying 4th root finding, as compared with GD model (3.7).

Example 3.2   By CTZD model (3.3) and DTZDK model (3.5)

Now, let us consider the following time-varying 4th root problem (written in the general form of a nonlinear equation):

$$x^4(t) - 3\cos(5t) - 7 = 0. \tag{3.9}$$

First, CTZD model (3.3) is exploited for online solution of (3.9) with the simulative results shown in Figure 3.2 (a). As seen from it, starting from randomly generated initial states in $[-3,3]$, neural states $x(t)$ of CTZD model (3.3) fit well with the theoretical 4th root $x^*(t)$ after a short time, which illustrates the efficacy of CTZD model (3.3) for the time-varying 4th root problem solving.

Then, DTZDK model (3.5) is exploited for online solution of (3.9) with the numerical results shown in Figure 3.2(b). Note that the time derivative of $a(t)$ can be obtained via a simple differentiation operation. In other words, $\dot{a}(t)$ is known (that is the reason why we choose such a discrete-time ZD model). From Figure 3.2(b), we see that neural states x_k of DTZDK model (3.5), starting from appropriate initial states $x(0)$ with step size $h = 0.8$, converge well to the theoretical time-varying 4th root within a few updates (or to say similarly, iterations).

Example 3.3   Choice of step size h

In this example, the following higher frequency problem is considered for choosing the suitable step size h in DTZDK model (3.5):

$$x^4(t) - \sin(7t) - 5 = 0. \tag{3.10}$$

With different step-size values (i.e., $h = 0.6, 0.8, 0.9, 0.95$, and 0.96), the numerical results of the maximal steady-state residual errors synthesized by DTZDK model (3.5) for (3.10) are shown in Table 3.1. The results show that, when step size h is increased from 0.6 to 0.95, the maximal steady-state residual error is decreasing; after that, by increasing step size h to 0.96, the maximal steady-state residual error starts to increase instead of continuing to decrease. In other words, similar to the observation of Example 2.2 in Chapter 2, the best step size h is again 0.95 or so.

Example 3.4   By DTZDK model (3.5) and DTZDU model (3.6)

As mentioned above, by employing the backward-difference rule to estimate the value of $\dot{a}(t)$, DTZDU model (3.6) is developed and generalized for online solution of the time-varying 4th root problem. In this example, to investigate the effectiveness of such a model, let us consider the following time-varying 4th root problem:

$$x^4(t) - \sin(20t) - 5 = 0. \tag{3.11}$$

For comparison, both DTZDK model (3.5) and DTZDU model (3.6) are exploited for solving the time-varying 4th root problem (3.11). Note that, different from the DTZDU model (3.6), DTZDK

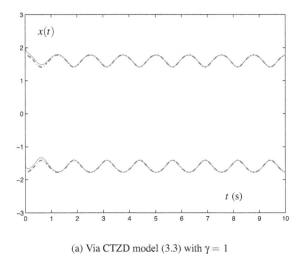

(a) Via CTZD model (3.3) with $\gamma = 1$

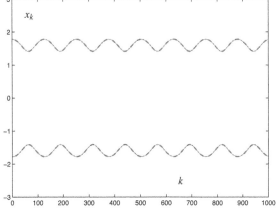

(b) Via DTZDK model (3.5) with $h = 0.8$

FIGURE 3.2 Finding the time-varying 4th root of (3.9) via CTZD model (3.3) and DTZDK model (3.5).

model (3.5) obtains $\dot{a}(t)$ by using the differentiation operation. As seen from Figure 3.3(a) and (b), with $\tau = 0.01$, the maximal steady-state residual errors of DTZDK model (3.5) and DTZDU model (3.6) are both quite small. This shows the effectiveness of DTZDU model (3.6) as compared with DTZDK model (3.5). Moreover, from Figure 3.3(b), (c), and (d), we see that the maximal steady-state residual error of DTZDU model (3.6) is reducing as τ decreases. More specifically, when τ is set as 0.0001, the maximal steady-state residual error is 2.027×10^{-6}. In summary, these sufficiently illustrate the effectiveness of DTZDU model (3.6).

Example 3.5 With time-derivative $\dot{a}(t)$ assumed unknown

For further investigation of DTZDU model (3.6), let us consider the following time-varying 4th root finding problem with $\dot{a}(t)$ unknown (or difficult to obtain):

$$x^4(t) - \sin(\cos(\sin(\cos(t)))) - 5 = 0. \tag{3.12}$$

TABLE 3.1   MSSREs of $|e_k|$ of DTZDK Model (3.5) Solving (3.10) with Different h Values

	h	τ	**Maximal steady-state residual error** (10^{-3})
	0.6	0.01	1.968
	0.8	0.01	1.477
$x^4(t) - \sin(7t) - 5$	0.9	0.01	1.312
	0.95	0.01	1.243 (best)
	0.96	0.01	1.486

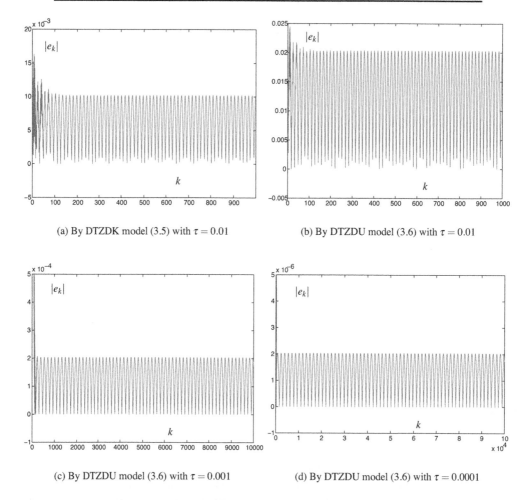

(a) By DTZDK model (3.5) with $\tau = 0.01$ (b) By DTZDU model (3.6) with $\tau = 0.01$

(c) By DTZDU model (3.6) with $\tau = 0.001$ (d) By DTZDU model (3.6) with $\tau = 0.0001$

FIGURE 3.3 Residual errors $|x_k^4 - \sin(20t_k) - 5|$ synthesized by CTZDK model (3.5) and DTZDU model (3.6) with different τ, where subfigure ranges are $10^3 \cdot [0,1] \times [-5,20] \cdot 10^{-3}$, $10^3 \cdot [0,1] \times [-5,25] \cdot 10^{-3}$, $10^4 \cdot [0,1] \times [-1,5] \cdot 10^{-4}$, and $10^5 \cdot [0,1] \times [-1,5] \cdot 10^{-6}$.

Evidently, DTZDK model (3.5) may not be proper for such a time-varying 4th root finding, since $\dot{a}(t)$ is difficult to obtain. Thus, it is preferable to exploit DTZTU model (3.6) for online solution of (3.12). As seen from Figure 3.4, the maximal steady-state residual errors are quite small, which substantiates the efficacy and precision of DTZDU model (3.6). In addition, Figure 3.4 illustrates again that the maximal steady-state residual error decreases in an $O(\tau^2)$ manner, as τ decreases.

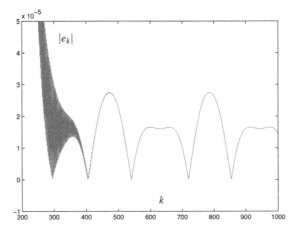

(a) By DTZDU model (3.6) with $\tau = 0.01$

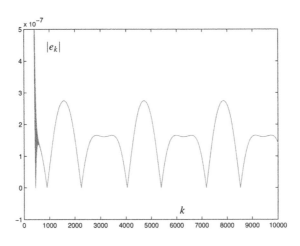

(b) By DTZDU model (3.6) with $\tau = 0.001$

FIGURE 3.4 Residual errors $|x_k^4 - \sin(\cos(\sin(\cos(t_k)))) - 5|$ synthesized by DTZDU model (3.6) with $h = 0.95$ and different τ, where subfigure ranges are $10^2 \cdot [2, 10] \times [-1, 5] \cdot 10^{-5}$ and $10^4 \cdot [0, 1] \times [-1, 5] \cdot 10^{-7}$.

3.5 Summary

In this chapter, a special class of neural dynamics termed zeroing dynamics (ZD) has been developed and investigated for the time-varying 4th root finding in real time. Such a ZD model has been elegantly introduced by defining an indefinite error function, instead of the nonnegative (or at least low-bounded) energy function usually associated with GD. In addition, we have generated the DTZD models for the online time-varying 4th root finding. Computer-simulation and numerical-experiment results have further illustrated the efficacy and precision of the CTZD model and its discrete-time models [i.e., DTZDK model (3.5) and DTZDU model (3.6)] for the time-varying 4th root finding.

Chapter 4

Time-Varying 5th Root Finding

Abstract

In this chapter, zeroing dynamics (ZD) is generalized to find the time-varying 5th root in real time or online. In addition, in order to get superior convergence and better precision of the ZD models, the power-sigmoid activation function is exploited. To solve the special case of the time-varying 5th root finding, i.e., the constant 5th root finding, we develop the simplified continuous-time ZD (S-CTZD) model and its discrete-time model (i.e., the S-DTZD model). Illustrative examples substantiate the efficacy of the ZD models for the time-varying and constant 5th roots finding.

4.1 Introduction

The solution of the time-varying 5th root in the form of $x^5(t) - a(t) = 0$ is considered as a fundamental and important problem arising in fields of computational mathematics and physics. For example, in [63], Panario and Thomson presented a method for computing the ρth root using a polynomial basis over finite fields of odd characteristic ρ (with $\rho \geqslant 5$), by taking advantage of a binomial reduction polynomial. As far as the authors know, there is a very little literature on this specific topic of the time-varying 5th root finding. Almost all of the reported methods are just for the constant 5th root finding, and may not be effective, precise, and efficient enough for the time-varying 5th root finding. In this chapter, based on the ZD design method, the ZD models are proposed, developed, and investigated for online solution of the time-varying 5th root. In order to achieve superior convergence and better precision, different activation functions are exploited in the proposed ZD models. In addition, the theoretical analysis and practical verification of the convergence properties of the proposed ZD models are presented. Note that, during the design procedure of ZD, the time-derivative information of the time-varying coefficient involved in the time-varying problem is fully utilized, and thus the resultant ZD model methodologically eliminates the lagging error (which is generated usually by the conventional algorithms, such as GD-type algorithms). More importantly, by making full use of the time-derivative information of the time-varying coefficient, the ZD model can achieve excellent convergence performance and solve the time-varying 5th root problem effectively, precisely, and efficiently.

4.2 ZD Models for Time-Varying Case

For further discussion, the ZD models are presented for the time-varying 5th root finding in this section. Now, let us consider the following time-varying 5th root problem (which, without loss of generality, is written in the form of a time-varying nonlinear equation):

$$x^5(t) - a(t) = 0, \ t \in [0, +\infty), \tag{4.1}$$

where $a(t) \in \mathbb{R}$ denotes a smoothly time-varying scalar, with its time derivative $\dot{a}(t) \in \mathbb{R}$ defined theoretically. They are assumed known or at least measurable in practice. Our object is to find $x(t) \in \mathbb{R}$ in real time t, such that the above time-varying nonlinear Equation (4.1) holds true at any time instant $t \in [0, +\infty)$. For presentation convenience, let $x^*(t) \in \mathbb{R}$ denotes the theoretical time-varying 5th root of $a(t)$.

4.2.1 CTZD Model

First, we define an indefinite error function by following Zhang *et al.*'s neural dynamic design method to monitor and control the time-varying 5th root finding process [102, 104, 117, 118, 122, 129, 134]:

$$e(t) = x^5(t) - a(t) \in \mathbb{R}.$$

Secondly, the time-derivative $\dot{e}(t)$ of error function $e(t)$ can be chosen and forced, such that $e(t)$ converges to zero preferably in an exponential manner. Specifically, we choose $\dot{e}(t)$ via the following ZD design formula [102, 104, 117, 118, 122, 129, 134]:

$$\frac{de(t)}{dt} = -\gamma\phi\big(e(t)\big), \tag{4.2}$$

where design parameter $\gamma > 0$, being the reciprocal of a capacitance parameter in the hardware implementation, should be set as large as the hardware would permit (e.g., in analog circuits or very large scale integration (VLSI)) or selected appropriately for simulative or experimental purposes. In addition, $\phi(\cdot) : \mathbb{R} \to \mathbb{R}$ denotes the general type of monotonically increasing odd activation function. In this chapter, we use again the linear activation function and the power-sigmoid activation function [please see the previous chapters (e.g., Subsections 2.3.2 and 3.2.2) for details if interested or necessary]. Note that different choices of design parameter γ and activation function $\phi(\cdot)$ affect the performance of the ZD models very much.

Third, expanding the above ZD design formula (4.2), we thus obtain

$$5x^4(t)\dot{x}(t) - \dot{a}(t) = -\gamma\phi\big(x^5(t) - a(t)\big),$$

which leads to the following CTZD model for the time-varying 5th root finding:

$$5x^4(t)\dot{x}(t) = \dot{a}(t) - \gamma\phi\big(x^5(t) - a(t)\big),$$

or, if $x(t) \neq 0$,

$$\dot{x}(t) = \frac{\dot{a}(t) - \gamma\phi\left(x^5(t) - a(t)\right)}{5x^4(t)}, \tag{4.3}$$

where $x(t)$, starting from a randomly generated nonzero initial condition $x(0) \in \mathbb{R}$, is the neural state corresponding to the theoretical 5th root $x^*(t)$ of (4.1). For the convergence properties of CTZD model (4.3), we have the following theorem.

Theorem 4 *Consider a smoothly time-varying nonzero scalar $a(t) \in \mathbb{R}$ in (4.1). If a monotonically increasing odd activation function $\phi(\cdot)$ is used, then neural state $x(t) \in \mathbb{R}$ of CTZD model (4.3), starting from randomly generated nonzero initial state $x(0) \in \mathbb{R}$, can converge to the theoretical time-varying 5th root $x^*(t)$ of $a(t)$.*

Proof Defining a Lyapunov function candidate $v(x(t),t) = (x^5(t) - a(t))^2/2 \geqslant 0$ for CTZD model (4.3), we derive its time derivative

$$\begin{aligned}
\dot{v}(x(t),t) &= \frac{\mathrm{d}v(x(t),t)}{\mathrm{d}t} \\
&= \left(5x^4(t)\dot{x}(t) - \dot{a}(t)\right)\left(x^5(t) - a(t)\right) \\
&= -\gamma\left(x^5(t) - a(t)\right)\phi\left(x^5(t) - a(t)\right).
\end{aligned}$$

As a monotonically increasing odd activation function is used, we have $\phi(-x^5(t) + a(t)) = -\phi(x^5(t) - a(t))$ and

$$\phi\left(x^5(t) - a(t)\right) \begin{cases} > 0, & \text{if } x^5(t) - a(t) > 0, \\ = 0, & \text{if } x^5(t) - a(t) = 0, \\ < 0, & \text{if } x^5(t) - a(t) < 0. \end{cases}$$

Therefore, the following results of equality and inequality hold true:

$$\left(x^5(t) - a(t)\right)\phi\left(x^5(t) - a(t)\right) \begin{cases} > 0, & \text{if } x^5(t) - a(t) \neq 0, \\ = 0, & \text{if } x^5(t) - a(t) = 0, \end{cases}$$

which guarantee the final negative-definiteness of $\dot{v}(x(t),t)$ with respect to error $e(t) = x^5(t) - a(t)$. It means that $\dot{v}(x(t),t) < 0$ for $x(t) \neq x^*(t)$, and $\dot{v}(x(t),t) = 0$ only for $x(t) = x^*(t)$. By Lyapunov theory, starting from randomly generated initial state $x(0) \in \mathbb{R}$, neural state $x(t)$ of CTZD model (4.3) can converge to the theoretical time-varying 5th root $x^*(t)$ of $a(t)$. To avoid the division-by-zero situation in computer simulation or hardware implementation of CTZD model (4.3), the nonzero $a(t)$ and $x(0)$ have to be required. The proof is thus complete. ∎

For comparative purposes, we also develop and exploit the GD model to solve the above problem (4.1). First, we define a scalar-valued square-based energy function $\varepsilon = (x^5(t) - a(t))^2/2$. Then, the following differential equation based on the negative-gradient information is obtained:

$$\frac{\mathrm{d}x(t)}{\mathrm{d}t} = -\gamma\frac{\partial\varepsilon}{\partial x} = -5\gamma x^4(t)\left(x^5(t) - a(t)\right).$$

From the above differential equation, we readily generalize the following nonlinear-activation form of the general GD model by using nonlinear activation function $\phi(\cdot)$:

$$\dot{x}(t) = -5\gamma x^4(t)\phi\left(x^5(t) - a(t)\right), \tag{4.4}$$

where $x(t) \in \mathbb{R}$, starting from randomly generated initial state $x(0) \in \mathbb{R}$, is the neural state corresponding to the theoretical time-varying 5th root $x^*(t)$ of (4.1).

4.2.2 DTZD Models

For potential hardware implementation on digital circuits of CTZD model (4.3), the discrete-time ZD (DTZD) models are developed and investigated for solving the time-varying 5th root problem (4.1) in this subsection.

Situation 1   DTZD model assuming $\dot{a}(t)$ known

To solve (4.1), we first refer to Euler forward-difference rule:

$$\dot{x}(t = k\tau) \approx \big(x((k+1)\tau) - x(k\tau)\big)/\tau.$$

Then, the DTZD model with $\dot{a}(t)$ known (i.e., the DTZDK model) for the time-varying $5th$ root finding is derived from CTZD model (4.3) as follows:

$$\frac{x_{k+1} - x_k}{\tau} = \frac{\dot{a}_k - \gamma\phi(x_k^5 - a_k)}{5x_k^4},$$

which is further written as

$$x_{k+1} = x_k - \frac{h\phi(x_k^5 - a_k) - \tau\dot{a}_k}{5x_k^4}. \tag{4.5}$$

Note that, when the linear activation function is exploited, DTZDK model (4.5) reduces to the following one:

$$x_{k+1} = x_k - \frac{h(x_k^5 - a_k) - \tau\dot{a}_k}{5x_k^4}. \tag{4.6}$$

Situation 2   DTZD model assuming $\dot{a}(t)$ unknown

In real-world applications, the analytical form of $\dot{a}(t)$ may be difficult to know, derive, or estimate. Thus, we investigate the discrete-time form of CTZD model (4.3) with $\dot{a}(t)$ unknown (i.e., the DTZDU model). So, as a simplest way, $\dot{a}(t)$ is approximated via Euler backward-difference rule, i.e.,

$$\dot{a}(t = k\tau) \approx \big(a(k\tau) - a((k-1)\tau)\big)/\tau.$$

Then, we similarly derive the following DTZDU model:

$$\frac{x_{k+1} - x_k}{\tau} = \frac{-\gamma\phi(x_k^5 - a_k)}{5x_k^4} + \frac{a_k - a_{k-1}}{5\tau x_k^4},$$

which is rewritten equivalently as

$$x_{k+1} = x_k - \frac{h\phi(x_k^5 - a_k)}{5x_k^4} + \frac{a_k - a_{k-1}}{5x_k^4}. \tag{4.7}$$

Similar to Situation 1, we obtain the following linearly-activated DTZDU model by exploiting the linear activation function in Equation (4.7):

$$x_{k+1} = x_k - \frac{h(x_k^5 - a_k)}{5x_k^4} + \frac{a_k - a_{k-1}}{5x_k^4}. \tag{4.8}$$

4.3 Simplified ZD Models for Constant Case and Newton Iteration

In this section, the simplified CTZD model and its discrete-time model are generalized for the constant 5th root finding.

4.3.1 S-CTZD Model

As a special case of the time-varying 5th root problem depicted in (4.1), the following constant 5th root problem is considered:

$$x^5(t) - a = 0, \tag{4.9}$$

where $a \in \mathbb{R}$ is a nonzero scalar constant. Since $\dot{a} = 0$ here, CTZD model (4.3) reduces to the following S-CTZD model for the constant 5th root finding:

$$\dot{x}(t) = \frac{-\gamma\phi\left(x^5(t) - a\right)}{5x^4(t)}. \tag{4.10}$$

For S-CTZD model (4.10), we have the following corollary on its convergence.

Corollary 2 *Consider a nonzero constant scalar $a \in \mathbb{R}$ depicted in (4.9). If a monotonically increasing odd activation function is used, then neural state $x(t) \in \mathbb{R}$ of S-CTZD (4.10), starting from randomly generated nonzero initial state $x(0) \in \mathbb{R}$, can converge to the theoretical 5th root x^* (or to say, $a^{1/5}$) of a.*

4.3.2 S-DTZD Model and Newton Iteration

Motivated by potential hardware implementation of S-CTZD model (4.10) exploited in the constant 5th root finding, we discretize S-CTZD model (4.10) by using Euler forward-difference rule. Then, the following S-DTZD model for the constant 5th root finding (4.9) is derived from S-CTZD model (4.10):

$$\frac{x_{k+1} - x_k}{\tau} = \frac{-\gamma\phi(x_k^5 - a)}{5x_k^4}, \tag{4.11}$$

which is further written as

$$x_{k+1} = x_k - \frac{h\phi(x_k^5 - a)}{5x_k^4}. \tag{4.12}$$

Look at S-DTZD model (4.12): If we utilize the linear activation function and set $h = 1$, S-DTZD model (4.12) reduces to

$$x_{k+1} = \frac{4x_k + a/x_k^4}{5}, \tag{4.13}$$

which is exactly Newton iteration for the constant 5th root finding. In other words, systematically, we discover a general form of Newton iteration for the constant 5th root finding by discretizing S-CTZD model (4.10). Specifically, Newton iteration is derived from CTZD model (4.3) by focusing on the constant 5th root problem (4.9) solving, discretizing S-CTZD model (4.10) via Euler forward difference, utilizing the linear activation function, and fixing the step size to be 1. This is regarded as a new effective explanation to Newton iteration for the constant 5th root finding.

For further research and for comparison, we generalize and develop Newton iteration to solve the time-varying 5th root problem (4.1) as

$$x_{k+1} = \frac{4x_k + a_k/x_k^4}{5}. \tag{4.14}$$

4.4   Illustrative Examples

The previous two sections have presented CTZD model (4.3), DTZD models (4.5)-(4.7), S-CTZD model (4.10), and S-DTZD model (4.12) for the online solution of the time-varying and constant scalar-valued 5th roots. In this section, illustrative examples, computer-simulation, and numerical-experiment results are provided for substantiating the efficacy of the ZD models with the aid of the suitable initial state $x(0)$, the linear activation function, and the power-sigmoid activation function with design parameters $\xi = 4$ and $p = 3$.

Example 4.1   By CTZD model (4.3) and GD model (4.4)

Now, let us consider the following time-varying 5th root problem:

$$x^5(t) - 2\sin(20t) - 1 = 0. \tag{4.15}$$

With design parameter $\gamma = 1$, CTZD model (4.3) is exploited for the real-time solution of (4.15). Specifically, as seen from Figure 4.1(a), starting from a randomly generated initial state in $[0,4]$, residual error $|e(t)| = |x^5(t) - \sin(20t) - 1|$ of CTZD model (4.3) converges to zero within a short time (i.e., 3 s or so). This means correspondingly that, after such a short time, neural state $x(t)$ of CTZD model (4.3) fits well with the theoretical 5th root $x^*(t)$. For comparative purposes, we exploit GD model (4.4) under the same conditions to solve the above same problem (4.15), and the simulative result is shown in Figure 4.1(b). The observation is that, using GD model (4.4), residual error $|e(t)|$ is oscillating and much larger than that of CTZD model (4.3). In other words, the solution of GD model (4.4) does not converge well to the time-varying theoretical solution of (4.15).

Example 4.2   Choice of step size h of DTZDK model (4.5)

In this example, we consider the following time-varying 5th root problem for choosing the suitable values of step size h in DTZDK model (4.5):

$$x^5(t) - 2\sin(2t) - 4 = 0. \tag{4.16}$$

The numerical results synthesized by DTZDK model (4.5) with different step-size values are shown in Table 4.1. As seen from the results, when step size h increases to 0.97 (or larger), the maximal steady-state residual error diverges. Thus, from the table, we conclude that, by increasing the value of step size h properly, the convergence performance of DTZDK model (4.5) is enhanced, and thus we choose $h \in [0.5, 0.96]$ for good convergence performance.

Example 4.3   By DTZDK model (4.5)

In this example, $\tau = 0.01$ and $h = 0.96$, we use DTZDK model (4.5) to solve the following two problems with different frequencies investigated and compared:

$$x^5(t) - 2\sin(2t) - 4 = 0, \tag{4.17}$$

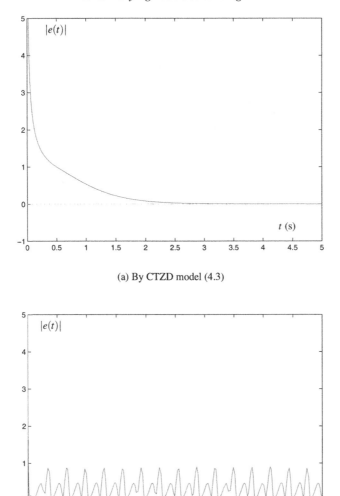

(a) By CTZD model (4.3)

(b) By GD model (4.4)

FIGURE 4.1 Residual errors $|e(t)| = |x^5(t) - \sin(20t) - 1|$ synthesized by CTZD model (4.3) and GD model (4.4) with $\gamma = 1$.

$$x^5(t) - 2\sin(20t) - 4 = 0. \tag{4.18}$$

Specifically, from Figure 4.2(a) and (b), we find that, when DTZDK model (4.5) is exploited to solve the low frequency problem (4.17), it has a better precision than that of online solution of the high frequency problem (4.18). In other words, it is much easier for DTZDK model (4.5) to obtain the more precise results in solving low-frequency time-varying problems rather than solving high-frequency time-varying problems. However, it is worth noting here that the precision of solving high-frequency time-varying problems can be increased effectively by decreasing the value of sampling gap τ, which is traditionally in an $O(\tau^2)$ manner.

TABLE 4.1   MSSREs of $|e_k|$ of DTZDK Model (4.5) Solving (4.16) with Different h Values

	h	τ	Maximal steady-state residual error (10^{-4})
	0.10	0.01	9.609
	0.30	0.01	3.212
	0.50	0.01	1.927
$x^5(t) - \sin(2t) - 4$	0.70	0.01	1.377
	0.90	0.01	1.070
	0.96	0.01	1.005 (best)
	0.97	0.01	divergence

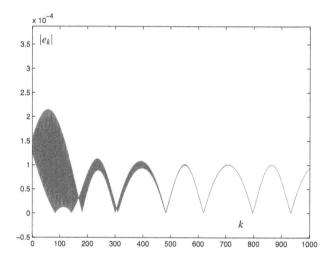

(a) Low frequency problem (4.17) solving with $k \in [0, 1000]$

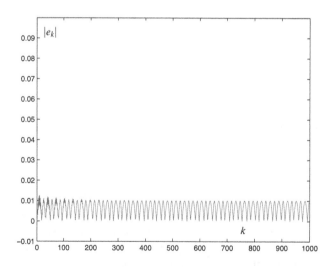

(b) High frequency problem (4.18) solving with $k \in [0, 1000]$

FIGURE 4.2 Solving the time-varying 5th root problems of different frequencies by DTZDK model (4.5), with labeled ordinate ranges of subfigures $[-0.5, 3.5] \cdot 10^{-4}$ and $[-1, 9] \cdot 10^{-2}$.

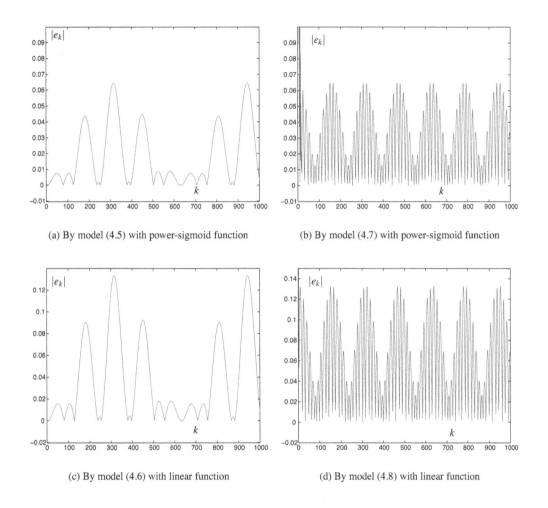

(a) By model (4.5) with power-sigmoid function

(b) By model (4.7) with power-sigmoid function

(c) By model (4.6) with linear function

(d) By model (4.8) with linear function

FIGURE 4.3 Finding the time-varying 5th root via DTZDK and DTZDU models, where labeled ranges of subfigures are $10^3 \cdot [0,1] \times [-1,9] \cdot 10^{-2}$, $10^3 \cdot [0,1] \times [-1,9] \cdot 10^{-2}$, $10^3 \cdot [0,1] \times [-2,12] \cdot 10^{-2}$, and $10^3 \cdot [0,1] \times [-2,14] \cdot 10^{-2}$.

Example 4.4 By DTZDK and DTZDU models

In this example, we use the following 5th root problem to investigate the effectiveness of DTZDK model (4.5) and DTZDU model (4.7):

$$x^5(t) - 3\sin(20t)\cos(2t) - 6 = 0. \tag{4.19}$$

For fair comparison, the two models use the same initial conditions and the same design parameters (e.g., $h = 0.9$ and $\tau = 0.01$). As seen from Figure 4.3(a) and (b), when the power-sigmoid activation function is used, the maximal steady-state residual errors of DTZDK model (4.5) and DTZDU model (4.7) are comparable, i.e., 6.45×10^{-2} and 6.46×10^{-2}, respectively. Note that, when the value of τ decreases, the maximal steady-state residual errors of the two models both become smaller. For example, when $\tau = 0.001$, the maximal steady-state residual errors of DTZDK model (4.5) and DTZDU model (4.7) are 6.426×10^{-4} and 6.462×10^{-4}, respectively. The numer-

TABLE 4.2   MSSREs of $|e_k|$ of DTZDU Model (4.7) Solving (4.20) with Different τ Values

	h	τ	Maximal steady-state residual error
	0.95	0.01	2.065×10^{-2}
$x^5(t) - \sin(3t)\cos\left(\sin(20t)\right) - 5$	0.95	0.001	2.051×10^{-4}
	0.95	0.0001	2.060×10^{-6}
	0.95	0.00001	2.062×10^{-8}

ical results of using the linear activation function are shown in Figure 4.3(c) and (d), where the maximal steady-state residual errors are slightly larger than those using the power-sigmoid activation function. The other observations and conclusions are similar to those using the power-sigmoid activation function. Simply put, both DTZDK and DTZDU models solve the time-varying 5th root problem effectively, and the solution precision can be improved greatly if τ decreases.

Example 4.5   Choice of τ of DTZDU model (4.7)

As presented, by using Euler backward-difference rule to estimate $\dot{a}(t)$, we develops DTZDU model (4.7). In Example 4.4, we have verified the effectiveness of DTZDU model (4.7). From the theory as well as the numerical-experiment results, we notice that the choice of τ plays an important role in determining the performance of DTZDU model (4.7). So, let us consider the following problem so as to investigate the role of τ more evidently:

$$x^5(t) - \sin(3t)\cos\left(\sin(20t)\right) - 5 = 0. \tag{4.20}$$

The corresponding numerical-experiment results are shown in Table 4.2. From the table, we discover that, when the value of τ decreases, the maximal steady-state residual error of DTZDU model (4.7) becomes much smaller [i.e., in an $O(\tau^2)$ manner]; and that the value of τ should be set appropriately small so as to satisfy the precision we need in practice as well as the real-time computation requirement.

Example 4.6   By DTZDU model (4.7) and Newton iteration (4.14)

Now and finally, let us consider the following time-varying 5th root problem:

$$x^5(t) - \sin\left(\cos(3t)\right)\cos\left(\cos(20t)\right) - 7 = 0. \tag{4.21}$$

In Examples 4.4 and 4.5, the effectiveness of DTZDU model (4.7) has been already verified. In this example, we further verify the superiority of DTZDU model (4.7) to Newton iteration (4.14) in terms of the time-varying 5th root finding. We set $\tau = 0.01$ and $h = 0.95$, and use the power-sigmoid activation function in DTZDU model (4.7). The numerical results are shown in the Figure 4.4. As seen from the figure, the maximal steady-state residual error of DTZDU model (4.7) is 1.738×10^{-2}; whereas that of Newton iteration (4.14) is 7.832×10^{-2}, more than 4.5 times larger than the former. These mean that DTZDU model (4.7) has very much superiority to Newton iteration (4.14) in terms of solving the time-varying 5th root problem.

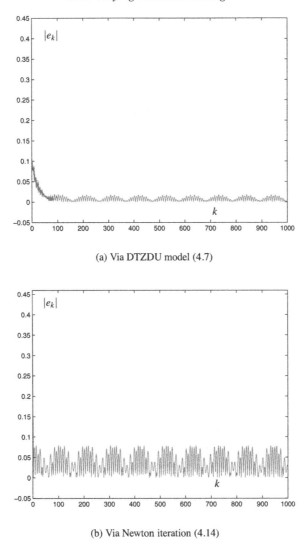

(a) Via DTZDU model (4.7)

(b) Via Newton iteration (4.14)

FIGURE 4.4 Finding the time-varying 5th root of (4.21) via DTZDU model (4.7) and Newton iteration (4.14).

4.5 Summary

In this chapter, by defining an indefinite error function, a class of neural dynamics (named zeroing dynamics, ZD) has been developed and investigated for the time-varying 5th root finding. For potential hardware implementation (based on digital circuits or digital computers) of CTZD model (4.3), we have developed and investigated the DTZD models. In addition, we have generated the simplified ZD models for the constant 5th root finding, and have discovered systematically that the S-DTZD model contains Newton iteration as a special case by using the linear activation function and setting the step size $h = 1$. Computer-simulation and numerical-experiment results have further substantiated the efficacy of the aforementioned ZD models for the time-varying and constant scalar-valued 5th roots finding.

Appendix: Extension to Time-Varying ρth Root Finding

Before ending this chapter, here and now, the design procedure of the ZD method is extended to the time-varying ρth root finding. Without loss of generality, the time-varying ρth root problem is generally described in the following form:

$$x^\rho(t) - a(t) = 0, \ t \in [0, +\infty), \tag{4.22}$$

where $\rho > 1 \in \mathbb{R}$ denotes a positive integer; and $a(t) \neq 0$ is a smoothly time-varying scalar, which, together with its time derivative $\dot{a}(t)$, is assumed to be known or at least measurable in practice. Our first objective is to find $x(t)$ in real time t such that the time-varying ρth root problem (4.22) holds true at any time instant $t \in [0, +\infty)$.

Therefore, the continuous-time ZD model for the time-varying ρth root finding is established via the following steps of the ZD design method.

Step 1. By the ZD design method [99, 102, 104, 117, 118, 122, 129, 134], we first define an indefinite error function, which is denoted by $e(t)$ and of which the time derivative is denoted by $\dot{e}(t)$. For the general root-finding problem discussed in this appendix, we define the following error function:

$$e(t) = x^\rho(t) - a(t). \tag{4.23}$$

Step 2. In order to force $e(t)$ to converge to zero, its time derivative $\dot{e}(t)$ is chosen via the following ZD design formula [99, 102, 104, 117, 118, 122, 129, 134]:

$$\dot{e}(t) = \frac{\mathrm{d}e(t)}{\mathrm{d}t} = -\gamma\phi\left(e(t)\right), \tag{4.24}$$

where design parameter $\gamma > 0 \in \mathbb{R}$ corresponds to the reciprocal of a capacitance parameter, and should be set as large as the hardware would permit or selected appropriately for simulative or experimental purposes. Besides, $\phi(\cdot)$ denotes a general monotonically increasing odd activation function, which can be linear or nonlinear. In addition, different choices of γ affect the performance of the resultant ZD model.

Step 3. By applying and expanding ZD design formula (4.24), the dynamic equation of the CTZD model for the time-varying ρth root finding is established finally:

$$\dot{e}(t) = \frac{\mathrm{d}e(t)}{\mathrm{d}t} = \rho x^{\rho-1}(t)\dot{x}(t) - \dot{a}(t) = -\gamma\phi\left(x^\rho(t) - a(t)\right),$$

which leads to the following CTZD model:

$$\rho x^{\rho-1}(t)\dot{x}(t) = \dot{a}(t) - \gamma\phi\left(x^\rho(t) - a(t)\right),$$

or, if $x(t) \neq 0$,

$$\dot{x}(t) = \frac{\dot{a}(t) - \gamma\phi\left(x^\rho(t) - a(t)\right)}{\rho x^{\rho-1}(t)}, \tag{4.25}$$

where $x(t) \in \mathbb{R}$, starting from a randomly generated nonzero initial condition $x(0) \in \mathbb{R}$, denotes the neural state corresponding to the theoretical solution $x^*(t)$ of (4.22).

Moreover, as our second objective, in order to transform CTZD model (4.25) to its corresponding discrete-time models for solving (4.22) online (i.e., with update index $k = 0, 1, 2, 3, \cdots$, and with sampling gap $\tau > 0$ being small enough), we refer to the following Euler forward-difference rule:

$$\dot{x}(t = k\tau) \approx (x((k+1)\tau) - x(k\tau))/\tau.$$

Therefore, the DTZD model with $\dot{a}(t)$ known (i.e., the DTZDK model) for the time-varying ρth root finding is derived from CTZD model (4.25) as

$$(x_{k+1} - x_k)/\tau = \frac{-\gamma\phi\left(x_k^\rho - a_k\right) + \dot{a}_k}{\rho x_k^{\rho-1}},$$

from which, with step size $h = \gamma\tau > 0$ selected appropriately, we further obtain

$$x_{k+1} = x_k - \frac{h\phi\left(x_k^\rho - a_k\right) - \tau\dot{a}_k}{\rho x_k^{\rho-1}}. \tag{4.26}$$

Similarly, we obtain the following DTZD model with $\dot{a}(t)$ unknown (i.e., the DTZDU model) for the time-varying ρth root finding by approximating $\dot{a}(t)$ using Euler backward difference rule:

$$x_{k+1} = x_k - \frac{h\phi(x_k^\rho - a_k)}{\rho x_k^{\rho-1}} + \frac{a_k - a_{k-1}}{\rho x_k^{\rho-1}}. \tag{4.27}$$

Being a topic of exercise, the corresponding verification (via computer simulations and numerical experiments) of the above ZD models for the time-varying ρth root finding are left for interested readers.

Part II

Nonlinear Equation Solving

Chapter 5

Time-Varying Nonlinear Equation Solving

Abstract

In this chapter, we generalize the zeroing dynamics model for solving online the general time-varying nonlinear equation in the form of $f(x(t),t) = 0 \in \mathbb{R}$. For comparative purposes, the gradient dynamics model is also employed for such a time-varying nonlinear equation solving. For achieving the superior convergence, different types of activation functions (linear, sigmoid, power functions, and power-sigmoid function) are investigated in this chapter. Computer-simulation results substantiate the theoretical analysis and efficacy of the ZD model for solving online the time-varying nonlinear equation.

5.1 Introduction

The solution of nonlinear equations in the form of $f(x) = 0$ is widely encountered in science and engineering fields. Many numerical algorithms have thus been presented and investigated [16,72,83]. In addition, the neural-dynamic approach is one of the important parallel-processing methods for solving equation and optimization problems [29]. For example, many studies have been reported on the real-time solution of algebraic equations, including matrix inversion and Sylvester equation [99–102, 104, 107, 114, 117, 118, 122, 128, 129, 134]. Generally speaking, most of the schemes as reported are related to the gradient-descent method or other methods intrinsically designed for constant problems solving. In other words, the coefficients of nonlinear equation $f(x) = 0$ are not the functions of time argument t (or simply put, not time-variant).

In this chapter, we develop and generalize Zhang *et al.*'s neural-dynamic design method (i.e., zeroing dynamics design method) to solve the time-varying nonlinear equation in the form of $f(x(t),t) = 0$ [133]. The resultant ZD model is elegantly introduced by defining an indefinite error function so as to make it decrease to zero exponentially. For comparative purposes, the conventional GD model is also investigated. In addition, different types of activation functions are investigated.

5.2 Problem Formulation and Solution Models

Our objective of this chapter is to find $x(t) \in \mathbb{R}$ in real time t, such that the following smoothly time-varying nonlinear equation holds true:

$$f(x(t),t) = 0 \in \mathbb{R}. \tag{5.1}$$

The existence of theoretical time-varying solution $x^*(t)$ at any time instant $t \in [0,+\infty)$ is assumed as a basis of discussion. In addition, some basic types of activation functions (such as linear, sigmoid, power, and power-sigmoid activation functions) are analyzed and discussed for the convergence of the ZD model.

5.2.1 ZD Model

Following Zhang *et al.*'s neural-dynamic design method, we first construct the following indefinite error function $e(t)$ so as to set up a ZD model to solve online the time-varying nonlinear Equation (5.1):

$$e(t) = f(x(t),t).$$

Secondly, the time-derivative of $e(t)$, i.e., $\dot{e}(t)$, can be chosen and forced mathematically, such that error function $e(t)$ converges to zero asymptotically and preferably exponentially. Specifically, we use the following ZD design formula [99–102, 104, 107, 114, 117, 118, 122, 128, 129, 134]:

$$\frac{de(t)}{dt} = -\gamma\phi\big(e(t)\big),$$

or equivalently, we have

$$\frac{df}{dt} = -\gamma\phi\Big(f\big(x(t),t\big)\Big), \tag{5.2}$$

where design parameter γ and activation function $\phi(\cdot)$ are defined the same as before.

Thirdly, expanding the above ZD design formula (5.2), we thus have

$$\frac{\partial f}{\partial x}\frac{dx}{dt} + \frac{\partial f}{\partial t} = -\gamma\phi\Big(f\big(x(t),t\big)\Big),$$

which leads to the following differential equation of the so-called ZD model:

$$\frac{\partial f}{\partial x}\dot{x}(t) = -\gamma\phi\Big(f\big(x(t),t\big)\Big) - \frac{\partial f}{\partial t} \quad \text{or} \quad \dot{x}(t) = -\Big(\gamma\phi\big(f(x(t),t)\big) + \frac{\partial f}{\partial t}\Big)\Big/\frac{\partial f}{\partial x}, \tag{5.3}$$

where $x(t)$, starting from a randomly generated initial condition $x(0) \in \mathbb{R}$, denotes the neural state corresponding to theoretical time-varying solution $x^*(t)$ of (5.1).

5.2.2 GD Model

As mentioned, the gradient-descent method is a conventional approach frequently used, e.g., to handle the constant situation of (5.1). According to the GD design approach, to solve time-varying nonlinear Equation (5.1), the gradient-descent method requires us to define a square-based energy function, such as $\mathscr{E}(t) = f^2(x(t),t)$. Then, a typical continuous-time adaptation rule based on the negative-gradient information leads to the following differential equation [which we term as a linearly-activated form of the GD model]:

$$\dot{x}(t) = \frac{dx}{dt} = -\frac{\gamma}{2}\frac{\partial\mathscr{E}}{\partial x} = -\gamma f\big(x(t),t\big)\frac{\partial f}{\partial x},$$

where γ is a positive design parameter used to scale the convergence rate; and $x(t)$, starting from a randomly generated initial condition $x(0) = x_0 \in \mathbb{R}$, denotes the neural state corresponding to theoretical solution $x^*(t)$. As an extension to the above design approach and under the inspiration of [104], we obtain the general nonlinearly activated form of the GD model (or to say, general GD model) by imposing nonlinear activation function $\phi(\cdot)$ as follows:

$$\dot{x}(t) = -\gamma \phi \left(f(x(t), t) \right) \frac{\partial f}{\partial x}. \tag{5.4}$$

In view of ZD model (5.3) and GD model (5.4), different choices for $\phi(\cdot)$ lead to different performance of the neural dynamics. Generally speaking, any monotonically increasing odd activation function $\phi(\cdot)$ can be used for the construction of the neural dynamics [132]. The following four basic types of activation functions are investigated in this chapter:

1) Linear activation function $\phi(e) = e$

2) Bipolar sigmoid activation function (with $\xi > 2$)

$$\phi(e) = \left(1 - \exp(-\xi e)\right) / \left(1 + \exp(-\xi e)\right)$$

3) Power activation function $\phi(e) = e^p$ with odd integer $p \geqslant 3$ (note that the linear activation function can be viewed as a special case of the power activation function with $p = 1$)

4) Power-sigmoid activation function (with $\xi \geqslant 1$ and $p \geqslant 3$)

$$\phi(e) = \begin{cases} e^p, & \text{if } |e| \geqslant 1, \\ \frac{1+\exp(-\xi)}{1-\exp(-\xi)} \frac{1-\exp(-\xi e)}{1+\exp(-\xi e)}, & \text{otherwise} \end{cases} \tag{5.5}$$

It is worth pointing out that other types of activation functions can be generated or extended based on the above basic types of activation functions.

5.3 Convergence Analysis

While the above section presents the general frameworks about the ZD and GD models for solving time-varying nonlinear equation $f(x(t), t) = 0$, more detailed design consideration and theoretical analysis are presented in this section. To analyze the convergence of the ZD model, we first introduce the following generalized definitions.

Definition 1 Neural dynamics is said to be globally convergent, if starting from any initial point taken in the whole associated Euclidean space, an every state trajectory converges to an equilibrium point x^* in a constant problem solving or a theoretical solution trajectory $x^*(t)$ in a time-varying problem solving.

Definition 2 A neural dynamics is said to be globally exponentially convergent, if every trajectory starting from any initial condition $x(t_0)$ satisfies

$$\|x(t) - x^*(t)\| \leqslant \eta \|x(t_0) - x^*(t)\| \exp(-\gamma(t - t_0)), \ \forall t \geqslant t_0 \geqslant 0,$$

where γ and η are positive constants (preferably independent of the initial state); $x^*(t)$ denotes an equilibrium point x^* in a constant problem solving or a theoretical solution trajectory $x^*(t)$ in a time-varying problem solving (potentially depending on the initial state); and symbol $\|\cdot\|$ denotes the norm (e.g., Euclidean norm) of a vector (which, in this chapter, denotes the absolute value of a scalar argument).

Therefore, we have the following theorems about ZD model (5.3).

Theorem 5 *Consider a solvable time-varying nonlinear equation depicted in (5.1), i.e., $f(x(t),t) = 0$. If a monotonically increasing odd function $\phi(\cdot)$ is used, neural state $x(t)$ of ZD model (5.3), starting from a randomly generated initial state $x(0) \in \mathbb{R}$, can converge to a theoretical time-varying solution $x^*(t)$ of $f(x(t),t) = 0$.*

Proof First, we define a Lyapunov function candidate $v(x(t),t) = f^2(x(t),t) \geqslant 0$ for ZD model (5.3), with its time derivative being

$$
\begin{aligned}
\dot{v}(x(t),t) &= \frac{dv(x(t),t)}{dt} = f(x(t),t)\left(\frac{\partial f}{\partial x}\frac{dx}{dt} + \frac{\partial f}{\partial t}\right) \\
&= -\gamma f(x(t),t)\phi\Big(f(x(t),t)\Big).
\end{aligned}
\tag{5.6}
$$

Because a monotonically increasing odd activation function is used in ZD model (5.3), we have $\phi(-f(x(t),t)) = -\phi(f(x(t),t))$ and

$$
\phi\Big(f(x(t),t)\Big)
\begin{cases}
> 0, & \text{if } f(x(t),t) > 0, \\
= 0, & \text{if } f(x(t),t) = 0, \\
< 0, & \text{if } f(x(t),t) < 0.
\end{cases}
$$

Then, we obtain

$$
f(x(t),t)\phi\Big(f(x(t),t)\Big)
\begin{cases}
> 0, & \text{if } f(x(t),t) \neq 0, \\
= 0, & \text{if } f(x(t),t) = 0,
\end{cases}
$$

which guarantees the final negative-definiteness of $\dot{v}(x(t),t)$. That is, $\dot{v}(x(t),t) < 0$ for $f(x(t),t) \neq 0$ [equivalently, $x(t) \neq x^*(t)$], and $\dot{v}(x(t),t) = 0$ for $f(x(t),t) = 0$ [equivalently, $x(t) = x^*(t)$]. By the Lyapunov theory, residual error $e(t) = f(x(t),t)$ can converge to zero; or, equivalently speaking, neural state $x(t)$ of ZD model (5.3) can converge to a theoretical solution $x^*(t)$ with $f(x^*(t),t) = 0$ starting from randomly generated initial state $x(0)$. The proof is thus complete.

Theorem 6 *In addition to Theorem 5, when neural state $x(t)$ of ZD model (5.3) converges to $x^*(t)$ from initial state $x(0)$ close enough to $x^*(0)$, ZD model (5.3) possesses the following properties.*

   1) If the linear activation function is used, then the exponential convergence with rate γ [in terms of residual error $e(t) = f(x(t),t) \to 0$] is achieved for ZD model (5.3).

   2) If the bipolar sigmoid activation function is used, then the superior convergence is achieved for error range $e(t) = f(x(t),t) \in [-\rho,\rho]$, $\exists \rho > 0$, as compared with the situation of using the linear activation function.

   3) If the power activation function is used, then the superior convergence is achieved for error ranges $(-\infty,-1)$ and $(1,\infty)$, as compared with the situation of using the linear activation function.

   4) If the power-sigmoid activation function is used, then the superior convergence is achieved for the whole error range $(-\infty,\infty)$, as compared with the situation of the linear activation function.

Proof Now we come to prove the additional convergence properties of ZD model (5.3) by considering the mentioned several types of activation functions $\phi(\cdot)$.

1) For the situation of using the linear activation function, it follows from (5.2) that $\mathrm{d}f(x(t),t)/\mathrm{d}t = -\gamma f(x(t),t)$, which yields $f(x(t),t) = \exp(-\gamma t)f(x(0),0)$. This proves the exponential convergence of ZD model (5.3) with rate γ in the sense of residual error $e(t) = f(x(t),t) \to 0$. Moreover, we show that neural state $x(t)$ of ZD model (5.3) exponentially converges to a theoretical time-varying solution $x^*(t)$ of time-varying nonlinear equation $f(x(t),t) = 0$ by the following procedure. With $x(0)$ close enough to $x^*(0)$, from the theorem of Taylor series expansion, we have

$$f\big(x(t),t\big) = f\big((x(t)-x^*(t)+x^*(t)),t\big)$$
$$= f\big(x^*(t),t\big) + \big(x(t)-x^*(t)\big)\frac{\partial f\big(x(t),t\big)}{\partial x}\bigg|_{x=x^*}$$
$$+ \frac{\big(x(t)-x^*(t)\big)^2}{2!}\frac{\partial f^{(2)}\big(x(t),t\big)}{\partial x^2}\bigg|_{x=x^*}$$
$$+\cdots+ \frac{\big(x(t)-x^*(t)\big)^n}{n!}\frac{\partial f^{(n)}\big(x(t),t\big)}{\partial x^n}\bigg|_{x=x^*}$$
$$+ \frac{\big(x(t)-x^*(t)\big)^{n+1}}{(n+1)!}\frac{\partial f^{(n+1)}\big(x(t),t\big)}{\partial x^{n+1}}\bigg|_{x=x^*} +\cdots.$$

Thus, in view of $f(x(t),t) = \exp(-\gamma t)f(x(0),0)$ and $f(x^*(t),t) = 0$, omitting the higher-order terms, we have

$$\big(x(t)-x^*(t)\big)\frac{\partial f\big(x(t),t\big)}{\partial x}\bigg|_{x=x^*} \approx \exp(-\gamma t)f\big(x(0),0\big),$$

which, if $\partial f(x(t),t)/\partial x|_{x(t)=x^*(t)} \neq 0$ and $\|\partial f(x(t),t)/\partial x|_{x(t)=x^*(t)}\| \geqslant \delta > 0$, yields

$$\big\|\big(x(t)-x^*(t)\big)\big\| \approx \exp(-\gamma t)\left\| f\big(x(0),0\big)/\left(\frac{\partial f\big(x(t),t\big)}{\partial x}\bigg|_{x=x^*}\right)\right\|$$
$$\leqslant \exp(-\gamma t)\big\| f\big(x(0),0\big)/\delta\big\|.$$

More generally, even if $\partial f^{(k)}(x(t),t)/\partial x^k|_{x(t)=x^*(t)} = 0$, $\forall k = 1,2,3,\cdots,(n-1)$, and $\partial f^{(n)}(x(t),t)/\partial x^n|_{x(t)=x^*(t)} \neq 0$ with $\|\partial f^{(n)}(x(t),t)/\partial x^n|_{x(t)=x^*(t)}\| \geqslant \delta > 0$, we still have

$$\big\|\big(x(t)-x^*(t)\big)\big\| \approx \exp(-\gamma t/n)\left\|\left(f\big(x(0),0\big)n!/\left(\frac{\partial f^{(n)}\big(x(t),t\big)}{\partial x^n}\bigg|_{x=x^*}\right)\right)^{\frac{1}{n}}\right\|$$
$$\leqslant \exp(-\gamma t/n)\left\|\big(f\big(x(0),0\big)n!/\delta\big)^{\frac{1}{n}}\right\|.$$

This implies that neural state $x(t)$ of ZD model (5.3) converges to theoretical solution $x^*(t)$ in an exponential manner.

2) When using function $\phi(e) = (1-\exp(-\xi e))/(1+\exp(-\xi e))$, we know that there exists an error range $e(t) = f(x(t),t) \in [-\rho,\rho]$ with $\rho > 0$, such that $\|\phi(f(x(t),t))\| > \|f(x(t),t)\|$ [104]. So, reviewing the proof of Theorem 5, we know that the superior convergence is achieved by ZD model (5.3) using the bipolar sigmoid activation function in such an error range, as compared with that using the linear activation function [132].

3) For the power activation function $\phi(e) = e^p$, the solution to (5.2) is

$$f\big(x(t),t\big) = f\big(x(0),0\big)\left((p-1)f^{p-1}\big(x(0),0\big)\gamma t + 1\right)^{-\frac{1}{p-1}}.$$

Besides, for $p = 3$, $e(t) = f(x(t),t) = f(x(0),0)/\sqrt{2f^2(x(0),0)\gamma t + 1}$. Evidently, as $t \to \infty$, $f(x(t),t) \to 0$. Reviewing the proof of Theorem 5, especially the Lyapunov function candidate $v(x(t),t) = f^2(x(t),t)/2$ and its time derivative $\dot{v}(x(t),t) = -\gamma f(x(t),t)\phi(f(x(t),t))$, in the error range $\|e(t)\| = \|f(x(t),t)\| \gg 1$, we have $f^{p+1}(x(t),t) \gg f^2(x(t),t) \gg \|f(x(t),t)\|$. In other words, the deceleration magnitude of the power activation function is much greater than that of the linear activation function. This implies that, when the power activation function is used, a much faster convergence is achieved by ZD model (5.3) for such error ranges in comparison with that using the linear activation function.

4) It follows from the above analysis that, to achieve the superior convergence, a high-performance neural dynamics can be developed by switching the power activation function to the sigmoid or linear activation function at the switching points $e(t) = f(x(t),t) = \pm 1$. Thus, if the power-sigmoid activation function is used with suitable design parameters ξ and p, the superior convergence is achieved by ZD model (5.3) for the whole error range $(-\infty, \infty)$, as compared with that using the linear activation function. $\qquad\square$

Remark 1 Nonlinearities always exist, which is one of the main motivations for us to investigate different activation functions. Even if the linear activation function is used, the nonlinear phenomena may still appear in its hardware implementation; e.g., in the form of saturation or inconsistency of the linear slope in analog implementation, and in the form of truncation and round-off errors in digital implementation. The investigation of different activation functions (such as the sigmoid function and the power function) gives us many more insights into positive and negative effects of nonlinearities existing in the implementation of the linear activation function. In addition, as shown in Theorem 6, the convergence of ZD model (5.3) using suitable types of nonlinear activation functions is evidently much faster than that using the linear activation function.

5.4   Illustrative Example

While the above two sections present ZD model (5.3) and GD model (5.4), this section substantiates them by showing the following simulative results and observations [which are all based on the power-sigmoid activation function depicted in (5.5) with $\xi = 4$ and $p = 3$].

For illustration and comparison purposes, both ZD model (5.3) and GD model (5.4) are exploited for solving online the time-varying nonlinear equation $f(x(t),t) = 0$ with the following specific example considered:

$$f(x(t),t) = x^2(t) - 2\sin(1.8t)x(t) + \sin^2(1.8t) - 1 = 0. \qquad (5.7)$$

The above equation is actually equal to $f(x(t),t) = (x(t) - \sin(1.8t) - 1)(x(t) - \sin(1.8t) + 1) = 0$, and its theoretical time-varying solutions are $x_1^*(t) = \sin(1.8t) + 1$ and $x_2^*(t) = \sin(1.8t) - 1$. The simulative results are illustrated in Figures 5.1 through 5.3. Specifically, with design parameter $\gamma = 1$, Figure 5.1 shows the comparison on the solution performance of ZD model (5.3) and GD model (5.4) during the solution processes of time-varying nonlinear Equation (5.7), where dash curves correspond to theoretical solutions $x^*(t)$, and solid curves correspond to neural states $x(t)$. As seen from Figure 5.1(a), neural states $x(t)$ of ZD model (5.3), starting from 50 randomly generated initial states in $[-4,4]$, converge to the theoretical solutions [i.e., $x_1^*(t)$ or $x_2^*(t)$]. This is because the time-derivative information of time-varying coefficients in (5.7) has been fully utilized in ZD model (5.3). In contrast, also starting from 50 randomly generated initial states in $[-4,4]$, neural states $x(t)$ of GD (5.4) do not fit well with the theoretical time-varying solution $x_1^*(t)$ or $x_2^*(t)$, which is

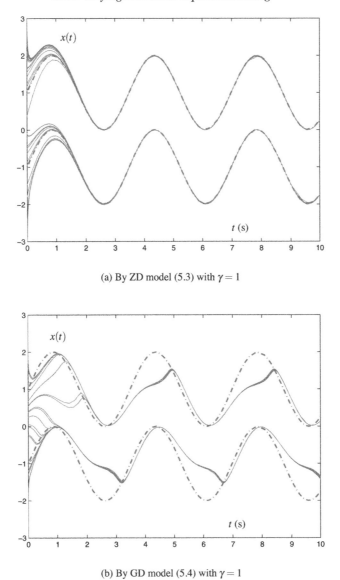

(a) By ZD model (5.3) with $\gamma = 1$

(b) By GD model (5.4) with $\gamma = 1$

FIGURE 5.1 Solution–performance comparison on ZD model (5.3) and GD model (5.4) for solving time-varying nonlinear Equation (5.7), where dash curves denote theoretical solutions.

shown clearly in Figure 5.1(b). Simply put, the steady-state residual errors of GD model (5.4) are considerably large. Furthermore, compared with the curves of theoretical solutions $x^*(t)$, the curves of GD states $x(t)$ have relatively serious distortion.

Moreover, in order to further investigate the convergence performance, we monitor the residual errors $|e(t)| = |x^2(t) - 2\sin(1.8t)x(t) + \sin^2(1.8t) - 1|$ during the equation-solving processes of both neural models. It is seen from Figure 5.2(a) and other simulative data that, by applying ZD model (5.3) to online solution of time-varying nonlinear Equation (5.7), its residual errors $|e(t)|$ converge to zero within 5 s. In contrast, as seen form Figure 5.2(b), by applying GD model (5.4) to online solution of time-varying nonlinear Equation (5.7) under the same conditions, its residual errors $|e(t)|$ are rather large and oscillating. More specifically, the maximal steady-state residual

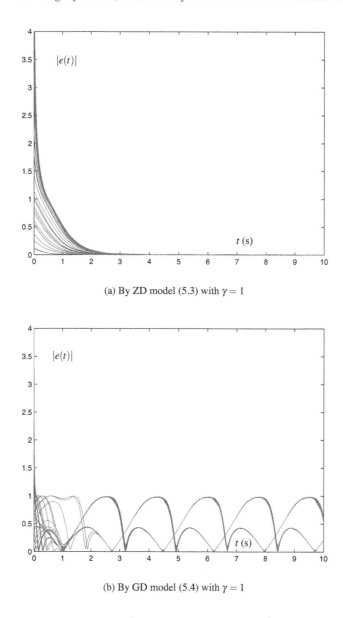

(a) By ZD model (5.3) with $\gamma = 1$

(b) By GD model (5.4) with $\gamma = 1$

FIGURE 5.2 Residual errors $|e(t)| = |x^2(t) - 2\sin(1.8t)x(t) + \sin^2(1.8t) - 1|$ synthesized by ZD model (5.3) and GD model (5.4) for solving time-varying nonlinear Equation (5.7).

errors are about 0.3146 and 0.0279 (as computed at $t = 100$ s), which respectively correspond to the situations of using design parameter $\gamma = 1$ and $\gamma = 10$. In addition to Figure 5.2(a), by using a large value of γ (e.g., $\gamma = 10$), the convergence performance of ZD model (5.3), as shown in Figure 5.3, is improved very much. More specifically, the maximal steady-state residual errors of the ZD model are about 1.0458×10^{-12} and 1.1990×10^{-13}, when design parameter $\gamma = 1$ and $\gamma = 10$. Note that such maximal steady-state residual errors should theoretically be zero but are numerically nonzero because of the finite-arithmetic simulation performed on finite-memory digital computers, with the spacing of floating point numbers being $2.220446049250313 \times 10^{-16}$ [57]. It is also worth pointing

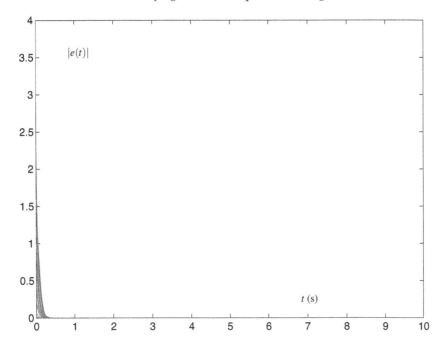

FIGURE 5.3 Residual errors $|e(t)| = |x^2 - 2\sin(1.8t)x + \sin^2(1.8t) - 1|$ synthesized by ZD model (5.3) with $\gamma = 10$ for solving time-varying nonlinear Equation (5.7).

out here that, as shown in Figures 5.2(a) and 5.3, as well as other simulative data, the convergence time of ZD model (5.3) is expedited from 5 s to 0.5 s and 0.05 s, as design parameter γ is increased from 1 to 10 and 100. These substantiate well that ZD model (5.3) has an exponential convergence property, which can be expedited effectively by increasing the value of design parameter γ.

In summary, the simulative results (i.e., Figures 5.1 through 5.3) have illustrated that ZD model (5.3) is more effective and efficient for solving in real time the time-varying nonlinear equations, as compared with GD model (5.4).

5.5 Summary

In this chapter, by defining an indefinite error function rather than the usually used nonnegative energy function, the ZD model has been developed and analyzed for solving the time-varying nonlinear equation in real time. In addition, it has been guaranteed that the residual error of the ZD model is decreasing to zero exponentially if the problem is solvable. The GD model has also been developed for comparison purposes. Theoretical and simulative results have both illustrated the efficacy and superiority of the ZD model (as compared with the GD model).

Chapter 6

Static Nonlinear Equation Solving

Abstract

A special class of neural dynamics (termed zeroing dynamics, ZD) is generalized, developed and investigated in this chapter for online solution of static nonlinear equation $f(x) = 0$ mostly for comparison and connection purposes. The gradient dynamics (GD) is thus also developed and exploited for solving online such a nonlinear equation. Conventionally and geometrically speaking, the GD model evolves along the surface descent direction (specifically, the tangent direction) of the square-based energy-function curve; but, how does the ZD model evolve? Trying to answer this question, the corresponding discrete-time ZD (DTZD) model is then developed and investigated. In terms of the constant nonlinear equation solving, Newton iteration is found once more to be a special case of the DTZD model. It is also discovered that, if a nonlinear equation possesses a local minimum point, the neural state of the CTZD model, starting from some initial value close to it, may move toward the local minimum point and then stop with warning information. In contrast, the neural state of the GD model falls into the local minimum point (with no warning). Furthermore, inspired by Wu's work, the improved CTZD models are proposed by defining two modified error functions, which handle such a local-minimum problem effectively.

6.1   Problem Formulation and Continuous-Time Models

Now, let us consider the nonlinear equation in a general form [112, 113, 127, 128],

$$f(x) = 0 \in \mathbb{R}, \tag{6.1}$$

where $f(\cdot) : \mathbb{R} \to \mathbb{R}$ is assumed continuously differentiable. Note that, traditionally and generally speaking, a constant problem [such as (6.1)] solving is theoretically proposed and practically generated via a short-time invariance hypothesis in an online manner from some time-varying problem [such as (5.1)] solving [2, 21, 27, 37, 84]. Our starting point now is to exploit ZD and GD models to find a root $x \in \mathbb{R}$, such that the above nonlinear Equation (6.1) holds true. For ease of presentation, let x^* denote such a theoretical solution (or to say, root, zero point) of nonlinear Equation (6.1).

6.1.1   CTZD Model

To monitor and control the solution process of nonlinear Equation (6.1), following Zhang *et al.*'s design method, we first define the indefinite error function $e(t) = f(x)$, where, evidently, if error function $e(t)$ converges to zero, then $x(t)$ converges to theoretical solution x^*.

Secondly, let the time derivative $\dot{e}(t)$ of error function $e(t)$ be chosen and forced mathematically, such that $e(t)$ converges to zero exponentially. Specifically, we choose $\dot{e}(t)$ via the following ZD design formula [112, 113, 127, 128]:

$$\frac{\mathrm{d}e(t)}{\mathrm{d}t} = -\gamma\phi\big(e(t)\big), \text{ i.e., } \frac{\mathrm{d}f(x)}{\mathrm{d}t} = -\gamma\phi\big(f(x)\big). \tag{6.2}$$

Thirdly, expanding the ZD design formula (6.2), we have the following differential equation of the CTZD model:

$$\frac{\partial f(x)}{\partial x}\frac{\mathrm{d}x}{\mathrm{d}t} = -\gamma\phi\big(f(x)\big), \text{ i.e., } f'(x)\dot{x} = -\gamma\phi\big(f(x)\big),$$

$$\text{or equivalently [if } f'(x) \neq 0], \ \dot{x}(t) = -\gamma\frac{\phi\big(f(x)\big)}{f'(x)}, \tag{6.3}$$

where $x(t)$, starting from randomly generated initial condition $x(0) \in \mathbb{R}$, denotes the neural state of CTZD model (6.3) corresponding to theoretical root x^* of (6.1). Note that, for better understanding the differences of constant and time-varying equations/problems solving, please see and compare Subsection 5.2.1.

In addition, for CTZD model (6.3) solving the nonlinear equation $f(x) = 0$, we have the following proposition about its convergence [112, 113, 127, 128, 134].

Proposition 5 *Consider a solvable nonlinear equation $f(x) = 0$. If monotonically increasing odd activation function $\phi(\cdot)$ is employed, then neural state $x(t)$ of CTZD model (6.3), starting from randomly generated initial condition $x(0) \in \mathbb{R}$, can converge to theoretical root x^* of nonlinear equation $f(x) = 0$ [of which the specific value of x^*, in the situation of not less than two zeros existing, depends on the sufficient closeness of initial state $x(0)$ to x^*].*

6.1.2   GD Model

In this subsection, for convenience of better and comparative reading, the design procedure of the GD models is presented as follows.

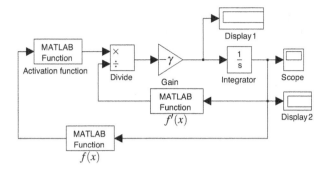

(a) Simulink-based model representation of CTZD (6.3)

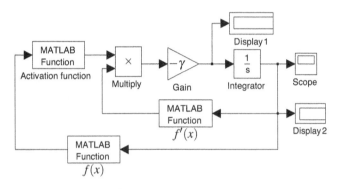

(b) Simulink-based model representation of GD (6.4)

FIGURE 6.1 Simulink-based model representations [127] about CTZD (6.3) and GD (6.4).

First, a square-based nonnegative energy function, such as $\mathscr{E}(x) = f^2(x)/2$, is constructed so that its minimum point is the root of the nonlinear equation $f(x) = 0$.

Secondly, the solution scheme is designed to evolve along a descent direction of this energy function until a minimal point is reached. The typical descent direction is the negative gradient of this energy function $\mathscr{E}(x)$, i.e., $-\partial \mathscr{E}(x)/\partial x$.

Thirdly, using the negative-gradient method to construct a neural-dynamic model for solving nonlinear equation $f(x) = 0$, we have the following neural-dynamic equation as the conventional linearly activated form of the GD model:

$$\frac{dx(t)}{dt} = -\gamma \frac{\partial \mathscr{E}(x)}{\partial x} = -\gamma f(x) \frac{\partial f(x)}{\partial x} = -\gamma f(x) f'(x).$$

Finally, as an extension of the above form of the GD model, we generalize a nonlinearly activated form of the GD model (i.e., the so-called general GD model) by exploiting nonlinear activation function $\phi(\cdot)$ as follows:

$$\dot{x}(t) = -\gamma \phi\big(f(x)\big) f'(x). \tag{6.4}$$

Moreover, according to the block diagrams of the neural-dynamic models depicted in [134], MATLAB® Simulink® based model representations of the CTZD and GD models are presented in Figure 6.1 [103, 124, 127]. The difference between CTZD model (6.3) and GD model (6.4) for solving this specific equation problem $f(x) = 0$ is interesting: CTZD model (6.3) has a division term of $f'(x)$, whereas GD model (6.4) has a multiplication term of $f'(x)$. In addition to the difference, in view of the fact that both CTZD and GD models can work effectively for this problem solving [134], people may ask why and how $f'(x)$ is needed, as well as how the performance of CTZD and GD models differs. In the following subsection, we try to answer the questions via illustrative examples.

6.1.3   Illustrative Examples

In the previous two subsections, we have presented the forms of the CTZD and GD models for zero (or to say, zero point, root) finding, together with the theoretical results of the CTZD model provided. In this subsection, three illustrative examples are presented to investigate and show the convergence performance of CTZD model (6.3) and GD model (6.4), which are based on the use of the power-sigmoid activation function with design parameters $\xi = 4$ and $p = 3$ [112,113,127,128,134].

Example 6.1 Let us consider the following nonlinear equation:

$$f(x) = (x-4)^{10}(x-1) = 0. \tag{6.5}$$

To check the solution correctness, the theoretical roots to the above nonlinear equation are given as $x_1^* = 4$ (a multiple root of order 10) and $x_2^* = 1$ (a simple root). Specifically, it is seen from the left subgraph of Figure 6.2 that, starting from 200 randomly generated initial states in $[-2,6]$, neural states $x(t)$ of CTZD model (6.3) converge to the theoretical root of nonlinear Equation (6.5), either $x_1^* = 4$ or $x_2^* = 1$. In contrast, by applying GD model (6.4), some less-correct solutions (i.e., different from $x_1^* = 4$ and $x_2^* = 1$) are generated, which is seen clearly in the right subgraph of Figure 6.2. In addition, it is seen from the figure that, starting from some initial states [e.g., $x(0) = 0.5$ here] close enough to the simple root $x^* = 1$, neural states $x(t)$ of both CTZD model (6.3) and GD model (6.4) converge to theoretical root $x^* = 1$. For the case of multiple root $x^* = 4$ of order 10, if the initial states [e.g., $x(0) = 2.5$ or $x(0) = 5.5$ here] are close enough to this multiple root, neural states $x(t)$ of CTZD model (6.3) converge well to theoretical root $x_1^* = 4$; but, under the same conditions, when applying GD model (6.4), neural states $x(t)$ of GD model (6.4) mostly converge to less correct solutions (or to say, approximate solutions with $0 < |x(40) - 4| < 1$).

  Moreover, we investigate the convergence performance of CTZD model (6.3) and GD model (6.4) for solving the nonlinear equation similar to (6.5) but with a multiple root of order less than 10 (specifically, 3 and 2); i.e., $f(x) = (x-4)^3(x-1) = 0$ and $f(x) = (x-4)^2(x-1) = 0$. The simulative

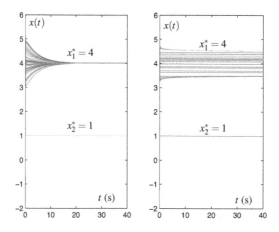

FIGURE 6.2 Online solution to $(x-4)^{10}(x-1) = 0$ by CTZD model (6.3) (left) and GD model (6.4) (right), with $\gamma = 1$.

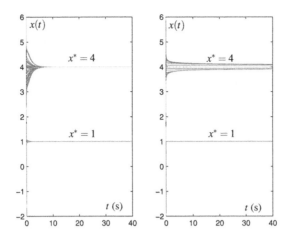

FIGURE 6.3 Online solution to $(x-4)^3(x-1) = 0$ by CTZD model (6.3) (left) and GD model (6.4) (right), with $\gamma = 1$.

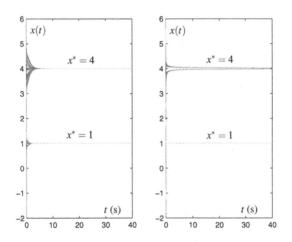

FIGURE 6.4 Online solution to $(x-4)^2(x-1) = 0$ by CTZD model (6.3) (left) and GD model (6.4) (right), with $\gamma = 1$.

results are presented in Figures 6.3 and 6.4. Specifically, starting from 200 initial states randomly generated in $[-2,6]$ and sufficiently close to simple root $x^* = 1$, neural states $x(t)$ of CTZD model (6.3) and GD model (6.4) converge well to such a theoretical root $x^* = 1$. For the case of the multiple root $x^* = 4$, neural states $x(t)$ of CTZD model (6.3) always converge well to such a theoretical root $x^* = 4$; but, similar to Figure 6.2, GD model (6.4) may still generate less correct (or approximate) solutions.

In summary, as seen from Figures 6.2 through 6.4, for the case of the multiple root (i.e., $x^* = 4$), neural states $x(t)$ of CTZD model (6.3) always converge accurately to the theoretical root regardless of its order. The efficacy is possibly owing to the division term of $f'(x)$ exploited in CTZD model

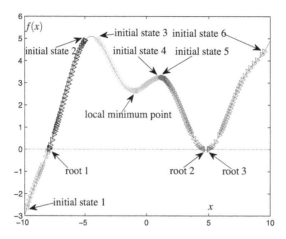

FIGURE 6.5 Convergence performance of neural states of CTZD model (6.3) for online solution of nonlinear Equation (6.6) starting from different initial states.

(6.3). In contrast, GD model (6.4) generates less correct results (or to say, approximate solutions). Furthermore, comparing the right subgraphs of Figures 6.2 through 6.4, we observe that, with the increase of the order of multiple root $x^* = 4$, the convergence performance of GD model (6.4) becomes (much) worse. This increasing inability, as we have investigated, is possibly owing to the multiplication term of $f'(x)$ exploited in GD model (6.4).

Example 6.2 Now, let us consider the following nonlinear equation:

$$f(x) = 0.01(x+7)(x-1)(x-8) + \sin x + 2.4 = 0,$$ (6.6)

with its three theoretical roots denoted by "root 1," "root 2" and "root 3" in Figure 6.5. Using CTZD model (6.3) to solve the above nonlinear Equation (6.6), as seen from Figure 6.5, starting from different initial states, neural states $x(t)$ converge to their corresponding theoretical roots. For example, if initial state $x(0)$ is -9.8 or -5.0 (corresponding to "initial state 1" or "initial state 2" in the figure), neural states $x(t)$ of CTZD model (6.3) converge to the "root 1," as depicted in Figure 6.5. Similarly, starting from initial state $x(0) = 1.0$ or 9.5 (corresponding to "initial state 5" and "initial state 6" in the figure), neural states $x(t)$ converge to "root 2" and "root 3," as depicted in Figure 6.5.

Correspondingly, let us see other simulation results, i.e., in Figure 6.6. Starting from 50 randomly generated initial states in $[-10, 10]$ and sufficiently close to the theoretical roots (i.e., "root 1," "root 2" and "root 3"), neural states $x(t)$ of both CTZD model (6.3) and GD model (6.4) converge well to them. But, for some initial states [e.g., $x(0) = -4$ or 0.8 corresponding to "initial state 3" or "initial state 4," respectively, in Figure 6.5], which are close to a local minimum point, GD model (6.4) generates a wrong solution; whereas, for this situation, neural states $x(t)$ of CTZD model (6.3) move toward the local minimum point and then stop with the following warning information.

```
Warning: Failure at t=1.479876e-005. Unable
to meet integration tolerances  without
reducing the step size below the smallest
value allowed (2.710505-020) at time t.
> In ode15s at 741 In znnvsgnnft at 20
```

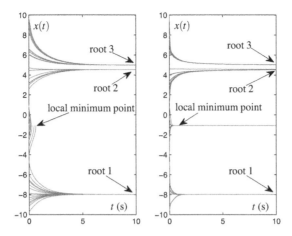

FIGURE 6.6 Online solution of $0.01(x+7)(x-1)(x-8)+\sin x+2.4=0$ by CTZD model (6.3) (left) and GD model (6.4) (right), with $\gamma=1$.

This warning reminds users to rechoose a suitable initial state with which to restart the CTZD-solution process for achieving the correct results. In contrast, GD model (6.4) yields a wrong solution [i.e., the local minimum point with $f'(x)=0$] in this situation, which may mislead the users into obtaining a "solution" of the nonlinear equation.

Example 6.3 Finally, let us consider the following nonlinear equation:

$$f(x) = \cos x + 3 = 0, \tag{6.7}$$

for another interesting observation. Evidently, the above nonlinear Equation (6.7) has no solution (i.e., no root, no zero). As seen from Figure 6.7, by applying CTZD model (6.3), neural states $x(t)$, starting from 20 initial states in $[0,10]$, move toward some values and then stop with a warning, as the above. In contrast, GD model (6.4) generates some misleading results again. The specific values involved, as the authors investigate, are related to the local minimum points satisfying $f'(x) = -\sin x = 0$; i.e., $x=0$ and $x=\pm\kappa\pi$ with $\kappa=1,2,3,\cdots$.

In summary, from the above three illustrative examples, we have the following observations and facts that CTZD model (6.3) is effective for solving nonlinear equation $f(x)=0$. In contrast, GD model (6.4) more probably yields wrong (or approximate) solutions for the case of a multiple root.

6.2 DTZD Models

For potential hardware implementation of CTZD model (6.3) for solving nonlinear Equation (6.1), we discretize it by using Euler forward-difference rule:

$$\dot{x}(t=k\tau) \approx (x_{k+1} - x_k)/\tau.$$

Then, the DTZD model is thus generated from CTZD model (6.3):

$$(x_{k+1} - x_k)/\tau = -\gamma\frac{\phi(f(x_k))}{f'(x_k)},$$

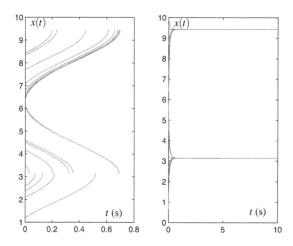

FIGURE 6.7 Online solution of CTZD model (6.3) (left) and GD model (6.4) (right) with $\gamma = 1$ for nonlinear equation $\cos x + 3 = 0$ which evidently has no theoretical root.

which is rewritten as

$$x_{k+1} = x_k - h\frac{\phi\left(f(x_k)\right)}{f'(x_k)}, \tag{6.8}$$

where parameter $h = \tau\gamma > 0$ denotes the step size that should be set appropriately.

6.2.1   Link and New Explanation to Newton Iteration

Now, let us look at DTZD model (6.8): If linear activation function $\phi(e) = e$ is used, DTZD model (6.8) reduces to

$$x_{k+1} = x_k - h\frac{f(x_k)}{f'(x_k)}; \tag{6.9}$$

and, if $h = 1$, DTZD model (6.9) further becomes

$$x_{k+1} = x_k - \frac{f(x_k)}{f'(x_k)}, \tag{6.10}$$

which is exactly Newton iteration presented in the textbooks and literature (see [25, 51, 57] and references therein) for solving nonlinear Equation (6.1). In other words, for Chapters 1 through 6 as a whole, we discover confirmedly that a general form of Newton iteration for solving nonlinear Equation (6.1) is derived by discretizing CTZD model (6.3). Evidently, the new effective explanation to Newton iteration is established; i.e., Newton iteration is one of the special cases of DTZD model (6.8) by focusing on the constant equation solving, discretizing it via Euler forward-difference rule, using a linear activation function and fixing the step size to be 1.

However, traditionally, Newton iteration for nonlinear Equation (6.1) solving has a standard textbook explanation; i.e., as mentioned, via Taylor expansion [25, 51, 57]. For readers' convenience as well as for completeness, such a derivation/construction/explanation procedure is repeated and compared below [25, 51, 57].

According to the Taylor expansion of nonlinear function $f(x)$ in (6.1) around the initial point of $x_0 \in \mathbb{R}$, we have

$$f(x) = f(x_0) + f'(x_0)(x - x_0) + f''(s)(x - x_0)^2/2, \tag{6.11}$$

where s is located somewhere between x_0 and x. Let $x = x^*$ in the above equation. Noting that $f(x^*) = 0$, we can get $0 = f(x^*) = f(x_0) + f'(x_0)(x^* - x_0) + f''(s)(x^* - x_0)^2/2$. As x_0 is close enough to x^*, the last term of the right-hand side of Equation (6.11) is so small (compared with the sum of the first two terms) that it can be neglected. In this way, the following equation is obtained: $0 \approx f(x_0) + f'(x_0)(x^* - x_0)$, which yields $x^* \approx x_0 - f(x_0)/f'(x_0)$. Letting x_{k+1} denote the next approximation of x^* and then substituting x_k for x_0 into $x^* \approx x_0 - f(x_0)/f'(x_0)$, we thus derive/construct/explain Newton iteration (6.10) [i.e., $x_{k+1} = x_k - f(x_k)/f'(x_k)$] approximately.

6.2.2 Theoretical Analysis

With the general form of DTZD model (6.8) and its link to Newton iteration presented, some design consideration and theoretical analysis are presented in this subsection by using the fixed-point iteration approach [128].

Theorem 7 *Assume that $f(x)$ of nonlinear Equation (6.1) is sufficiently continuously differentiable on $[x^* - \delta, x^* + \delta]$, with x^* being a root of order q (if $q = 1$, x^* is a simple root). When step size h of DTZD model (6.8) satisfies $0 < h < 2q/\phi'(0)$, there generally exists such a positive number δ, such that the sequence $\{x_k\}$ generated by DTZD model (6.8) converges to x^* for any initial state $x_0 \in [x^* - \delta, x^* + \delta]$.*

Proof Following the proof approach of [25, 51, 57], about DTZD model (6.8), we know that the fixed-point iteration function is $g(x) = x - h\phi(f(x))/f'(x)$. By applying the fixed-point iteration approach [128], the key of the proof is to analyze $g'(x)$:

$$
\begin{aligned}
g'(x) &= 1 - h\frac{\phi'(f(x))(f'(x))^2 - \phi(f(x))f''(x)}{(f'(x))^2} \\
&= 1 - h\phi'(f(x)) + h\frac{\phi(f(x))f''(x)}{(f'(x))^2}.
\end{aligned}
\tag{6.12}
$$

1) If x^* is a simple root, we have $f(x^*) = 0$ and $f'(x^*) \neq 0$. Substituting x^* into (6.12), in view of $\phi(f(x^*)) = 0$ [as $\phi(\cdot)$ is odd], we have

$$g'(x^*) = 1 - h\phi'(0).$$

So, when $0 < h < 2/\phi'(0)$, the above expression $1 - h\phi'(0)$ satisfies $-1 < 1 - h\phi'(0) < 1$, i.e., $|g'(x^*)| < 1$. As $g'(x)$ is continuous at and around the point x^*, it is possible to find a $\delta > 0$, such that $|g'(x)| < 1$ for any $x \in [x^* - \delta, x^* + \delta]$. According to the fixed-point iteration approach [128], it is then proved that, if initial state $x_0 \in [x^* - \delta, x^* + \delta]$, the sequence $\{x_k\}$ ($k = 0, 1, 2, \cdots$) resulting from the fixed-point iteration function $g(x)$ [i.e., generated by DTZD model (6.8)] converges to x^* [which is a simple root of nonlinear equation $f(x) = 0$].

2) Now, let us investigate the situation that x^* is a multiple root of order $q \geqslant 2$; i.e., $f(x^*) = 0$, $f'(x^*) = 0, \cdots, f^{(q-1)}(x^*) = 0$, but $f^{(q)}(x^*) \neq 0$. So, we denote $f(x)$ as $f(x) = (x - x^*)^q m(x)$ with $m(x^*) \neq 0$. Now let us analyze $g'(x)$ in (6.12). Then we have

$$g'(x^*) = \lim_{x \to x^*} g'(x) = 1 - h\phi'(0) + h\lim_{x \to x^*}\frac{\phi(f(x))f''(x)}{(f'(x))^2},$$

of which the last term can be derived (via L'Hopital's rule) as

$$\lim_{x \to x^*} \frac{\phi\big(f(x)\big)f''(x)}{\big(f'(x)\big)^2} = \lim_{x \to x^*} \frac{\phi\big(f(x)\big)\big((x-x^*)^q m(x)\big)''}{\Big(\big((x-x^*)^q m(x)\big)'\Big)^2} \quad [5pt]$$

$$= \lim_{x \to x^*} \phi\big(f(x)\big)\Bigg[\frac{q(q-1)(x-x^*)^{q-2} m(x)}{\Big(q(x-x^*)^{q-1} m(x) + (x-x^*)^q m'(x)\Big)^2}$$

$$+ \frac{2q(x-x^*)^{q-1} m'(x) + (x-x^*)^q m''(x)}{\Big(q(x-x^*)^{q-1} m(x) + (x-x^*)^q m'(x)\Big)^2}\Bigg]$$

$$= \lim_{x \to x^*} \frac{\phi\big(f(x)\big)\Big(q-1+2\frac{(x-x^*)m'(x)}{m(x)} + \frac{(x-x^*)^2 m''(x)}{qm(x)}\Big)}{q(x-x^*)^q m(x)\Big(1+(x-x^*)\frac{m'(x)}{qm(x)}\Big)^2}$$

$$= \Bigg(\lim_{x \to x^*} \frac{\phi\big(f(x)\big)}{q(x-x^*)^q m(x)}\Bigg)(q-1)$$

$$= (q-1)\lim_{x \to x^*} \frac{\big(\phi\big(f(x)\big)\big)'}{\big(q(x-x^*)^q m(x)\big)'}$$

$$= (q-1)\lim_{x \to x^*}\Bigg[\frac{\phi'\big(f(x)\big)q(x-x^*)^{q-1} m(x)}{q^2(x-x^*)^{q-1} m(x) + q(x-x^*)^q m'(x)}$$

$$+ \frac{\phi'\big(f(x)\big)(x-x^*)^q m'(x)}{q^2(x-x^*)^{q-1} m(x) + q(x-x^*)^q m'(x)}\Bigg]$$

$$= (q-1)\lim_{x \to x^*} \frac{\phi'\big(f(x)\big)\big(qm(x) + (x-x^*)m'(x)\big)}{q^2 m(x) + q(x-x^*)m'(x)}$$

$$= (q-1)\lim_{x \to x^*} \frac{\phi'\big(f(x)\big)}{q} = \frac{(q-1)\phi'(0)}{q}.$$

Thus, we have $g'(x^*) = 1 - h\phi'(0) + h(q-1)\phi'(0)/q = 1 - h\phi'(0)/q$. So, similar to the above analysis on the simple-root situation, in this q-order multiple-root situation, if step size h is chosen, such that $0 < h < 2q/\phi'(0)$, then $|g'(x^*)| < 1$. As $g'(x)$ is continuous at and around the point x^*, it is possible to find a $\delta > 0$, such that $|g'(x)| < 1$ for any $x \in [x^* - \delta, x^* + \delta]$ as well. According to the fixed-point iteration approach [128], it is proved again that, if initial state $x_0 \in [x^* - \delta, x^* + \delta]$, the sequence $\{x_k\}$ ($k = 0, 1, 2, \cdots$) resulting from the fixed-point iteration function $g(x)$ [i.e., generated by DTZD model (6.8)] converges to x^* (being a multiple root in the second situation).

By summarizing the above situations, the proof is complete.

It is worth pointing out that, if step size h of DTZD model (6.8) satisfies $0 < h < 2q/\phi'(0)$, there generally exists an initial state x_0 close enough to such a root, which ensures the convergence of DTZD model (6.8). Besides, the values of step size h and initial state x_0 need to be selected appropriately to ensure (and better expedite) the convergence: Large values of step size may make DTZD model (6.8) converge fast to a root of (6.1), but too large values of step size may result in divergence. The (optimal) choosing of initial state x_0 is also an interesting issue.

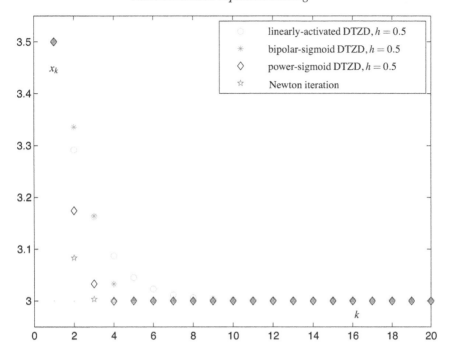

FIGURE 6.8 Convergence performance of DTZD model (6.8) using different activation functions with, $h = 0.5$ and Newton iteration solving $(x-3)(x-1) = 0$ for $x_1^* = 3$.

6.2.3 Illustrative Examples

In this subsection, numerical-experiment results and observations are provided for substantiating the theoretical analysis, the link (to Newton iteration) and efficacy of DTZD model (6.8) with the aid of suitable initial state x_0, activation function $\phi(\cdot)$, and step size h.

Example 6.4 Now, let us consider the following nonlinear equation:

$$f(x) = (x-3)(x-1) = 0. \tag{6.13}$$

Evidently, the theoretical roots of nonlinear Equation (6.13) are written as $x_1^* = 3$ and $x_2^* = 1$ (both of which are simple roots) for the verification of DTZD model (6.8). By applying DTZD model (6.8) to solve the above Equation (6.13), numerical-experiment results are shown in Figure 6.8. Note that, in this example, we choose initial state $x_0 = 3.5$ close to $x_1^* = 3$, and DTZD model (6.8) converges to $x_1^* = 3$, in addition to setting step size $h = 0.5$. From the figure, we find that, after a sufficient number of iterations, DTZD model (6.8) always converges to theoretical root $x_1^* = 3$ even when different activation functions are used. In addition, when $h = 0.5$, the fastest convergence (within about 4 iterations) is achieved by using the power-sigmoid activation function. Besides, Newton iteration is compared, of which the superior performance is depicted via pentagrams in Figure 6.8.

For the same nonlinear Equation (6.13) solving, it is worth investigating the convergence properties of DTZD model (6.8) with different values of step size h. The numerical results are shown in Figure 6.9, where the power-sigmoid activation function is exploited with $\xi = 4$ and $p = 3$. In this situation, by Theorem 7, $\phi'(0) = \xi(1 + \exp(-\xi))/(2(1 - \exp(-\xi))) = 2.0746$, and thus h should satisfy $0 < h < 2/\phi'(0) = 0.9640$ (see also Subsection 2.3.2). Let us choose the initial state $x_0 = 3.5$, and compare among different step-size values (e.g., $h = 0.05, 0.1, 0.6$ and 1). As seen from Figure 6.9, the convergence speed can be expedited by appropriately increasing the value of step size h. For example, in the situation of $h = 0.6$, theoretical root $x_1^* = 3$ is reached by the power-sigmoid

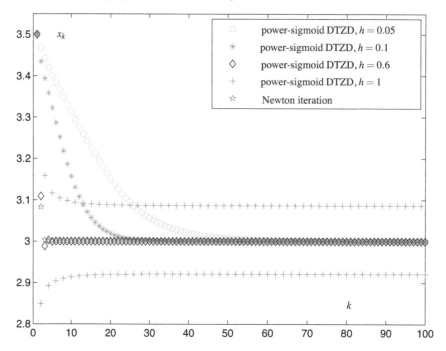

FIGURE 6.9 Convergence performance of DTZD model (6.8) using power-sigmoid activation function, with different h and Newton iteration solving $(x-3)(x-1) = 0$ for $x_1^* = 3$.

activation DTZD model (or to say, power-sigmoid DTZD model) (6.8) within about 4 iterations; whereas, in the situations of $h = 0.1$ and $h = 0.05$, root $x_2^* = 3$ is reached with about 30 and 70 iterations, respectively. Furthermore, too large values of h may lead to cyclic oscillations or even divergence, e.g., when $h = 1 > 0.9640$, as illustrated in Figure 6.9.

Example 6.5 Let us consider the following nonlinear equation:

$$f(x) = (x-4)^5(x+2) = 0. \tag{6.14}$$

Different from (6.13), this nonlinear Equation (6.14) theoretically has a multiple root $x_1^* = 4$ of order 5 (i.e., $q = 5$ here) and a simple root $x_2^* = -2$. By starting from initial state $x_0 = 4.5$, we examine the effectiveness of DTZD model (6.8) on solving for the multiple root $x_1^* = 4$. As we employ the power-sigmoid activation function with $\xi = 4$ and $p = 3$, in this situation, by Theorem 7, h should satisfy

$$0 < h < 2q/\phi'(0) = 10/2.0746 = 4.8202.$$

Thus, setting step size $h = 0.8$, as shown in Figure 6.10, we find that DTZD model (6.8) converges to $x_2^* = 4$ within a fewer iterations as compared with traditional Newton iteration (6.10). This illustrates the superior effectiveness of DTZD model (6.8) in multiple roots finding, in addition to the important link to Newton iteration.

6.3   Comparison between CTZD Model and Newton Iteration

In this section, we compare CTZD model (6.3) and Newton Iteration (6.10), both of which are now exploited for online solution of static nonlinear Equation (6.1).

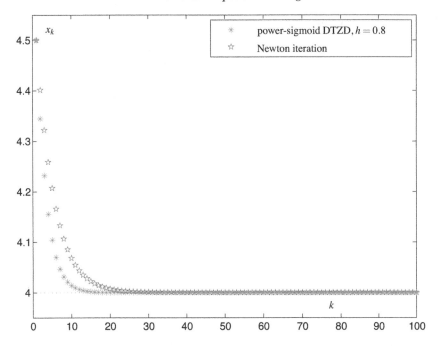

FIGURE 6.10 Convergence performance of DTZD model (6.8) using power-sigmoid activation function, with $h = 0.8$ and Newton iteration solving $(x-4)^5(x+2) = 0$ for $x_1^* = 4$.

6.3.1 Theoretical Comparison

In addition to the main difference being CTZD model for time-varying problems solving and Newton iteration for constant problems solving, other differences of the two schemes (or to say, models, systems) lie in the following facts.

1) CTZD model (6.3) is a continuous-time method, of which the design method is based on the elimination of an indefinite error function. In contrast, Newton iteration (6.10) is a discrete-time (iterative) method, which is traditionally thought to be obtained from Taylor series expansion [or newly thought in our research to be derived by discretizing CTZD model (6.3) with some specific conditions].

2) The convergence behavior of CTZD model (6.3) is that neural state $x(t)$, starting from an initial condition, traverses every point of the curve until the root is found. In contrast, the convergence behavior of Newton iteration (6.10) is that the current point jumps to the next point on the function curve by finding the intersection between the X-axis and the straight line tangent to the curve of $y = f(x)$ at the current point.

3) The theoretical analysis on the convergence of CTZD model (6.3) is based on the well-known Lyapunov theory or preferably the theory of ordinary differential equations (ODEs) [132]. Differing from that of CTZD model, the standard theoretical analysis on the convergence of Newton iteration (6.10) is based on the well-known fixed-point theorem and related knowledge.

4) The convergence speed of CTZD model (6.3) can be expedited (theoretically, arbitrarily faster) by increasing the value of γ and using a suitable activation function. In contrast, the convergence rate for Newton iteration (6.10) is fixed, which is quadratic for the simple root finding and linear for the multiple root finding [57].

6.3.2   Illustrative Examples

In this subsection, by performing numerical comparison in different situations, computer-simulation and numerical-experiment results are provided to show the characteristics of CTZD model (6.3) and Newton iteration (6.10). Though Newton iteration (6.10) can be obtained from CTZD model (6.3), their convergence behaviors are quite different from each other, which evidently results in different advantages of CTZD model (6.3) and Newton iteration (6.10) for solving the nonlinear equation problem. To investigate and compare the efficacy of such two models, let us consider the following four illustrative examples.

Example 6.6   With x defined in some range

It is widely encountered in science and engineering problems that the variable x in nonlinear function $f(x)$ is defined only in a finite range. For illustration and comparison, let us consider the following nonlinear equation:

$$f(x) = \cos(x) = 0, \text{ with } x \in [0, \pi]. \tag{6.15}$$

Evidently, the only theoretical root in the defining range $[0, \pi]$ is $x^* = \pi/2$. Both CTZD model (6.3) and Newton iteration (6.10) are then exploited to solve the above nonlinear Equation (6.15). By using CTZD model (6.3) with $\gamma = 1$, we see from Figure 6.11(a) that, starting from initial state $x(0) = 3$, neural state $x(t)$ converges to the desired theoretical root $x^* = \pi/2 \approx 1.57$ after a short time (e.g., 3 s or so). But, for the same initial state $x_0 = 3$, as shown in Figure 6.11(b), state x_k of Newton iteration (6.10) converges to a different root $-3\pi/2 \approx -4.71$, which is out of range $[0, \pi]$. Actually, the Newton iteration state calculated from the first iteration $x_1 = x_0 - f(x_0)/f'(x_0) = -4.015$ has already been out of range $[0, \pi]$. Simply put, CTZD model (6.3) generates a correct solution [i.e., the so-called "correct" state in Figure 6.11(a)], whereas Newton iteration (6.10) generates a wrong solution [i.e., the so-called "wrong" state Figure 6.11(b)].

So, as seen from the simulation results, CTZD model (6.3) has superior effectiveness to Newton iteration (6.10) in the situation of x defined in some range, which is consistent with their theoretical convergence behaviors discussed in Subsection 6.3.1. In addition, the initial condition of Newton iteration (6.10) should be chosen close enough to the desired theoretical root; otherwise, it may converge to some other root. Moreover, as seen from Figure 6.12, the absolute value of the slope $f'(x_0)$ should not be small; otherwise [e.g., the slope $f'(x_0)$ of (6.15) is about -0.14], the tangent line of function $f(x)$ at point x_0 is nearly horizontal, and the first intersection x_1 is far away from x_0, leading to the consequence $\{x_k\}$ running out of the defining range $[0, \pi]$.

Example 6.7   Divergent oscillation of Newton iteration

According to the fixed-point theorem given in [51], Newton iteration (6.10) may oscillate and diverge when $|g'(x)| \geqslant 1$ on an interval containing root x^*. For illustration and comparison, let us consider the following nonlinear equation:

$$f(x) = \arctan(x) = 0. \tag{6.16}$$

Evidently, one theoretical root is $x^* = 0$. Using Newton iteration (6.10) to solve nonlinear Equation (6.16), we obtain $g'(x) = -2x \cdot \arctan(x)$. Specifically, as seen from Figure 6.13(a), the initial state is set as $x_0 = 1.45$, and then the sequence $\{x_k\}$ generated by Newton iteration (6.10) is divergently oscillating (with $x_1 = -1.55, x_2 = 1.85, x_3 = -2.89, \cdots$), i.e., the so-called "wrong" (or to say, divergent) state trajectory of Newton iteration (6.10). In contrast, as shown in Figure 6.13(b), neural state $x(t)$ of CTZD model (6.3), starting from the same initial state, converges to theoretical root $x^* = 0$ after a short time, i.e., the so-called "correct" (or to say, convergent) state trajectory of CTZD

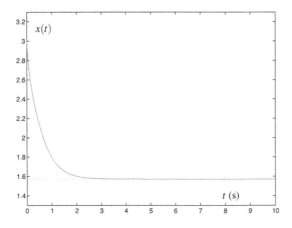

(a) "Correct" state trajectory of CTZD model (6.3)

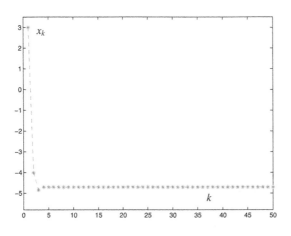

(b) "Wrong" state trajectory of Newton iteration (6.10)

FIGURE 6.11 Online solution of nonlinear Equation (6.15) by CTZD model (6.3), with $\gamma = 1$ and by Newton iteration (6.10), starting from the same initial state $x(0) = x_0 = 3$.

model (6.3). These illustrate the superior effectiveness of CTZD model (6.3) [as compared with Newton iteration (6.10)], in addition to the important link between them.

Example 6.8 Convergence speed

To compare the convergence speed of CTZD model (6.3) and Newton iteration (6.10), let us consider the following nonlinear equation:

$$f(x) = (x-3)^5(x+1) = 0. \tag{6.17}$$

Evidently, this nonlinear Equation (6.17) theoretically has a multiple root $x_1^* = 3$ of order 5 and a simple root $x_2^* = -1$. As shown in Figure 6.14(a), starting from initial state $x(0) = 3.5$, neural state $x(t)$ of CTZD model (6.3), with $\gamma = 10$, converges to the multiple root $x_1^* = 3$ (about 1.4 s) much faster than that of CTZD model (6.3), with $\gamma = 1$ (about 14 s). Furthermore, if $\gamma = 100$, the convergence time is 0.14 s. Thus, it is substantiated that CTZD model (6.3) has an exponential convergence

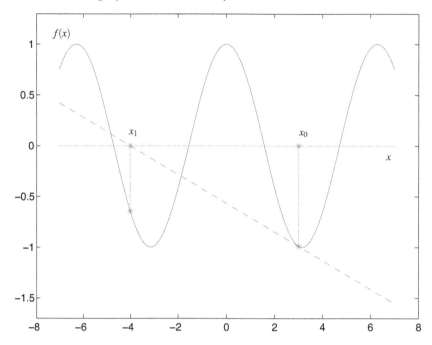

FIGURE 6.12 Graphical interpretation of the failure of Newton iteration (6.10) when solving nonlinear Equation (6.15).

property, and the convergence speed can be expedited effectively by increasing the value of design parameter γ. In contrast, for the same initial state, as seen from Figure 6.14(b), the convergence rate of Newton iteration (6.10) is fixed, which takes about 25 iterations to reach the theoretical solution. Besides, for root $x_1^* = 3$, the convergence rate for Newton iteration (6.10) is linear as it is a multiple root.

Example 6.9 Nonlinear equation containing local minimum

Let us consider the situation of the nonlinear equation containing a local minimum. According to the different convergence behaviors theoretically discussed in Subsection 6.3.1, neural state $x(t)$ of CTZD model (6.3) may move toward the local minimum point and then stop; in contrast, the state x_k of Newton iteration (6.10) may converge to the theoretical root within a few iterations. For further illustration and comparison, let us consider the following nonlinear equation:

$$f(x) = x^3 - x - 3 = 0. \tag{6.18}$$

The theoretical root we consider is $x^* = 1.6717$. As shown in Figure 6.15(a) and (b), starting from initial state $x(0) = 0.5$, neural state $x(t)$ of CTZD model (6.3), with $\gamma = 1$, cannot converge to the theoretical root x^*; instead, it stops at the point $x = -0.5774$, which is the local minimum of nonlinear Equation (6.18). In contrast, starting from the same initial state, the state x_k of Newton iteration (6.10) converges to the theoretical root within a few iterations. These illustrate that Newton iteration (6.10) can have a superior effectiveness property when nonlinear function $f(x)$ possesses a local minimum, as compared with CTZD model (6.3). Note that, as observed from related simulation results, when we choose the initial state $x_0 = 0.6$ or a bigger value, neural state $x(t)$ of CTZD model (6.3) with $\gamma = 1$ converges to the theoretical root. Thus, if nonlinear function $f(x)$ possesses a local minimum, the initial state of CTZD model (6.3) has to be set close enough to the theoretical root so as for state $x(t)$ to avoid falling into the basin of attraction of the local minimum of $f(x)$.

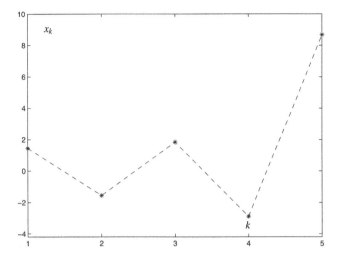

(a) "Wrong" state trajectory of Newton iteration (6.10)

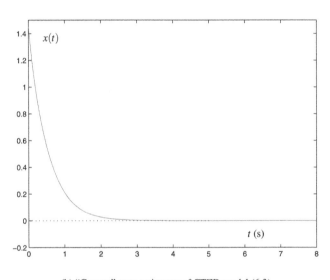

(b) "Correct" state trajectory of CTZD model (6.3)

FIGURE 6.13 Online solution of (6.16) by Newton iteration (6.10) and CTZD model (6.3), with $\gamma = 1$, starting from the same initial state $x_0 = x(0) = 1.45$.

6.4 Further Discussion to Avoid Local Minimum

In this section, two CTZD models with modified error functions (MEFs) are proposed for solving the nonlinear Equation (6.1) containing some local minimum point(s).

6.4.1 CTZD Model with the First MEF

To solve the nonlinear Equation (6.1) containing at least a local minimum point, following Zhang *et al.*'s neural-dynamic design method (i.e., zeroing dynamics design method), we first define the

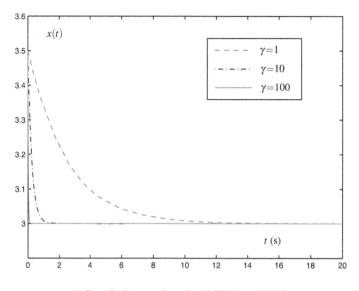

(a) Expedited state trajectories of CTZD model (6.3)

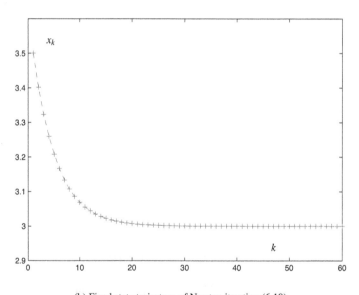

(b) Fixed state trajectory of Newton iteration (6.10)

FIGURE 6.14 Online solution of (6.17) by CTZD model (6.3), with different γ and by Newton iteration (6.10), starting from the same initial state $x(0) = x_0 = 3.5$.

following MEF, which is also an indefinite error function for proposing a CTZD model (i.e., a so-called CTZD model based on the first MEF, FMEF-based CTZD):

$$e(t) = \exp(\mu x)f(x), \text{ with } \mu > 0, \tag{6.19}$$

which differs from the error function defined explicitly or implicitly in the previous three sections [i.e., $f(x)$ itself]. Note that, if we force $e(t)$ converge to zero, in view of $\exp(\mu x) \neq 0$, then $f(x)$ converges to zero as well and neural state $x(t)$ converges to a root of nonlinear Equation (6.1).

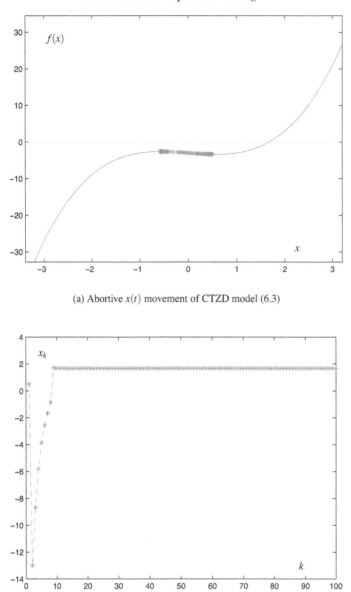

(a) Abortive $x(t)$ movement of CTZD model (6.3)

(b) Successful trajectory of Newton iteration (6.10)

FIGURE 6.15 Convergence performance of CTZD model (6.3), with $\gamma = 1$ and Newton iteration (6.10) for online solution of nonlinear Equation (6.18), starting from the same initial state $x(0) = x_0 = 0.5$.

Secondly, the time derivative $\dot{e}(t)$ of error function $e(t)$ should be chosen, such that $e(t)$ exponentially converges to zero. Specifically, $\dot{e}(t)$ is chosen via the following Zhang *et al.*'s design formula (i.e., ZD design formula) [99–102, 104, 107, 114, 117, 118, 122, 128, 129, 134]:

$$\frac{\mathrm{d}e(t)}{\mathrm{d}t} = -\gamma\phi\big(e(t)\big), \tag{6.20}$$

or to say,

$$\frac{\mathrm{d}\left(\exp(\mu x)f(x)\right)}{\mathrm{d}t} = -\gamma\phi\left(\exp(\mu x)f(x)\right). \tag{6.21}$$

Thirdly, expanding the above ZD design formula (6.21), with time $t \in [0, +\infty)$ and $f'(x) = \partial f(x)/\partial x$, we obtain the following derivation result:

$$\frac{\partial\left(\exp(\mu x)f(x)\right)}{\partial x}\frac{\mathrm{d}x}{\mathrm{d}t} = -\gamma\phi\left(\exp(\mu x)f(x)\right),$$

and then the FMEF-based CTZD model:

$$\exp(\mu x)(\mu f(x) + f'(x))\dot{x} = -\gamma\phi\left(\exp(\mu x)f(x)\right),$$

or, depicted in explicit dynamics,

$$\dot{x}(t) = -\gamma\frac{\phi\left(\exp(\mu x)f(x)\right)}{\exp(\mu x)(\mu f(x) + f'(x))}, \tag{6.22}$$

where $x(t)$ denotes the neural state of the FMEF-based CTZD model (6.22) corresponding to theoretical root x^* of nonlinear Equation (6.1).

Theorem 8 *Consider a solvable nonlinear equation $f(x) = 0$, where function $f(\cdot)$ is continuously differentiable (i.e., with at least the first-order derivative at some interval containing x^*). If a monotonically increasing odd activation function $\phi(\cdot)$ is used, then neural state x of the FMEF-based CTZD model (6.22), starting from any initial state $x(0) = x_0$ close enough to x^*, converges to the theoretical root x^* of nonlinear equation $f(x) = 0$ [note that the specific value of x^*, in the situation of not less than two roots existing, depends on the sufficient closeness of initial state x_0].*

Proof We define a Lyapunov function candidate $v(x) = \exp(2\mu x)f^2(x)/2$ for the FMEF-based CTZD model (6.22). In fact, we have

1) $v(x^*) = 0$, and $v(x) > 0$ for any $x \neq x^*$;

2) its time derivative

$$\begin{aligned}
\dot{v}(x) &= \frac{\mathrm{d}v(x)}{\mathrm{d}t} = \frac{\partial v(x)}{\partial x}\frac{\mathrm{d}x}{\mathrm{d}t} \\
&= \mu\exp(2\mu x)f^2(x)\frac{\mathrm{d}x}{\mathrm{d}t} + \exp(2\mu x)f(x)f'(x)\frac{\mathrm{d}x}{\mathrm{d}t} \\
&= \exp(2\mu x)f(x)(\mu f(x) + f'(x))\frac{\mathrm{d}x}{\mathrm{d}t} \\
&= -\gamma\exp(\mu x)f(x)\phi(\exp(\mu x)f(x)).
\end{aligned} \tag{6.23}$$

As the monotonically increasing odd activation function is used, we have $\phi(-f(x)) = -\phi(f(x))$. Besides, $\exp(\mu x)$ is always greater than zero. Therefore, we obtain

$$\phi\left(\exp(\mu x)f(x)\right)\begin{cases} > 0, & \text{if } f(x) > 0, \\ = 0, & \text{if } f(x) = 0, \\ < 0, & \text{if } f(x) < 0, \end{cases}$$

and

$$\exp(\mu x)f(x)\phi\left(\exp(\mu x)f(x)\right)\begin{cases} > 0, & \text{if } f(x) \neq 0, \\ = 0, & \text{if } f(x) = 0, \end{cases}$$

which guarantees the final negative-definiteness of $\dot{v}(x)$. That is, in terms of $f(x)$, we have $\dot{v}(x) < 0$ for $f(x) \neq 0$ [equivalently, $x \neq x^*$], and $\dot{v}(x) = 0$ for $f(x) = 0$ [equivalently, $x = x^*$]. By the Lyapunov theory, neural state x of FMEF-based CTZD model (6.22) can converge to a theoretical solution x^* with $f(x^*) = 0$. Equivalently, $e(t) = \exp(\mu x) f(x)$ can converge to zero. Moreover, at the local minima [with $f'(x) = 0$], the denominator of the right-hand side of FMEF-based CTZD model (6.22) is not zero. Hence, neural state $x(t)$ of FMEF-based CTZD model (6.22) can escape from the local minima and converge to a theoretical root nearby. Note that, for comparison, neural state $x(t)$ of the conventional CTZD model cannot escape from the local minima because its related denominator becomes zero. The proof is now complete.

It is worth pointing out that different convergence performance of FMEF-based CTZD model (6.22) can be achieved by using different design parameter γ and different activation function $\phi(\cdot)$. In general, any monotonically increasing odd function can be chosen as the activation function of the FMEF-based CTZD model. In addition, by reviewing the FMEF-based CTZD model (6.22), if we utilize linear activation function $\phi(e) = e$, the FMEF-based CTZD model (6.22) reduces to

$$\dot{x}(t) = -\gamma \frac{f(x)}{\mu f(x) + f'(x)}.$$

6.4.2 CTZD Model with the Second MEF

For comparison and investigation purposes, we define another modified error function for proposing a new CTZD model (i.e., a so-called CTZD model based on the second MEF, for short, SMEF-based CTZD) to solve nonlinear Equation (6.1):

$$e(t) = f(x)/x^2 \text{ for } x \neq 0 \text{ and } f(0) \neq 0. \tag{6.24}$$

Note that, when $x \neq 0$, if we force $e(t)$ converge to zero, in view of $1/x^2 \neq 0$, then $f(x)$ converges to zero as well and state x converges to a root of (6.1).

Similar to the design of the above FMEF-based CTZD model (6.22), the time derivative $\dot{e}(t)$ of error function $e(t)$ in (6.24) is chosen so that $e(t)$ converges to zero; simply, via the Zhang *et al.*'s design formula:

$$\frac{de(t)}{dt} = -\gamma \phi\left(e(t)\right),$$

or equivalently,

$$\frac{d\left(f(x)/x^2\right)}{dt} = -\gamma \phi\left(f(x)/x^2\right). \tag{6.25}$$

In (6.25), design parameter γ and activation function $\phi(\cdot)$ are defined the same as those in the FMEF-based CTZD model (6.22). Besides, from (6.25), it follows that

$$\dot{x}(t) = -\gamma \frac{\phi\left(f(x)/x^2\right) x^3}{-2f(x) + x f'(x)}, \tag{6.26}$$

where x is the state and output of the SMEF-based CTZD model (6.26) corresponding to theoretical root x^* of nonlinear Equation (6.1). In addition, if we utilize linear activation function $\phi(e) = e$, the SMEF-based CTZD model (6.26) reduces to

$$\dot{x}(t) = -\gamma \frac{x f(x)}{-2f(x) + x f'(x)}.$$

Similar to Theorem 8, we have the following direct theoretical result (i.e., a corollary) on the convergence of the SMEF-based CTZD model (6.26).

Corollary 3 *Consider a solvable nonlinear equation $f(x) = 0$ [with $f(0) \neq 0$], where function $f(\cdot)$ is continuously differentiable (i.e., with at least the first-order derivative at some interval containing x^*). If monotonically increasing odd activation function $\phi(\cdot)$ is used, then neural state x of the SMEF-based CTZD model (6.26), starting from any nonzero initial state x_0 close enough to x^*, converges to the theoretical root x^* of nonlinear equation $f(x) = 0$.*

6.4.3   Illustrative Example

The previous two subsections have proposed the CTZD models with two modified error functions [i.e., the FMEF-based CTZD model (6.22) and SMEF-based CTZD model (6.26)] for solving the nonlinear equation containing at least a local minimum point. In this subsection, computer-simulation results and observations are provided for substantiating the efficacy of the two new CTZD models.

To investigate the effectiveness of the FMEF-based CTZD model (6.22) and SMEF-based CTZD model (6.26), let us consider the following nonlinear equation:

$$f(x) = 0.01(x+7)(x-1)(x-8) + \sin x + 2.4 = 0, \qquad (6.27)$$

with its three theoretical roots denoted by "root 1," "root 2," and "root 3" in Figure 6.16. Both of CTZD models (6.22) and (6.26) are exploited for solving the above nonlinear Equation (6.27), which includes a local minimum point $x \approx -1.08$. Using the FMEF-based CTZD model (6.22) with $\mu = 1$ and $\gamma = 10$, we see from Figure 6.16(a) that, starting from four illustrative initial states, neural states $x(t)$ all converge to their corresponding theoretical roots.

1) For example, if initial state $x(0)$ is -8.9 (corresponding to "initial state 1" in the figure), neural state $x(t)$ of the FMEF-based CTZD model (6.22) converges to the "root 1," as depicted in Figure 6.16(a).

2) Similarly, starting from initial states $x(0) = 3.15$ and 9.5 (corresponding to "initial state 3" and "initial state 4," respectively, in the figure), neural states $x(t)$ of the FMEF-based CTZD model (6.22) converge to "root 2" and "root 3," respectively, which is also shown in Figure 6.16(a).

3) But, for some initial state [e.g., $x(0) = 2.9$ corresponding to "initial state 2" in Figure 6.16(a)], which is close to the local minimum point, the neural state of the previously proposed CTZD model (6.3) moves toward the local minimum point and then stops with a warning information (see also Examples 6.2 and 6.9); whereas neural state of GD model (6.4) converges to the local minimum point without any warning. On the other hand, in this situation, starting from the "initial state 2," neural state $x(t)$ of the FMEF-based CTZD model (6.22) escapes from the local minimum point and converges to "root 1" successfully and correctly, which is shown in Figure 6.16(a).

Furthermore, to investigate and verify confirmedly the convergence performance of the FMEF-based CTZD model (6.22), as shown in Figure 6.16(b), starting from 80 randomly generated initial states, neural states $x(t)$ of the FMEF-based CTZD model (6.22), with $\mu = 1$ and $\gamma = 10$, all converge well to the theoretical roots (i.e., "root 1," "root 2," and "root 3") within a short time (e.g., 0.6 s or so).

In addition, the simulative results of the SMEF-based CTZD model (6.26) for solving nonlinear Equation (6.27) are illustrated in Figure 6.16(c) and (d). Similar to the situation of the FMEF-based CTZD model (6.22), starting from four illustrative initial states, neural states $x(t)$ of the SMEF-based CTZD model (6.26) all converge to their corresponding theoretical roots (of which the state-trajectory subfigure is omitted due to result similarity). It is worth mentioning here that, especially starting from "initial state 2," neural state $x(t)$ of the SMEF-based CTZD model (6.26) escapes from the local minimum point and converges to "root 1" successfully. To further investigate

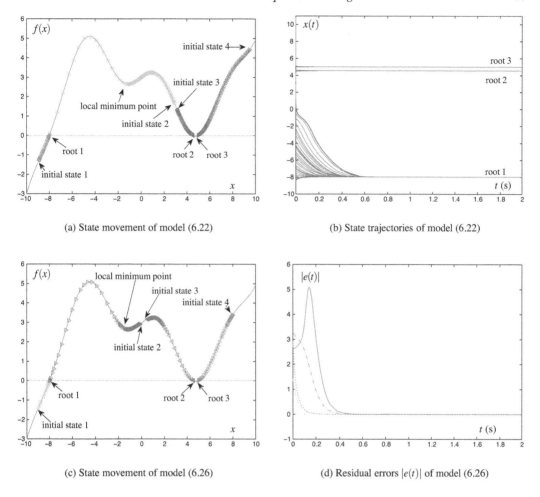

(a) State movement of model (6.22)

(b) State trajectories of model (6.22)

(c) State movement of model (6.26)

(d) Residual errors $|e(t)|$ of model (6.26)

FIGURE 6.16 Convergence performance of CTZD models (6.22) and (6.26) solving nonlinear Equation (6.27), where labeled ranges of subfigures are $[-10, 10] \times [-3, 6]$, $[0, 2] \times [-10, 10]$, $[-10, 10] \times [-3, 6]$, and $[0, 2] \times [-1, 6]$.

and verify the convergence performance, as shown in Figure 6.16(d), the corresponding residual errors $|e(t)| = |0.01(x(t) + 7)(x(t) - 1)(x(t) - 8) + \sin x(t) + 2.4|$ are monitored during the problem solving processes of the SMEF-based CTZD model (6.26), starting from the four initial states. From Figure 6.16(d), we see that the residual errors all converge to zero within around 0.6 s.

Based on the above simulative results, we draw the conclusion that both of the FMEF-based CTZD model (6.22) and SMEF-based CTZD model (6.26) escape from the local minimum point and converge to a theoretical root nearby successfully. These also illustrate the effectiveness of the proposed new CTZD models with modified error functions on solving nonlinear equation problem (6.27). Note that using modified (or to say, different) error functions as the starting point of the ZD model design is an interesting research direction.

In order to investigate insightfully the convergence performance of the FMEF-based CTZD model (6.22) and SMEF-based CTZD model (6.26) for solving the nonlinear Equation (6.27), including a local minimum point, the function graphs of $f(x) = 0.01(x + 7)(x - 1)(x - 8) + \sin x + 2.4$ and $\exp(\mu x) f(x)$ are plotted in Figure 6.17, and the function graphs of $f(x)/x^2$ are plotted in Figure 6.18.

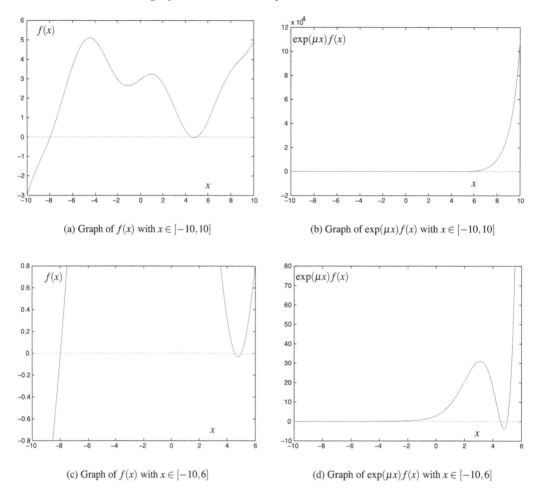

(a) Graph of $f(x)$ with $x \in [-10,10]$

(b) Graph of $\exp(\mu x)f(x)$ with $x \in [-10,10]$

(c) Graph of $f(x)$ with $x \in [-10,6]$

(d) Graph of $\exp(\mu x)f(x)$ with $x \in [-10,6]$

FIGURE 6.17 Function graphs [112] of $f(x)$ and $\exp(\mu x)f(x)$ explaining the efficacy of CTZD model (6.22) on solving nonlinear Equation (6.27), where ordinate ranges of subfigures are $[-3,6]$, $[-2,12] \cdot 10^4$, $[-0.8,0.8]$, and $[-10,80]$.

1) Specifically, as seen from Figure 6.17, compared with the situation of $f(x)$, the local minimum point $x \approx -1.08$ disappears in the curve of modified error function $\exp(\mu x)f(x)$, while the zero points (or to say, root points) of $f(x)$ do not change in $\exp(\mu x)f(x)$. These therefore guarantee that the solution process of the FMEF-based CTZD model (6.22) escapes from "the local minimum point" and converges to the theoretical root successfully.

2) Moreover, as seen from Figure 6.17(a) and (c), as well as Figure 6.18, the local minimum point $x \approx -1.08$ of $f(x)$ also disappears in the curve of error function $f(x)/x^2$, while the zero points of $f(x)$ do not change in $f(x)/x^2$. These, therefore, guarantee as well that the solution process of the SMEF-based CTZD model (6.26) escapes from "the local minimum point" and converges to the theoretical root successfully.

This function analysis explains the efficacy of the FMEF-based CTZD model (6.22) and SMEF-based CTZD model (6.26) on solving nonlinear Equation (6.27), which includes a local minimum point $x \approx -1.08$.

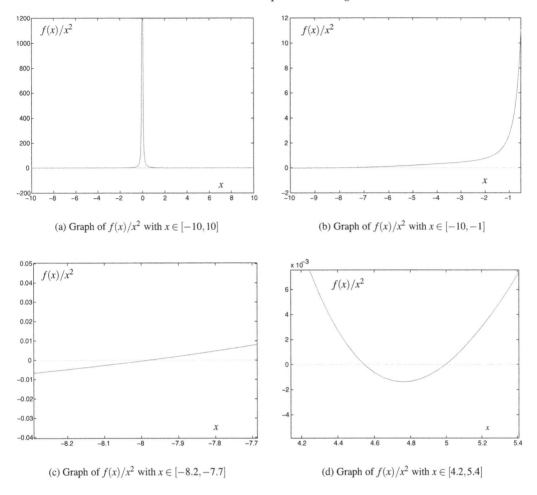

(a) Graph of $f(x)/x^2$ with $x \in [-10, 10]$

(b) Graph of $f(x)/x^2$ with $x \in [-10, -1]$

(c) Graph of $f(x)/x^2$ with $x \in [-8.2, -7.7]$

(d) Graph of $f(x)/x^2$ with $x \in [4.2, 5.4]$

FIGURE 6.18 Function graphs of $f(x)/x^2$ explaining the efficacy of CTZD model (6.26) on solving (6.27), where labeled ranges of subfigures are $[-10, 10] \times [2, 12] \cdot 10^2$, $[-10, -1] \times [-2, 12]$, $[-8.2, -7.7] \times [-4, 5] \cdot 10^{-2}$, and $[4.2, 5.4] \times [-4, 6] \cdot 10^{-3}$.

Before ending this section, it is worth mentioning that, similar to Section 6.2, the discrete-time models of the FMEF-based CTZD model (6.22) and SMEF-based CTZD model (6.26) can also be developed for solving online the nonlinear Equation (6.27), including a local minimum point. The corresponding numerical experiments are left to interested readers to complete as a topic of exercise.

6.5 Summary

For solving nonlinear equations, a special class of neural dynamics has been generalized, developed, modeled, and compared in this chapter by following Zhang *et al.*'s method. Different from the conventional GD model, the resultant CTZD model has been elegantly derived from an indefinite error function. Following the previous studies on the GD model and the CTZD model [134], we have further investigated and compared the convergence performance of such two models. As an extension and generalization of our previous work, the DTZD model has been developed from the CTZD model for solving the static nonlinear equations in the discrete manner. Moreover, we have discovered that the DTZD model contains Newton iteration as a special case in some sense, which has further presented another new reasonable explanation/interpretation about the theoretical origination of the Newton iteration. In addition, we have improved the CTZD model by defining two modified error functions and generated new neural-dynamic forms to overcome the local-minimum problem. Both theoretical analysis and simulative/numerical results have illustrated the efficacy and superiority of the ZD models for static nonlinear equations solving. At the end of the chapter, it is also worth mentioning that possible links between the ZD method and other dynamics methods (such as dynamic relaxation method [84] and Davidenko-type method [37]), as well as comparisons with other numerical methods (such as [2,21,27]), are left to interested readers to complete as topics of exercise.

Chapter 7

System of Nonlinear Equations Solving

Abstract

In this chapter, a general CTZD model is proposed and investigated for solving the system of time-varying nonlinear equations. The solution error of such a CTZD model can large-scale exponentially converge to zero, which is analyzed with different activation functions considered. In addition, such a CTZD model is discretized for potential digital hardware implementation. This produces the DTZDK and DTZDU models according to the criterion whether the time-derivative information is known or not. The DTZD models are further improved by using Broyden method, and thus DTZDK-B and DTZDU-B models are generated. The efficacy of such ZD models is verified via simulative and numerical examples.

7.1   Problem Formulation and CTZD Model

In this chapter, much more complex nonlinear equations are further considered and investigated. Specifically, let us consider the following solvable system of time-varying nonlinear Equations [123]:

$$\mathbf{f}(\mathbf{x}(t),t) = 0 \in \mathbb{R}^n, \ t \in [0,+\infty), \tag{7.1}$$

where $\mathbf{f}(\cdot) : \mathbb{R}^n \to \mathbb{R}^n$ denotes the differentiable nonlinear mapping function. Our main objective is to find $\mathbf{x}(t) \in \mathbb{R}^n$ in real time t, such that the above system of smoothly time-varying nonlinear equations depicted in (7.1) holds true.

7.1.1   CTZD Model

To monitor and control the solving process of the system of time-varying nonlinear equations, we define the following vector-valued indefinite error function:

$$\mathbf{e}(t) = \mathbf{f}(\mathbf{x}(t),t) = [f_1(\mathbf{x}(t),t), f_2(\mathbf{x}(t),t), \cdots, f_n(\mathbf{x}(t),t)]^{\mathrm{T}}, \tag{7.2}$$

where the jth element of $\mathbf{e}(t)$, $e_j(t) = f_j(\mathbf{x}(t),t)$, with $j = 1,2,\cdots,n$. Evidently, if $\mathbf{e}(t)$ converges to zero, then $\mathbf{x}(t)$ converges to the theoretical solution $\mathbf{x}^*(t)$.

Then, to make every element of $\mathbf{e}(t)$ converge to zero, a suggested ZD design formula is adopted as follows [99–102, 104, 107, 114, 117, 118, 122, 128, 129, 134]:

$$\frac{d\mathbf{e}(t)}{dt} = -\gamma\Phi(\mathbf{e}(t)), \text{ i.e., } \frac{d\mathbf{f}(\mathbf{x}(t),t)}{dt} = -\gamma\Phi(\mathbf{f}(\mathbf{x}(t),t)), \tag{7.3}$$

where design parameter $\gamma > 0$, being the reciprocal of a capacitance parameter, is used to scale the convergence rate of the ZD model, which should be set as large as the hardware would permit, or selected appropriately for simulative or experimental purposes. Besides, $\Phi(\cdot) : \mathbb{R}^n \to \mathbb{R}^n$ denotes a vector array of activation functions; and $\phi(\cdot)$ is used to denote the element of $\Phi(\cdot)$, i.e., a specific activation function. Generally, any monotonically increasing odd activation function $\phi(\cdot)$ can be used for constructing the neural-dynamic model with convergence expedited. Note again that different choices of activation function $\phi(\cdot)$ lead to different convergence performance. In this chapter, the following two types of $\phi(\cdot)$ are used:

1) Linear activation function $\phi(e_j) = e_j$

2) Power-sum activation function $\phi(e_j) = \sum_{\kappa=1}^{N} e_j^{2\kappa-1}$ with integer $N \geqslant 2$

Finally, by expanding ZD design formula (7.3), the following differential equation of the CTZD model is obtained as

$$J(\mathbf{x}(t),t)\dot{\mathbf{x}}(t) = -\gamma\Phi(\mathbf{f}(\mathbf{x}(t),t)) - \frac{\partial\mathbf{f}(\mathbf{x}(t),t)}{\partial t}, \tag{7.4}$$

where

$$J(\mathbf{x}(t),t) = \frac{\partial\mathbf{f}}{\partial\mathbf{x}} = \begin{bmatrix} \frac{\partial f_1}{\partial x_1} & \frac{\partial f_1}{\partial x_2} & \cdots & \frac{\partial f_1}{\partial x_n} \\ \frac{\partial f_2}{\partial x_1} & \frac{\partial f_2}{\partial x_2} & \cdots & \frac{\partial f_2}{\partial x_n} \\ \vdots & \vdots & \ddots & \vdots \\ \frac{\partial f_n}{\partial x_1} & \frac{\partial f_n}{\partial x_2} & \cdots & \frac{\partial f_n}{\partial x_n} \end{bmatrix} \in \mathbb{R}^{n\times n}, \quad \frac{\partial\mathbf{f}}{\partial t} = \begin{bmatrix} \frac{\partial f_1}{\partial t} \\ \frac{\partial f_2}{\partial t} \\ \vdots \\ \frac{\partial f_n}{\partial t} \end{bmatrix} \in \mathbb{R}^n.$$

If Jacobian matrix $J(\mathbf{x}(t),t)$ is nonsingular (in this chapter, we only consider this situation), the above dynamics is rewritten as below:

$$\dot{\mathbf{x}}(t) = -J^{-1}(\mathbf{x}(t),t)\left(\gamma\Phi(\mathbf{f}(\mathbf{x}(t),t)) + \frac{\partial\mathbf{f}(\mathbf{x}(t),t)}{\partial t}\right), \tag{7.5}$$

where $\mathbf{x}(t)$, starting from randomly generated initial condition $\mathbf{x}(0) \in \mathbb{R}^n$, denotes the neural state (or to say, state vector) corresponding to theoretical solution $\mathbf{x}^*(t)$ of the system of time-varying nonlinear Equation (7.1).

7.1.2 Theoretical Analysis

In this subsection, detailed design consideration and theoretical analysis of CTZD model (7.4) for solving the system of time-varying nonlinear Equations (7.1) are presented as follows.

Theorem 9 *Consider a solvable system of time-varying nonlinear Equations (7.1). If monotonically increasing odd activation function array $\Phi(\cdot)$ is employed, CTZD model (7.4) for solving (7.1) is large-scale asymptotically stable.*

Proof We define a Lyapunov function candidate as follows:

$$v(\mathbf{x}(t),t) = \frac{1}{2}\|\mathbf{f}(\mathbf{x}(t),t)\|_2^2 = \frac{1}{2}\mathbf{f}^{\mathrm{T}}(\mathbf{x}(t),t)\mathbf{f}(\mathbf{x}(t),t) = \frac{1}{2}\sum_{j=1}^{n} f_j^2 \geqslant 0, \tag{7.6}$$

which guarantees the positive-definiteness of $v(\mathbf{x}(t),t)$; i.e., $v(\mathbf{x}(t),t) > 0$ for any $f_j(\mathbf{x}(t),t) \neq 0$, and $v(\mathbf{x}(t),t) = 0$ only for all $f_j(\mathbf{x}(t),t) = 0$, with $j = 1, 2, \cdots, n$. Thus, along the system trajectory of CTZD model (7.4), time derivative $\dot{v}(\mathbf{x}(t),t)$ is

$$\dot{v}(\mathbf{x}(t),t) = \frac{\mathrm{d}v(\mathbf{x}(t),t)}{\mathrm{d}t} = \sum_{j=1}^{n} f_j \frac{\mathrm{d}f_j}{\mathrm{d}t} = -\gamma \sum_{j=1}^{n} f_j(\mathbf{x}(t),t)\phi(f_j(\mathbf{x}(t),t)). \tag{7.7}$$

The element of $\Phi(\cdot)$ [i.e., $\phi(\cdot)$] is a monotonically increasing odd function, thus $\phi(-f_j(\mathbf{x}(t),t)) = -\phi(f_j(\mathbf{x}(t),t))$, and

$$\phi(f_j(\mathbf{x}(t),t)) \begin{cases} > 0, & \text{if } f_j(\mathbf{x}(t),t) > 0, \\ = 0, & \text{if } f_j(\mathbf{x}(t),t) = 0, \\ < 0, & \text{if } f_j(\mathbf{x}(t),t) < 0. \end{cases} \tag{7.8}$$

Hence, we have

$$f_j(\mathbf{x}(t),t)\phi(f_j(\mathbf{x}(t),t)) \begin{cases} > 0, & \text{if } f_j(\mathbf{x}(t),t) \neq 0, \\ = 0, & \text{if } f_j(\mathbf{x}(t),t) = 0, \end{cases} \tag{7.9}$$

which guarantees the final negative-definiteness of $\dot{v}(\mathbf{x}(t),t)$; i.e., $\dot{v}(\mathbf{x}(t),t) < 0$ for $\|\mathbf{f}(\mathbf{x}(t),t)\|_2 \neq 0$ [i.e., for any $f_j(\mathbf{x}(t),t) \neq 0$, with $j = 1, 2\cdots, n$] and $\dot{v}(\mathbf{x}(t),t) = 0$ for $\|\mathbf{f}(\mathbf{x}(t),t)\|_2 = 0$ [i.e., for all $f_j(\mathbf{x}(t),t) = 0$, with $j = 1, 2\cdots, n$]. Apparently, Lyapunov function candidate (7.6) is positive definite [with respect to $\mathbf{f}(\mathbf{x}(t),t)$], and its time derivative (7.7) is negative definite, which satisfy the requirement of the Lyapunov theory. In addition, if $\|\mathbf{f}(\mathbf{x}(t),t)\|_2 \to \infty$, Lyapunov function candidate (7.6) tends to infinity. Thus, CTZD model (7.4) for solving (7.1) is large-scale asymptotically stable (or to say, large-scale convergent). Please also see and compare with Definitions 1 and 2 in Section 5.3 of Chapter 5 (i.e., the connection and difference between the two terms "global" and "large-scale"). Note that, when $e_j(t) = f_j(\mathbf{x}(t),t)$ with $j = 1, 2\cdots, n$ converges to zero, neural state $\mathbf{x}(t)$ of CTZD model (7.4), starting from randomly generated initial state $\mathbf{x}(0)$, converges to the theoretical solution $\mathbf{x}^*(t)$. The proof is now complete.

Theorem 10 *In addition to Theorem 9, CTZD model (7.4) possesses the following properties.*

1) *Using the linear activation function array, CTZD model (7.4) possesses exponential convergence with rate γ and in two ways proved [i.e., residual error $e_j(t) = f_j(\mathbf{x}(t),t) \to 0$ for any $j \in \{1, 2, \cdots, n\}$, and neural state $\mathbf{x}(t) \to \mathbf{x}^*(t)$].*

2) *Using the power-sum activation function array, CTZD model (7.4) possesses superior convergence (i.e., better than the γ-rate exponential convergence), as compared with that using the linear activation function array.*

Proof The convergence analysis of CTZD model (7.4) activated by the above two types of activation functions is presented as below.

1) If CTZD model is activated by the linear activation function array, then we obtain $\mathrm{d}\mathbf{f}(\mathbf{x}(t),t)/\mathrm{d}t = -\gamma\mathbf{f}(\mathbf{x}(t),t)$ from Equation (7.3), and its analytical solution

$$\mathbf{f}(\mathbf{x}(t),t) = \exp(-\gamma t)\mathbf{f}(\mathbf{x}(0),0),$$

which equals to $e_j(t) = f_j(\mathbf{x}(t),t) = \exp(-\gamma t)f_j(\mathbf{x}(0),0)$ with $j = 1, 2, \cdots, n$. This proves that CTZD model (7.4) has exponential convergence with rate γ in terms of the consistent situation of any $e_j(t) = f_j(\mathbf{x}(t),t) \to 0$.

On the other hand, CTZD model (7.4) possesses the same convergence rate γ in terms of neural state $\mathbf{x}(t) \to \mathbf{x}^*(t)$ proved via the following procedure. Let $\tilde{\mathbf{x}}(t) \in \mathbb{R}^n$ denote the difference between the neural state solution $\mathbf{x}(t)$ of CTZD model (7.4) and the theoretical solution $\mathbf{x}^*(t)$ of Equation (7.1), i.e.,

$$\tilde{\mathbf{x}}(t) = \mathbf{x}(t) - \mathbf{x}^*(t) = [\tilde{x}_1(t), \tilde{x}_2(t), \cdots, \tilde{x}_n(t)]^T \in \mathbb{R}^n.$$

In view of $\mathbf{f}(\mathbf{x}^*(t), t) = \mathbf{f}([x_1^*(t), x_2^*(t), \cdots, x_n^*(t)]^T, t) = 0$, according to the aforementioned definition,

$$\mathbf{f}(\mathbf{x}(t), t) = \mathbf{f}([x_1(t), x_2(t), \cdots, x_n(t)]^T, t)$$
$$= \mathbf{f}([x_1^*(t) + \tilde{x}_1(t), x_2^*(t) + \tilde{x}_2(t), \cdots, x_n^*(t) + \tilde{x}_n(t)]^T, t).$$

By Taylor series expansion, omitting higher-order terms, we have

$$\mathbf{f}(\mathbf{x}(t), t) \approx \mathbf{f}(\mathbf{x}^*(t), t) + \sum_{j=1}^n \frac{\partial}{\partial x_j} \mathbf{f}(\mathbf{x}^*(t), t)(x_j(t) - x_j^*(t))$$

$$= \begin{bmatrix} \frac{\partial}{\partial x_1} f_1(\mathbf{x}^*(t), t) & \frac{\partial}{\partial x_2} f_1(\mathbf{x}^*(t), t) & \cdots & \frac{\partial}{\partial x_n} f_1(\mathbf{x}^*(t), t) \\ \frac{\partial}{\partial x_1} f_2(\mathbf{x}^*(t), t) & \frac{\partial}{\partial x_2} f_2(\mathbf{x}^*(t), t) & \cdots & \frac{\partial}{\partial x_n} f_2(\mathbf{x}^*(t), t) \\ \vdots & \vdots & \ddots & \vdots \\ \frac{\partial}{\partial x_1} f_n(\mathbf{x}^*(t), t) & \frac{\partial}{\partial x_2} f_n(\mathbf{x}^*(t), t) & \cdots & \frac{\partial}{\partial x_n} f_n(\mathbf{x}^*(t), t) \end{bmatrix} \begin{bmatrix} \tilde{x}_1(t) \\ \tilde{x}_2(t) \\ \vdots \\ \tilde{x}_n(t) \end{bmatrix}$$

$$= J(\mathbf{x}^*(t), t)\tilde{\mathbf{x}}(t).$$

Thus, we have $\mathbf{f}(\mathbf{x}(t), t) = \exp(-\gamma t)\mathbf{f}(\mathbf{x}(0), 0) \approx J(\mathbf{x}^*(t), t)\tilde{\mathbf{x}}(t)$. It is worth pointing out that, owing to the matrix $J(\mathbf{x}(t), t)$ being nonsingular, $\tilde{\mathbf{x}}(t) \approx \exp(-\gamma t)J^{-1}(\mathbf{x}^*(t), t)\mathbf{f}(\mathbf{x}(0), 0)$, and thus we have $\|\tilde{\mathbf{x}}(t)\|_2 \approx \exp(-\gamma t)\|J^{-1}(\mathbf{x}^*(t), t)\mathbf{f}(\mathbf{x}(0), 0)\|_2$. According to the invertibility condition [104, 132], for nonsingular matrix J with $\|J\|_F \leqslant \beta < \infty$ [where $\|\cdot\|_F$ denotes the F-norm (or to say, Frobenius norm) of a matrix argument], there exists positive real number φ, such that $\|J^{-1}\|_F \leqslant \varphi < \infty$. That is, $\|J^{-1}(\mathbf{x}^*(t), t)\|_F \leqslant \varphi < \infty$. With $\|\mathbf{f}(\mathbf{x}(0), 0)\|_2 \leqslant \alpha < \infty$, we finally have

$$\|\tilde{\mathbf{x}}(t)\|_2 \leqslant \exp(-\gamma t)\|J^{-1}(\mathbf{x}^*(t), t)\|_F \|\mathbf{f}(\mathbf{x}(0), 0)\|_2 \leqslant \alpha\varphi\exp(-\gamma t),$$

which shows again the exponential convergence of CTZD model (7.4) in terms of neural state $\mathbf{x}(t) \to \mathbf{x}^*(t)$.

2) If CTZD model (7.4) is activated by the power-sum activation function array, then we obtain $\mathrm{d}f_j(\mathbf{x}(t), t)/\mathrm{d}t = -\gamma \sum_{\kappa=1}^N f_j^{2\kappa-1}(\mathbf{x}(t), t)$ from Equation (7.3). Specifically, for $N = 2$, $\mathrm{d}f_j(\mathbf{x}(t), t)/\mathrm{d}t = -\gamma(f_j^3(\mathbf{x}(t), t) + f_j(\mathbf{x}(t), t))$, and its analytical solution [123] is written as

$$f_j(\mathbf{x}(t), t) = \frac{\mathrm{sgn}(f_j(\mathbf{x}(0), 0))\exp(-\gamma t)}{\sqrt{1 + f_j^{-2}(\mathbf{x}(0), 0) - \exp(-2\gamma t)}}, \tag{7.10}$$

where $\mathrm{sgn}(\cdot)$ denotes the signum function (or to say, operator) of an input scalar, which outputs 1, 0 or -1 corresponding to a positive, zero or negative input, respectively. Evidently, as $t \to \infty$, $f_j(\mathbf{x}(t), t) \to 0$, and $\|\mathbf{f}(\mathbf{x}(t), t)\|_2 = \sqrt{\sum_{j=1}^n f_j^2(\mathbf{x}(t), t)} \to 0$. Besides, from the proof of Theorem 9, for $v(\mathbf{x}(t), t) = \|\mathbf{f}(\mathbf{x}(t), t)\|_2^2/2$ and its time derivative $\dot{v}(\mathbf{x}(t), t) = -\gamma \sum_{j=1}^n f_j(\mathbf{x}(t), t)\phi(f_j(\mathbf{x}(t), t))$, we know that the following inequality holds: $f_j(\mathbf{x}(t), t)\phi(f_j(\mathbf{x}(t), t)) = \sum_{\kappa=1}^N f_j^{2\kappa}(\mathbf{x}(t), t) > f_j^2(\mathbf{x}(t), t)$ in the whole error range $e_j(t) = f_j(\mathbf{x}(t), t) \in (-\infty, +\infty)$. This implies that, when using the power-sum activation function array (or to say, power-sum activation functions), a much faster convergence is achieved by CTZD model (7.4) in such an error range in comparison with the situation of using the linear activation function array (or to say, linear activation functions).

7.1.3 Illustrative Example

In this subsection, via the following illustrative example, we simulatively substantiate the effectiveness of CTZD model (7.4) for solving the system of time-varying nonlinear equations depicted in (7.1).

Example 7.1 With $\mathbf{x}(t) = [x_1(t), x_2(t), x_3(t), x_4(t)]^T$, let us consider the following system of time-varying nonlinear equations:

$$\mathbf{f}(\mathbf{x}(t), t) = \begin{bmatrix} \ln x_1(t) - 1/(t+1) \\ x_1(t)x_2(t) - \exp(1/(t+1))\sin t \\ x_1^2(t) - (\sin t)x_2(t) + x_3(t) - 2 \\ x_1^2(t) - x_2^2(t) + x_3(t) + x_4(t) - t \end{bmatrix} = 0, \text{ where } x_1(t) > 0. \qquad (7.11)$$

The Jacobian matrix $J(\mathbf{x}(t), t)$ and the time-varying theoretical solution $\mathbf{x}^*(t)$ (which is given only for checking the correctness of the CTZD model solution) are written respectively as

$$J(\mathbf{x}(t), t) = \begin{bmatrix} 1/x_1(t) & 0 & 0 & 0 \\ x_2(t) & x_1(t) & 0 & 0 \\ 2x_1(t) & -\sin t & 1 & 0 \\ 2x_1(t) & -2x_2(t) & 1 & 1 \end{bmatrix} \text{ and } \mathbf{x}^*(t) = \begin{bmatrix} \exp(1/(t+1)) \\ \sin t \\ 2 - \exp(2/(t+1)) + \sin^2 t \\ t - 2 \end{bmatrix}.$$

The corresponding computer-simulation results of CTZD model (7.4) with different values of γ for solving (7.11) are illustrated in Figures 7.1 through 7.4, as well as in Table 7.1. It is worth mentioning that all neural states $x(t)$ start from initial states $\mathbf{x}(0) = [c/2, 4 - c, -8 + c, -6 + c]^T$, where $c = 1, 2, \cdots, 8$. Specifically, Figures 7.1 and 7.2 illustrate neural states $\mathbf{x}(t)$ of CTZD model (7.4) using different activation functions (i.e., linear and power-sum activation functions) over time interval $[0, 10]$ s. As seen from Figure 7.1, starting from initial states $\mathbf{x}(0)$, neural states $\mathbf{x}(t)$ of CTZD model (7.4) all converge to theoretical solution $\mathbf{x}^*(t)$. Moreover, the convergence of CTZD model (7.4) is expedited effectively by increasing the value of γ, which is seen from Figure 7.2. In other words, the larger the value of γ is, the faster the neural state $\mathbf{x}(t)$ of CTZD model (7.4) converges to $\mathbf{x}^*(t)$. Besides, the solution errors $\|\mathbf{x}(t) - \mathbf{x}^*(t)\|_2$ of CTZD model (7.4) with different values of γ and different activation functions are shown in Figures 7.3 and 7.4. Both of the figures further illustrate that the convergence of solution errors is expedited effectively with γ increased. In addition, the convergence time of CTZD model (7.4) using the power-sum activation function array is shorter than that using the linear activation function array.

Additionally, to achieve the solution precision $\|\mathbf{x}(t) - \mathbf{x}^*(t)\|_2 < 0.01$, the average convergence time of CTZD model (7.4) with different activation functions and different values of N and γ for solving the system of time-varying nonlinear Equations (7.11) is shown in Table 7.1. With the linear activation function array used [i.e., $\mathbf{f}(\mathbf{x}(t), t) = \exp(-\gamma t)\mathbf{f}(\mathbf{x}(0), 0)$], the average convergence time decreases roughly by 10 times, as the value of γ increases by 10 times. With the power-sum activation function array used, the average convergence time is obviously shorter than that using the linear activation function array. For example, when $\gamma = 1$, the average convergence time becomes shorter, with N getting larger. Note that, when $\gamma = 10$ or even larger, the average convergence time is not much different, even with N getting larger. The main reason of this phenomenon is that, with N becoming larger, the tiny change of $\mathbf{x}^5$, $\mathbf{x}^7$ or even $\mathbf{x}^9$ of power-sum activation functions does not make much difference (or to say, contribution) to the decrease of the convergence time, which is already very small. Besides, for $\gamma = 1000$, to achieve the solution precision (i.e., < 0.01), the average convergence time of CTZD model (7.4) is smaller than 0.01 s. Thus, CTZD model (7.4) with a suitable value of γ can be used to solve the system of time-varying nonlinear equations in real time.

In summary, the above simulative results illustrate the efficacy of CTZD model (7.4) for solving the system of time-varying nonlinear equations.

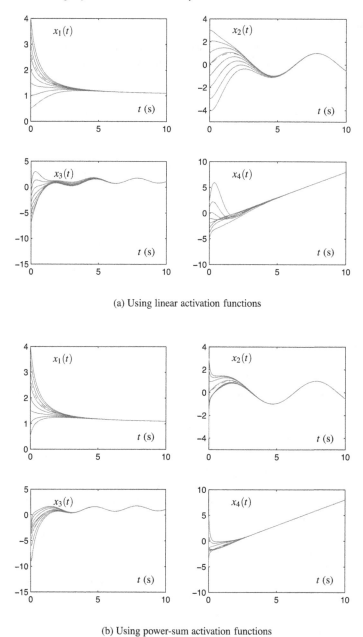

(a) Using linear activation functions

(b) Using power-sum activation functions

FIGURE 7.1 Neural states $\mathbf{x}(t)$ of CTZD model (7.4) using different activation functions with $N = 3$ and $\gamma = 1$ for solving system of time-varying nonlinear Equations (7.11).

7.2 Discrete-Time Models

To obtain the discrete-time models of CTZD model (7.4) for solving online the system of time-varying nonlinear equations, CTZD model (7.5) is further considered as a research basis. That is,

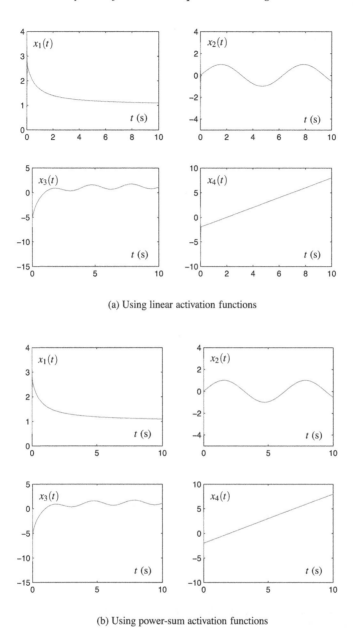

(a) Using linear activation functions

(b) Using power-sum activation functions

FIGURE 7.2 Neural states $\mathbf{x}(t)$ of CTZD model (7.4) using different activation functions with $N = 3$ and $\gamma = 100$ for solving system of time-varying nonlinear Equations (7.11).

denoting $\partial \mathbf{f}(\mathbf{x}(t),t)/\partial t = \mathbf{f}_t'(\mathbf{x}(t),t)$, defining the function $\mathbf{g}(\mathbf{x}(t),t)$ as

$$\mathbf{g}(\mathbf{x}(t),t) = -J^{-1}(\mathbf{x}(t),t)(\gamma \mathbf{f}(\mathbf{x}(t),t) + \mathbf{f}_t'(\mathbf{x}(t),t)),$$

and using Euler forward difference, from (7.5) we obtain (with the linear activation function array used for the analytical convenience):

$$\mathbf{x}_{k+1} = \mathbf{x}_k + \tau \mathbf{g}(\mathbf{x}_k,t_k) \quad \text{and} \quad t_{k+1} = t_k + \tau,$$

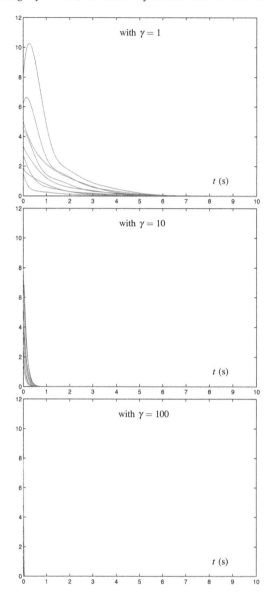

FIGURE 7.3 Solution errors $\|\mathbf{x}(t) - \mathbf{x}^*(t)\|_2$ of CTZD model (7.4) using linear activation functions with different γ, where subfigure ranges are all $[0, 10] \times [0, 12]$.

where $k = 0, 1, 2, \cdots$ denotes the update index (or to say, sampling index), which discretizes CTZD model (7.5). Then the DTZDK model is obtained as follows [i.e., with time-derivative $\mathbf{f}'_t(\mathbf{x}_k, t_k)$ assumed to be known]:

$$\mathbf{x}_{k+1} = \mathbf{x}_k - J^{-1}(\mathbf{x}_k, t_k)\big(h\mathbf{f}(\mathbf{x}_k, t_k) + \tau\mathbf{f}'_t(\mathbf{x}_k, t_k)\big), \qquad (7.12)$$

where $h = \tau\gamma > 0$ denotes the step size. For the situation of the time-derivative $\mathbf{f}'_t(\mathbf{x}_k, t_k)$ unknown, we replace $\mathbf{f}'_t(\mathbf{x}_k, t_k)$ in DTZDK model (7.12) by

$$\big(\mathbf{f}(\mathbf{x}_k, t_k) - \mathbf{f}(\mathbf{x}_k, t_{k-1})\big)/\tau$$

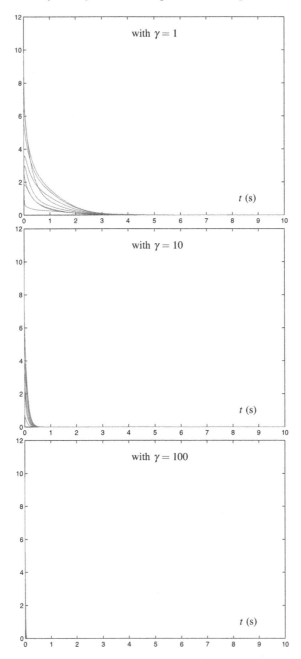

FIGURE 7.4 Solution errors $\|\mathbf{x}(t) - \mathbf{x}^*(t)\|_2$ of CTZD model (7.4) using power-sum activation functions with $N = 3$ and different γ, where subfigure ranges are all $[0, 10] \times [0, 12]$.

and obtain the DTZDU model as follows:

$$\mathbf{x}_{k+1} = \mathbf{x}_k + J^{-1}(\mathbf{x}_k, t_k)\big((1+h)\mathbf{f}(\mathbf{x}_k, t_k) - \mathbf{f}(\mathbf{x}_k, t_{k-1})\big). \tag{7.13}$$

TABLE 7.1 Average Convergence Time of CTZD Model (7.4) Using Different Activation Functions with Different Parameter Values to Achieve Precision $\|\mathbf{x}(t) - \mathbf{x}^*(t)\|_2 < 0.01$

γ	Linear activation	Power-sum activation			
		$N = 2$	$N = 3$	$N = 4$	$N = 5$
1	6.5944500	4.8321125	4.6814375	4.6669500	4.6001875
10	0.6352750	0.5546125	0.5436375	0.5480125	0.5538875
100	0.0659000	0.0599750	0.0601500	0.0603875	0.0606625
1000	0.0066500	0.0060250	0.0060375	0.0061125	0.0061000

7.2.1 Broyden Method

For solving a system of static nonlinear equations depicted in $\mathbf{f}(\mathbf{x}) = 0 \in \mathbb{R}^n$ with $\mathbf{x} \in \mathbb{R}^n$, Newton iteration [57] is developed as a special case of DTZDK model (7.12):

$$\mathbf{x}_{k+1} = \mathbf{x}_k - J^{-1}(\mathbf{x}_k)\mathbf{f}(\mathbf{x}_k), \tag{7.14}$$

where $J(\mathbf{x})$ is the Jacobian matrix of $\mathbf{f}(\mathbf{x})$. It is easy to derive the link between Newton iteration and the ZD models, which is thus omitted here. Note that they both suffer from the disadvantages of computing the inverse of Jacobian matrix, which may be time consuming. Broyden [7] suggested using a successively updated matrix B_k to approximate $J^{-1}(\mathbf{x}_k)$ at every iteration, and the update formula of B_k is

$$B_k = B_{k-1} + \frac{(\mathbf{p}_{k-1} - B_{k-1}\mathbf{q}_{k-1})\mathbf{p}_{k-1}^{\mathrm{T}}B_{k-1}}{\mathbf{p}_{k-1}^{\mathrm{T}}B_{k-1}\mathbf{q}_{k-1}},$$

where two auxiliary vectors are used: $\mathbf{p}_{k-1} = \mathbf{x}_k - \mathbf{x}_{k-1}$ and $\mathbf{q}_{k-1} = \mathbf{f}(\mathbf{x}_k) - \mathbf{f}(\mathbf{x}_{k-1})$. Thus, Equation (7.14) becomes

$$\mathbf{x}_{k+1} = \mathbf{x}_k - B_k\mathbf{f}(\mathbf{x}_k),$$

which significantly reduces the amount of calculation.

7.2.2 DTZD Models Aided with Broyden Method

Similar to Newton iteration, the DTZDK and DTZDU models require computing the inverse of Jacobian matrix $J(\mathbf{x}_k, t_k)$ at every step. Therefore we try to improve such two DTZD models using the Broyden method [121]. That is, $J^{-1}(\mathbf{x}_k, t_k)$ is approximated by B_k, where the update formula of B_k is

$$B_k = B_{k-1} + \frac{(\mathbf{p}_{k-1} - B_{k-1}\mathbf{q}_{k-1})\mathbf{p}_{k-1}^{\mathrm{T}}B_{k-1}}{\mathbf{p}_{k-1}^{\mathrm{T}}B_{k-1}\mathbf{q}_{k-1}}, \tag{7.15}$$

with $\mathbf{p}_{k-1} = \mathbf{x}_k - \mathbf{x}_{k-1}$ and $\mathbf{q}_{k-1} = \mathbf{f}(\mathbf{x}_k, t_k) - \mathbf{f}(\mathbf{x}_{k-1}, t_k)$. Replacing $J^{-1}(\mathbf{x}_k, t_k)$ by B_k in DTZDK model (7.12) yields the DTZDK-B model (i.e., the DTZDK model aided with the Broyden method):

$$\mathbf{x}_{k+1} = \mathbf{x}_k + B_k(h\mathbf{f}(\mathbf{x}_k, t_k) + \tau\mathbf{f}_t'(\mathbf{x}_k, t_k)). \tag{7.16}$$

Doing the same in DTZDU model (7.13), we obtain the DTZDU-B model (i.e., the DTZDU model aided with the Broyden method):

$$\mathbf{x}_{k+1} = \mathbf{x}_k + B_k\big((1+h)\mathbf{f}(\mathbf{x}_k, t_k) - \mathbf{f}(\mathbf{x}_k, t_{k-1})\big). \tag{7.17}$$

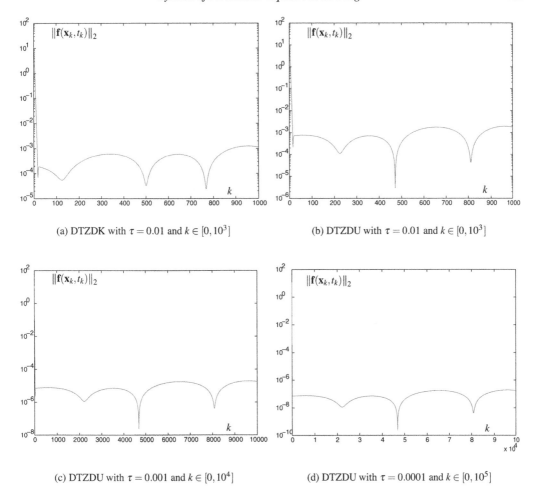

(a) DTZDK with $\tau = 0.01$ and $k \in [0, 10^3]$

(b) DTZDU with $\tau = 0.01$ and $k \in [0, 10^3]$

(c) DTZDU with $\tau = 0.001$ and $k \in [0, 10^4]$

(d) DTZDU with $\tau = 0.0001$ and $k \in [0, 10^5]$

FIGURE 7.5 Residual errors $\|\mathbf{f}(\mathbf{x}_k, t_k)\|_2$ of DTZDK model (7.12) and DTZDU model (7.13) for solving system of time-varying nonlinear Equations (7.18), with different τ, where ordinate ranges of subfigures are $[10^{-5}, 10^2]$, $[10^{-6}, 10^2]$, $[10^{-8}, 10^2]$, and $[10^{-10}, 10^2]$.

7.2.3 Illustrative Examples

In this subsection, we numerically illustrate the efficacy of DTZDK model (7.12), DTZDU model (7.13), DTZDK-B model (7.16), and DTZDU-B model (7.17). In addition, we simply use $h = 0.5$, and only investigate the effect of τ on the models.

Example 7.2 Let us consider and solve the following system of time-varying nonlinear equations:

$$\mathbf{f}(\mathbf{x}(t), t) = \begin{bmatrix} x_2(t) - \sqrt[3]{t+1} \\ x_1(t) - x_2^3(t) + \exp(x_3(t)) - \cos t \\ x_1(t)x_2^3(t) - 2t\exp(x_3(t)) - (\cos t - 2t)(t+1) \end{bmatrix} = 0. \tag{7.18}$$

On one hand, using DTZDK and DTZDU models, the residual errors $\|\mathbf{f}(\mathbf{x}_k, t_k)\|_2$ for solving the system are shown in Figure 7.5. As seen from the figure, DTZDK model (7.12) is better than DTZDU model (7.13) when using the same value of τ. The maximum steady-state residual errors of DTZDU model (7.13), with $\tau = 0.01$, $\tau = 0.001$ and $\tau = 0.0001$, are of the order of 10^{-3}, 10^{-5},

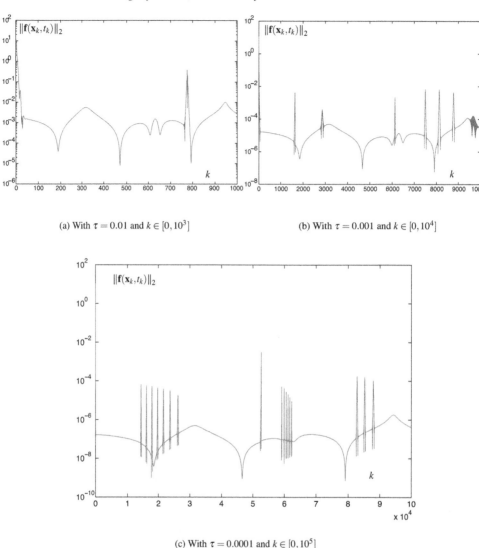

(a) With $\tau = 0.01$ and $k \in [0, 10^3]$ (b) With $\tau = 0.001$ and $k \in [0, 10^4]$

(c) With $\tau = 0.0001$ and $k \in [0, 10^5]$

FIGURE 7.6 Residual errors $\|\mathbf{f}(\mathbf{x}_k, t_k)\|_2$ of DTZDU-B model (7.17) for solving system of time-varying nonlinear Equations (7.18) with different τ, where ordinate ranges of subfigures are $[10^{-6}, 10^2]$, $[10^{-8}, 10^2]$, and $[10^{-10}, 10^2]$.

and 10^{-7}, respectively, showing an $O(\tau^2)$ pattern. Besides, note that all ordinates (or to say, vertical coordinates) of the numerical-experiment results (i.e., figures) are plotted in log scale.

On the other hand, using the DTZDU-B model as the improved DTZD model, its first big advantage is the speed. As we need not calculate the Jacobian inverse at every step, instead, by doing simple matrix-vector calculation according to the update formula (7.15), much time is saved. Further analysis of the speed improvement can be found at [7]. Therefore, in the following content, we mainly illustrate the improved models' precision (or to say, accuracy if implemented/coded on digital computers as numerical algorithms), which is measured in the form of residual error $\|\mathbf{f}(\mathbf{x}_k, t_k)\|_2$ as before. Using DTZDU-B model (7.17) to solve (7.18), the numerical results are shown in Figure 7.6. Note that there are some peaks in the figure (introduced by Broyden method), but most of the residual errors of DTZDU-B model (7.17) here are close to those of DTZDU model (7.13)

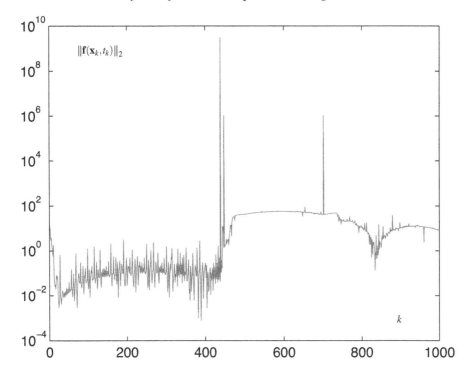

FIGURE 7.7 Residual error of "pure" Broyden method, with $\tau = 0.01$ solving (7.18).

[i.e., those in Figure 7.5(b) through (d)]. Similarly, an $O(\tau^2)$ pattern (or to say, manner) is observed roughly; i.e., when τ is reduced by 10 times, the residual error $\|\mathbf{f}(\mathbf{x}_k, t_k)\|_2$ of DTZDU-B model (7.17) at the nonpeak interval of index k generally reduce by 100 times. It is worth noting that, because of the similarity of the results and figures to the above ones, the corresponding numerical experiments of DTZDK-B model (7.16) are not presented (but left to interested readers to complete as a topic of exercise).

Before ending the example, it is worth comparing with the "pure" Broyden method (which is actually a variant of Newton iteration). Note that the improved DTZD-B (i.e., DTZDK-B and DTZDU-B) models can also be viewed as the improvements of the original Broyden method, which extend the ability of the Broyden method to solve the system of time-varying nonlinear equations (STVNE). For comparative purposes, now let us apply the original Broyden method (i.e., the "pure" Broyden method) directly to solving the STVNE problem (7.18) without combining ZD techniques. Thus, the updated formula for B_k is (7.15) with the same $\mathbf{p}_{k-1}$ but with $\mathbf{q}_{k-1} = \mathbf{f}(\mathbf{x}_k, t_k) - \mathbf{f}(\mathbf{x}_{k-1}, t_{k-1})$, and the updated formula for $\mathbf{x}_k$ is $\mathbf{x}_{k+1} = \mathbf{x}_k - B_k \mathbf{f}(\mathbf{x}_k, t_k)$. As Figure 7.7 shows, the "pure" Broyden method for solving the STVNE problem is not good enough, because it does not make use of the time-derivative information of the system. From the comparison between DTZDU-B [i.e., corresponding to Figure 7.6(a)] and the "pure" Broyden method [i.e., corresponding to Figure 7.7], it is evident that the Broyden-method aided DTZD models are effective improvements on the Broyden method for solving the STVNE.

Example 7.3 Another advantage of the improved models over the unimproved ones is that the improved models can easily cope with the systems of which the Jacobian may become singular during

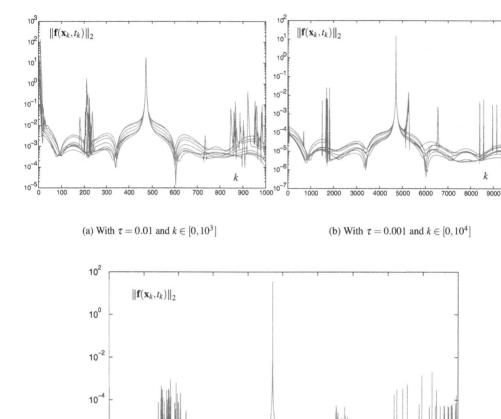

(a) With $\tau = 0.01$ and $k \in [0, 10^3]$

(b) With $\tau = 0.001$ and $k \in [0, 10^4]$

(c) With $\tau = 0.0001$ and $k \in [0, 10^5]$

FIGURE 7.8 Residual errors $\|\mathbf{f}(\mathbf{x}_k, t_k)\|_2$ of DTZDU-B model (7.17) for solving system of time-varying nonlinear Equations (7.19), with different τ and with 8 different initial states, where ordinate ranges of subfigures are $[10^{-5}, 10^3]$, $[10^{-7}, 10^2]$, and $[10^{-10}, 10^2]$.

time evolution. Thus, let us consider the following system of time-varying nonlinear equations:

$$\mathbf{f}(\mathbf{x}(t), t) = \begin{bmatrix} x_1(t) - 1/(t+1) \\ x_1^2(t) + x_2(t) - \exp(1/(t+1)) \sin t \\ x_1^2(t) - x_2(t) \sin t + x_3(t) - 2 \\ x_1^2(t) - x_2^2(t) + x_3(t) + \cos(t/3) x_4(t) + 2 + 2 \sin t \end{bmatrix} = 0. \qquad (7.19)$$

The corresponding Jacobian matrix of (7.19) is

$$J(\mathbf{x}(t),t) = \begin{bmatrix} 1 & 0 & 0 & 0 \\ 2x_1(t) & 1 & 0 & 0 \\ 2x_1(t) & -\sin t & 1 & 0 \\ 2x_1(t) & 2x_2(t) & 1 & \cos(t/3) \end{bmatrix},$$

which is singular at $t = 3\pi/2 \approx 4.7$ s. The numerical-experiment results of DTZDU-B model (7.17) solving (7.19) are shown in Figure 7.8. Note that, in this numerical experiment, we use 8 different initial states x_0 to have more general results, from which we see that, no matter what τ is and what the initial state is, the residual errors $\|\mathbf{f}(\mathbf{x}_k,t_k)\|_2$ at the singular time-instant $t = 3\pi/2$ are of similar order, while the other index interval shows the $O(\tau^2)$ pattern similar to those of the previous experiments (i.e., Figure 7.6). It is worth noting again that, because of the similarity of the results and figures to the above ones, the corresponding numerical experiments of DTZDK-B model (7.16) are not presented (but left to interested readers to complete as a topic of exercise).

7.3 Summary

In this chapter, a general CTZD model has been proposed and investigated for solving the system of time-varying nonlinear equations. The solution error of such a CTZD model can exponentially converge to zero, which has been analyzed with different activation functions. In addition, such a CTZD model has been discretized for potential hardware implementation. This has produced the DTZDK and DTZDU models according to the criterion whether the time-derivative information is known or not. Furthermore, the DTZD models have further been improved by using the Broyden method, and thus the DTZDK-B and DTZDU-B models have been generated. The efficacy of the ZD models has been verified via the computer-simulation and numerical-experiment results.

Part III

Matrix Inversion

Chapter 8

ZD Models and Newton Iteration

Abstract

In this chapter, we develop and investigate a discrete-time zeroing dynamics (DTZD) model for matrix inversion, which is depicted by a system of difference equations. Comparing with Newton iteration, we find and confirm that the DTZD model incorporates Newton iteration as a special case. Noticing this relationship, we perform numerical comparison on different situations of using the DTZD model and Newton iteration for matrix inversion. Different types of activation functions and different step-size values are examined for the superior convergence and better stability of the DTZD model. Numerical examples illustrate the efficacy of the DTZD model and Newton iteration for online matrix inversion.

8.1 Introduction

The problem of matrix inversion is considered to be one of the fundamental problems widely encountered in science and engineering. It is usually an essential part of many solutions; e.g., statistics [115], as preliminary steps for optimization [100], signal processing [80], electromagnetic systems [71], robotic control [81, 96], and physics [85]. In many engineering applications, the online matrix inversion is desired.

Since the mid-1980s, efforts have been directed toward computational aspects of fast matrix inversion, and many algorithms have thus been proposed [23, 85, 88, 96]. Due to the in-depth research in artificial neural networks, various dynamic and analog solvers have been developed, investigated, and implemented on specific architectures [54, 80]. Suited for analog very large scale integration (VLSI) implementation [8] and based on its potential high-speed parallel processing, the neural-dynamic approach is now regarded as a powerful alternative for online problems solving [8, 54, 81].

Recently, a special class of neural dynamics (termed zeroing dynamics, ZD) has been proposed by Zhang *et al.* for online matrix inversion [99, 102, 104, 118]. Such a continuous-time ZD (CTZD) model is designed based on a matrix-valued indefinite error function, instead of a scalar-valued energy function associated with gradient dynamics [54]. In this chapter, we develop and investigate a DTZD model for online matrix inversion [118]. General nonlinear activation functions and variable learning step size can also be used for such a model. When the linear activation functions (or to say, linear activation function array) and step size $h = 1$ are used, the DTZD model reduces exactly to Newton iteration for matrix inversion. Noticing the relationship between the two approaches, we perform numerical comparison on different situations of using the DTZD model and Newton iteration for matrix inversion. Different types of activation functions and different step-size values are examined for superior convergence and better stability of the DTZD model [118].

8.2   ZD Models

To solve for a matrix inverse, the ZD design method is usually based on the definition equation, $AX - I = XA - I = 0$, where $A \in \mathbb{R}^{n \times n}$ is assumed to be square and nonsingular, and $I \in \mathbb{R}^{n \times n}$ is the identity matrix.

8.2.1   CTZD Model

Following Zhang *et al.*'s neural-dynamic design method (i.e., zeroing dynamics design method), to monitor and control the matrix-inversion process, we define the following matrix-valued indefinite error function (instead of scalar-valued energy function $\|AX(t) - I\|_{\mathrm{F}}^2 / 2$ associated with gradient dynamics):

$$E(t) = AX(t) - I.$$

To make $E(t)$ converge to zero (specifically, every entry $e_{ij}(t)$ is convergent to zero, $\forall \ i, j = 1, 2, \cdots, n$), the error function's time-derivative $\dot{E}(t)$ is chosen via the following ZD design formula [99–102, 104, 107, 114, 117, 118, 122, 128, 129, 134]:

$$\frac{\mathrm{d}E(t)}{\mathrm{d}t} = -\Gamma \Phi\big(E(t)\big), \tag{8.1}$$

where $\Gamma \in \mathbb{R}^{n \times n}$ is a positive-definite matrix used to scale the convergence rate of the inversion process, and $\Phi(\cdot) : \mathbb{R}^{n \times n} \to \mathbb{R}^{n \times n}$ denotes an activation function matrix-mapping (or to say, matrix-array, array). In general, any monotonically increasing odd activation function $\phi(\cdot)$, being the ijth element of matrix mapping $\Phi(\cdot)$, can be used for the construction of such a CTZD model. To illustrate the main ideas, four types of activation functions $\phi(\cdot)$ are investigated in this chapter as follows:

1) Linear activation function $\phi(e_{ij}) = e_{ij}$

2) Bipolar sigmoid activation function (with $\xi > 2$)

$$\phi(e_{ij}) = \frac{1 - \exp(-\xi e_{ij})}{1 + \exp(-\xi e_{ij})}$$

3) Power activation function $\phi(e_{ij}) = e_{ij}^p$ with odd integer $p \geqslant 3$

4) Power-sigmoid activation function

$$\phi(e_{ij}) = \begin{cases} e_{ij}^p, & \text{if } |e_{ij}| \geqslant 1 \\ \frac{1+\exp(-\xi)}{1-\exp(-\xi)} \frac{1-\exp(-\xi e_{ij})}{1+\exp(-\xi e_{ij})}, & \text{otherwise} \end{cases}$$

with suitable design parameters $\xi \geqslant 1$ and $p \geqslant 3$

New activation functions can thus be generated based on these basic types.

Expanding ZD design formula (8.1) leads to the following implicit dynamic equation of the CTZD model:

$$A\dot{X}(t) = -\Gamma\Phi(AX(t) - I),$$

where $X(t)$, starting from an initial state $X(0) = X_0 \in \mathbb{R}^{n \times n}$, denotes the state matrix corresponding to the theoretical inverse of matrix A.

Similar to usual neural-dynamic approaches, the matrix-valued design parameter $\Gamma \in \mathbb{R}^{n \times n}$ in the above CTZD model, being a set of the reciprocals of capacitance parameters, should be set as large as the hardware would permit (e.g., in analog circuits or VLSI [8]) or selected appropriately for simulative or experimental purposes. To keep the differential equations well conditioned, it is desired to keep Γ well conditioned. That is, the eigenvalues of Γ should be in the same scale. For simplicity, we assume $\Gamma = \gamma I$ with scalar design parameter $\gamma > 0$. As a consequence, we keep every $e_{ij}(t)$ ($\forall i, j = 1, \cdots, n$) converge at the same rate and almost at the same time, which also simplifies the neural-dynamics design and analysis. The CTZD model is thus in the following specific form:

$$A\dot{X}(t) = -\gamma\Phi(AX(t) - I). \tag{8.2}$$

Evidently, the above CTZD model is depicted in implicit dynamics, instead of explicit dynamics usually associated with gradient dynamics [8, 54, 80]. In addition, for CTZD model (8.2), the following theoretical results are obtained [118].

Lemma 1 *Given nonsingular matrix $A \in \mathbb{R}^{n \times n}$, if monotonically increasing odd function array $\Phi(\cdot)$ is used, then state matrix $X(t)$ of CTZD model (8.2), starting from any initial state $X_0 \in \mathbb{R}^{n \times n}$, converges to theoretical inverse A^{-1} of matrix A. In addition, CTZD model (8.2) possesses the following properties.*

1) If the linear activation function array is used, then the global exponential convergence with rate γ is achieved for (8.2).

2) If the bipolar sigmoid activation function array is used, then the superior convergence is achieved for (8.2) for error range $[-\delta, \delta]$, $\exists \delta > 0$, as compared with the situation of using the linear activation function array.

3) If the power activation function array is used, then the superior convergence is achieved for (8.2) for error ranges $(-\infty, -1)$ and $(1, +\infty)$, as compared with the situation of using the linear activation function array.

4) If the power-sigmoid activation function array is used, then the superior convergence is achieved for the whole error range $(-\infty, +\infty)$, as compared with the situation of using the linear activation function array.

8.2.2 DTZD Model

For potential hardware implementation of CTZD model (8.2) based on digital circuits or computers, we discretize CTZD model (8.2) by using Euler forward-difference rule, $\dot{X}(t) \approx (X_{k+1} - X_k)/\tau$, where $\tau > 0$ denotes the sampling gap, and X_k denotes the kth iteration (or to say, sampling, update) of $X(t = k\tau)$ with $k = 1, 2, \cdots$. We thus generate the discrete-time model of CTZD model (8.2) (i.e., the DTZD model) as follows:

$$AX_{k+1} = AX_k - h\Phi(AX_k - I), \tag{8.3}$$

where $h = \tau\gamma > 0$ is the step size that should be selected appropriately for the convergence to theoretical inverse A^{-1}. Because of A being invertible, DTZD model (8.3) is rewritten in the following intermediate form:

$$X_{k+1} = X_k - hA^{-1}\Phi(AX_k - I).$$

In view of the convergence of $X(t)$ to A^{-1} in (8.2) [in other words, $X(t) \approx A^{-1}$ for time t being large enough], we simply replace the A^{-1} term of the above intermediate DTZD form with the X_k term. This replacement yields the following explicit difference equation of DTZD model for matrix inversion:

$$X_{k+1} = X_k - hX_k\Phi(AX_k - I). \tag{8.4}$$

As far was we know, in terms of Equation (8.4), different choices of h and $\Phi(\cdot)$ also lead to different performance of such a DTZD model.

8.2.3 Link to Newton Iteration

Now, look at DTZD model (8.4). When we use the array of linear activation function $\phi(e_{ij}) = e_{ij}$, DTZD model (8.4) reduces to

$$X_{k+1} = X_k - hX_k(AX_k - I). \tag{8.5}$$

For $h = 1$, the above Equation (8.5) further becomes

$$X_{k+1} = X_k - X_k(AX_k - I) = 2X_k - X_kAX_k, \tag{8.6}$$

which is exactly Newton iteration for matrix inversion. In other words, we discover that the general form of Newton iteration for matrix inversion is given by DTZD model (8.4). It is worth mentioning that Newton iteration (8.6) is strongly stable in the sense that it computes a good approximation of A^{-1} even if the computation is performed with finite precision. In the ensuing sections, we show that DTZD model (8.4) [including Newton iteration (8.6) as a special case] is strongly stable as well, provided that activation function array $\Phi(\cdot)$ and step size h are appropriately chosen.

8.3 Choices of Initial State X_0

In this section, we address the quadratic convergence of DTZD model (8.6), which starts from initial state $X_0 = \alpha A^T$, with $\alpha = 2/\text{tr}(AA^T)$ for general nonsingular matrix A, or preferably starts from initial state $X_0 = \alpha I$, with $\alpha = 2/\text{tr}(A)$ for positive-definite or negative-definite matrix A. At the end of this section, we apply such choices of initial state X_0 to the general situation of DTZD model (8.4). Now, we present such main results via the following lemma and theorems [118], especially the most important Theorem 14 and its subsequent remarks of this section.

Lemma 2 *If $A \in \mathbb{R}^{n \times n}$ is nonsingular, then AA^T is positive definite and has only real positive eigenvalues.*

Theorem 11 *Given nonsingular matrix* $A \in \mathbb{R}^{n \times n}$, *let* $\lambda_{\max}(AA^{\mathrm{T}})$ *and* $\lambda_{\min}(AA^{\mathrm{T}})$ *represent the largest and smallest eigenvalues of* AA^{T}, *respectively. If* $0 < \alpha < \beta = 2/(\lambda_{\max}(AA^{\mathrm{T}}) + \lambda_{\min}(AA^{\mathrm{T}}))$, *then*

$$1 - \frac{2}{\mathrm{cond}_2^2(A) + 1} = \|I - \beta AA^{\mathrm{T}}\|_2 < \|I - \alpha AA^{\mathrm{T}}\|_2 < 1, \tag{8.7}$$

where the two-norm of matrix A, $\|A\|_2 = \sqrt{\lambda_{\max}(AA^{\mathrm{T}})}$, *is determined by the square root of the largest eigenvalue of* AA^{T} *(in other words, the largest singular value of* A*). In addition, the two-norm based condition-number of matrix* A, $\mathrm{cond}_2(A) = \|A\|_2 \|A^{-1}\|_2 = \sqrt{\lambda_{\max}(AA^{\mathrm{T}})/\lambda_{\min}(AA^{\mathrm{T}})} \geqslant 1$ *(in other words, the ratio of the largest singular value of* A *to its smallest one).*

Proof We first start from the proof of the rightmost part of Equation (8.7). Note that $0 < \alpha < \beta = 2/(\lambda_{\max}(AA^{\mathrm{T}}) + \lambda_{\min}(AA^{\mathrm{T}}))$, as assumed. For ease of writing, we abbreviate here $\lambda_{\max}(AA^{\mathrm{T}})$ and $\lambda_{\min}(AA^{\mathrm{T}})$ as $\lambda_{\max}$ and $\lambda_{\min}$, respectively.

Let $\mu_i \in \mathbb{R}$ and $\xi \in \mathbb{R}^n$ denote, respectively, the ith eigenvalue and its corresponding eigenvector of $I - \alpha AA^{\mathrm{T}}$, for $i = 1, 2, \cdots, n$. That is, $(I - \alpha AA^{\mathrm{T}}) \xi = \xi - \alpha AA^{\mathrm{T}} \xi = \mu_i \xi$ holds true. We thus have $AA^{\mathrm{T}} \xi = ((1 - \mu_i)/\alpha) \xi$, which defines that $\lambda_i = (1 - \mu_i)/\alpha$ is an eigenvalue (or, exactly speaking, the ith eigenvalue) of AA^{T}, while ξ is its corresponding eigenvector of AA^{T}. In view of A being nonsingular and Lemma 2, we thus have $\lambda_{\max} \geqslant \lambda_i \geqslant \lambda_{\min} > 0$, for any $i \in \{1, 2, \cdots, n\}$.

It is known that, for symmetric matrix $I - \alpha AA^{\mathrm{T}}$,

$$\|I - \alpha AA^{\mathrm{T}}\|_2 = \max_{1 \leqslant i \leqslant n} |\mu_i| = \max_{1 \leqslant i \leqslant n} (|1 - \alpha \lambda_i|). \tag{8.8}$$

For the purpose of analyzing the above equation, it follows from $0 < \alpha < \beta = 2/(\lambda_{\max} + \lambda_{\min})$ that,

$$0 < \alpha \lambda_{\max} + \alpha \lambda_{\min} < 2,$$
$$\alpha \lambda_{\min} - 1 < 1 - \alpha \lambda_{\max} \leqslant 1 - \alpha \lambda_{\min} < 1,$$
$$|1 - \alpha \lambda_{\max}| \leqslant 1 - \alpha \lambda_{\min} < 1.$$

From the above inequalities and Equation (8.8), we have

$$\|I - \alpha AA^{\mathrm{T}}\|_2 = \max_{1 \leqslant i \leqslant n} (|1 - \alpha \lambda_i|) = \max(|1 - \alpha \lambda_{\min}|, |1 - \alpha \lambda_{\max}|) = 1 - \alpha \lambda_{\min} < 1,$$

which thus completes the proof of the rightmost part of (8.7).

Secondly, following the above result, we complete readily the proof of the left part of Equation (8.7) by considering $0 < \alpha < \beta = 2/(\lambda_{\max} + \lambda_{\min})$:

$$\|I - \alpha AA^{\mathrm{T}}\|_2 = 1 - \alpha \lambda_{\min} > 1 - \beta \lambda_{\min} = \|I - \beta AA^{\mathrm{T}}\|_2$$
$$= 1 - \frac{2\lambda_{\min}}{\lambda_{\max} + \lambda_{\min}} = 1 - \frac{2}{\lambda_{\max}/\lambda_{\min} + 1}$$
$$= 1 - \frac{2}{\mathrm{cond}_2^2(A) + 1}.$$

The proof of Theorem 11 is thus complete.

Theorem 12 *Given nonsingular matrix* $A \in \mathbb{R}^{n \times n}$ *and* $\alpha = 2/\mathrm{tr}(AA^{\mathrm{T}})$, *we have*

$$1 - \frac{2}{\mathrm{cond}_2^2(A) + n - 1} \leqslant \|\alpha AA^{\mathrm{T}} - I\|_2 \leqslant 1 - \frac{2}{(n-1)\mathrm{cond}_2^2(A) + 1}. \tag{8.9}$$

Proof By the notation used in the above, $\lambda_i > 0$ denotes the ith eigenvalue of AA^T, while $\lambda_{max} > 0$ and $\lambda_{min} > 0$ denotes, respectively, the largest and the smallest eigenvalues of AA^T. We know $\text{tr}(AA^T) = \sum_{i=1}^n \lambda_i \geqslant \lambda_{max} + \lambda_{min}$, and thus $0 < \alpha = 2/\text{tr}(AA^T) \leqslant 2/(\lambda_{max} + \lambda_{min})$.

It follows from the proof of Theorem 11 that

$$\|\alpha AA^T - I\|_2 = 1 - \alpha\lambda_{min} = 1 - \frac{2\lambda_{min}}{\text{tr}(AA^T)} < 1. \tag{8.10}$$

For the purpose of analyzing the above equation, we derive the lower bound of $\text{tr}(AA^T)/\lambda_{min}$ as

$$\frac{\text{tr}(AA^T)}{\lambda_{min}} = \sum_{\kappa=1}^n \frac{\lambda_\kappa}{\lambda_{min}} \geqslant \frac{\lambda_{max}}{\lambda_{min}} + \sum_{i=1}^{n-1} \frac{\lambda_{min}}{\lambda_{min}} = \text{cond}_2^2(A) + n - 1$$

and the upper bound of $\text{tr}(AA^T)/\lambda_{min}$ as

$$\frac{\text{tr}(AA^T)}{\lambda_{min}} = \sum_{\kappa=1}^n \frac{\lambda_\kappa}{\lambda_{min}} \leqslant \sum_{i=1}^{n-1} \frac{\lambda_{max}}{\lambda_{min}} + \frac{\lambda_{min}}{\lambda_{min}} = (n-1)\text{cond}_2^2(A) + 1.$$

Thus, from the above two bounds, we have

$$\frac{2}{(n-1)\text{cond}_2^2(A) + 1} \leqslant \frac{2\lambda_{min}}{\text{tr}(AA^T)} \leqslant \frac{2}{\text{cond}_2^2(A) + n - 1},$$

which, together with (8.10), gives rise to Equation (8.9) and thus completes the proof of Theorem 12.

Theorem 13 *For positive-definite or negative-definite matrix $A \in \mathbb{R}^{n\times n}$, given $\alpha = 2/\text{tr}(A)$, we have*

$$1 - \frac{2}{\text{cond}_2(A) + n - 1} \leqslant \|\alpha A - I\|_2 \leqslant 1 - \frac{2}{(n-1)\text{cond}_2(A) + 1}. \tag{8.11}$$

Proof Here, for positive-definite matrix A (rather than the general nonsingular A discussed previously), we use $\lambda_i > 0$ to denote the ith eigenvalue of A, and use $\lambda_{max} > 0$ and $\lambda_{min} > 0$ to denote, respectively, the largest and the smallest eigenvalues of A. Please beware of the notational difference between this proof and the proofs of Theorems 11 and 12.

We know $\text{tr}(A) = \sum_{i=1}^n \lambda_i \geqslant \lambda_{max} + \lambda_{min}$, and thus $0 < \alpha = 2/\text{tr}(A) \leqslant 2/(\lambda_{max} + \lambda_{min})$. It follows from the proof of Theorem 11 that

$$\|\alpha A - I\|_2 = 1 - \alpha\lambda_{min} = 1 - \frac{2\lambda_{min}}{\text{tr}(A)} < 1. \tag{8.12}$$

Similar to the proof of Theorem 12, we have

$$\frac{2}{(n-1)\text{cond}_2(A) + 1} \leqslant \frac{2\lambda_{min}}{\text{tr}(A)} \leqslant \frac{2}{\text{cond}_2(A) + n - 1},$$

which, together with (8.12), gives rise to Equation (8.11) and thus completes the proof of Theorem 13.

Theorem 14 *For general nonsingular matrix $A \in \mathbb{R}^{n\times n}$, we choose initial state $X_0 = \alpha A^T$ with $\alpha = 2/\text{tr}(AA^T)$ for DTZD model (8.6) (namely, Newton iteration) which, starting with $\|AX_0 - I\|_2 < 1$, quadratically converges to theoretical inverse A^{-1}. In addition, for positive-definite or negative-definite matrix A, we choose initial state $X_0 = \alpha I$ with $\alpha = 2/\text{tr}(A)$ for (8.6) which, starting with a smaller value of $\|AX_0 - I\|_2 < 1$, quadratically converges to theoretical inverse A^{-1}.*

Proof Let $E_k = AX_k - I$ denote the residual error matrix at the kth iteration during the matrix-inverse computation. Premultiplying Equation (8.6) by A yields

$$AX_{k+1} = AX_k - AX_k(AX_k - I).$$

Residual error E_{k+1} is thus written as

$$
\begin{aligned}
E_{k+1} &= AX_{k+1} - I = AX_k - AX_k(AX_k - I) - I \\
&= (AX_k - I) - AX_k(AX_k - I) \\
&= E_k - (E_k + I)(E_k) = -E_k^2 \\
&= -E_0^{2^{k+1}} = -(AX_0 - I)^{2^{k+1}},
\end{aligned}
\tag{8.13}
$$

which implies that the sequence $\{E_k\}$ quadratically converges to zero, provided that initial state X_0 is chosen, such that $\|AX_0 - I\|_2 < 1$. Simple manipulations yield $X_k - A^{-1} = -A^{-1}(AX_0 - I)^{2^k} \to 0$, which implies that the sequence $\{X_k\}$ quadratically converges to A^{-1} as well, if X_0 is chosen, such that $\|AX_0 - I\|_2 < 1$.

Now, we come to choose initial state X_0 satisfying $\|AX_0 - I\|_2 < 1$. It follows from Theorem 11 that $X_0 = \alpha A^T$ with $0 < \alpha \leqslant 2/(\lambda_{\max}(AA^T) + \lambda_{\min}(AA^T))$ is a reasonably good initial state, which ensures $\lim_{k \to \infty} X_k = A^{-1}$. The larger the value of α is, the better the initial state X_0 is, in the sense that $\|AX_0 - I\|_2$ is smaller.

However, to avoid the computation of $\lambda_{\min}(AA^T)$ and $\lambda_{\max}(AA^T)$ (which is usually time-consuming), for general nonsingular matrix A, it is more practical to exploit Theorem 12 and choose $X_0 = \alpha A^T$ with $\alpha = 2/\operatorname{tr}(AA^T)$, as such an initial state ensures $\lim_{k \to \infty} X_k = A^{-1}$.

In addition, for positive-definite or negative-definite matrix A, we exploit Theorem 13 to choose a better initial state $X_0 = 2I/\operatorname{tr}(A)$ that starts DTZD model (8.6) with a smaller value of $\|AX_0 - I\|_2$ and ensures $\lim_{k \to \infty} X_k = A^{-1}$. This is in view of the following inequalities:

$$1 \leqslant \operatorname{cond}_2(A) \leqslant \operatorname{cond}_2^2(A),$$

$$1 - \frac{2}{\operatorname{cond}_2(A) + n - 1} \leqslant 1 - \frac{2}{\operatorname{cond}_2^2(A) + n - 1},$$

$$1 - \frac{2}{(n-1)\operatorname{cond}_2(A) + 1} \leqslant 1 - \frac{2}{(n-1)\operatorname{cond}_2^2(A) + 1}.$$

The proof is thus complete.

Remark 2 If the largest and smallest eigenvalues of a matrix are known or can be computed readily, then we choose initial state $X_0 = 2A^T/(\lambda_{\max}(AA^T) + \lambda_{\min}(AA^T))$ for such a general nonsingular matrix A, or preferably choose initial state $X_0 = 2I/(\lambda_{\max}(A) + \lambda_{\min}(A))$ for such a positive-definite or negative-definite matrix A. These choices have smaller values of $\|AX_0 - I\|_2$, as compared with Theorem 14. For example, corresponding to the two cases in the first sentence of this remark, in the former case, $\|AX_0 - I\|_2 = 1 - 2/(\operatorname{cond}_2^2(A) + 1)$, while, in the latter case, $\|AX_0 - I\|_2 = 1 - 2/(\operatorname{cond}_2(A) + 1)$. However, it is well known that matrix eigenvalues are usually difficult to compute, and thus Theorem 14 may be a more practical way to choose X_0.

Remark 3 With respect to different definitions of matrix norms, there are many other choices of α in the design of initial state $X_0 = \alpha A^T$ for general nonsingular matrix A. For example, we set $\alpha = 1/\|A\|_1\|A\|_\infty$. In addition, in the design of initial state $X_0 = \alpha I$ for positive-definite matrix A, we set $\alpha = 1/\|A\|_1$, which generates $\|AX_0 - I\|_2 \leqslant 1 - 1/(\sqrt{n}\operatorname{cond}_2(A))$. Similarly, in the design of initial state $X_0 = \alpha I$ for negative-definite matrix A, we set $\alpha = -1/\|A\|_1$. However, for circuit-implementation purposes, we prefer to use $\alpha = 2/\operatorname{tr}(AA^T) = 2/(\sum_{i=1}^n \sum_{j=1}^n a_{ij}^2)$ [or $\alpha = 2/\operatorname{tr}(A) =$

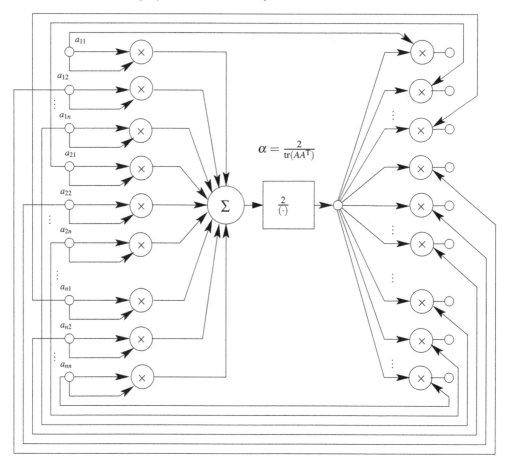

FIGURE 8.1 Block diagram for generating initial state $X_0 = 2A^T / \mathrm{tr}(AA^T)$.

$2/(\sum_{i=1}^n a_{ii})$ for positive-definite or negative-definite matrix A], which can be implemented more easily than other norm-based choices. The lower complexity of circuit implementation of trace-based X_0 motivates Theorem 14. Furthermore, Figure 8.1 presents the block diagram for generating initial state $X_0 = 2A^T / \mathrm{tr}(AA^T)$ by using multiplication circuits, summation circuits, and a division circuit.

Before ending this section, it is worth mentioning that the above results can be applied to the general situation of DTZD model (8.4), which includes the linear model (8.5) as a special case. Specifically speaking, to start the general DTZD model (8.4), initial state X_0 can take the form $X_0 = \alpha A^T$, with $\alpha = 2/\mathrm{tr}(AA^T)$ for inverting the general nonsingular matrix A, or preferably take the form $X_0 = \alpha I$ with $\alpha = 2/\mathrm{tr}(A)$ for inverting the positive-definite or negative-definite matrix A. The analysis will be given in the ensuing section.

8.4   Choices of Step Size h

When we exploit DTZD model (8.4) to solve for a matrix inverse, after initial state X_0 is chosen appropriately as in the preceding section, the next important issue to be discussed here is the appropriate choices of step size h. In the following two subsections, with respect to different activation functions $\phi(\cdot)$, we present, respectively, the rough bounds and optimal values of step size h, which are analyzed for the convergence and superior convergence of DTZD model (8.4) for the matrix-inversion purpose.

8.4.1 Bounds of Step Size h

In this subsection, we address the bounds of step size h, which guarantee the convergence of DTZD model (8.5) (namely, the linear model) starting from the initial state X_0 given in Section 8.3. Specifically, the bounds of step size h in (8.5) are $0 < h \leqslant 1$. At the end of this subsection, we extend the analysis of step-size bounds to the general situation of DTZD model (8.4), which starts from the initial state X_0 given in Section 8.3 as well.

Lemma 3 *Given nonsingular matrix $A \in \mathbb{R}^{n \times n}$, for unitary matrices U and V satisfying $U^{\mathrm{T}}U = I$ and $V^{\mathrm{T}}V = I$, respectively, we have $\|UA\|_2 = \|AV\|_2 = \|UAV\|_2 = \|A\|_2$ [118].*

Theorem 15 *Given nonsingular matrix $A \in \mathbb{R}^{n \times n}$, if step size h satisfies $0 < h \leqslant 1$, then DTZD model (8.5), starting from initial state $X_0 = \alpha A^T$ with $\alpha = 2/\mathrm{tr}(AA^{\mathrm{T}})$, converges to theoretical inverse A^{-1}. In addition, for positive-definite or negative-definite matrix A, if step size h satisfies $0 < h \leqslant 1$, then DTZD model (8.5), starting from initial state $X_0 = \alpha I$ with $\alpha = 2/\mathrm{tr}(A)$, converges to theoretical inverse A^{-1}.*

Proof Let $E_k = AX_k - I$ denote the residual error matrix at the kth iteration during the matrix-inverse computation. Premultiplying Equation (8.5) by A yields

$$AX_{k+1} = AX_k - hAX_k(AX_k - I).$$

Residual error E_{k+1} is thus written as

$$
\begin{aligned}
E_{k+1} = AX_{k+1} - I &= AX_k - hAX_k(AX_k - I) - I \\
&= (AX_k - I) - hAX_k(AX_k - I) \\
&= E_k - h(E_k + I)E_k.
\end{aligned}
\tag{8.14}
$$

Evidently, error matrix $E_0 = AX_0 - I$ is symmetric (i.e., $E_0^{\mathrm{T}} = E_0$), because $X_0 = \alpha A^{\mathrm{T}}$, with $\alpha = 2/\mathrm{tr}(AA^{\mathrm{T}})$ for general nonsingular matrix A, or $X_0 = \alpha I$ with $\alpha = 2/\mathrm{tr}(A)$ for positive-definite or negative-definite matrix A. Then, according to Equation (8.14), we know that the next-iteration error-matrix E_1 is symmetric as well:

$$
\begin{aligned}
E_1^{\mathrm{T}} = (E_0 - h(E_0 + I)E_0)^{\mathrm{T}} &= E_0^{\mathrm{T}} - hE_0^{\mathrm{T}}(E_0 + I)^{\mathrm{T}} \\
&= E_0^{\mathrm{T}} - hE_0^{\mathrm{T}}(E_0^{\mathrm{T}} + I^{\mathrm{T}}) = E_0 - hE_0(E_0 + I) \\
&= E_0 - h(E_0^2 + E_0) = E_0 - h(E_0 + I)E_0 \\
&= E_1.
\end{aligned}
$$

Similarly, error matrix E_k is proved to be symmetric, for any $k = 0, 1, 2, \cdots$ Thus, E_k is diagonalized by the following transformation: $E_k = U\Lambda_k U^{\mathrm{T}}$, where $U \in \mathbb{R}^{n \times n}$ is a unitary matrix with $U^{\mathrm{T}} = U^{-1}$, and matrix Λ_k is diagonal. It follows from Equation (8.14) and the above diagonal transformation that

$$
\begin{aligned}
E_{k+1} &= U\Lambda_k U^{\mathrm{T}} - h(U\Lambda_k U^{\mathrm{T}} + I)U\Lambda_k U^{\mathrm{T}} \\
&= U\Lambda_k U^{\mathrm{T}} - h(U\Lambda_k^2 U^{\mathrm{T}} + U\Lambda_k U^{\mathrm{T}}) \\
&= U\Lambda_k U^{\mathrm{T}} - hU(\Lambda_k^2 + \Lambda_k)U^{\mathrm{T}} \\
&= U(\Lambda_k - h(\Lambda_k^2 + \Lambda_k))U^{\mathrm{T}}.
\end{aligned}
\tag{8.15}
$$

By defining $\Lambda_{k+1} = \Lambda_k - h(\Lambda_k^2 + \Lambda_k)$, error matrix E_{k+1} becomes $E_{k+1} = U\Lambda_{k+1}U^{\mathrm{T}}$. It follows from Lemma 3 that $\|E_{k+1}\|_2 = \|U\Lambda_{k+1}U^{\mathrm{T}}\|_2 \leqslant \|E_k\|_2 = \|U\Lambda_k U^{\mathrm{T}}\|_2$ if and only if $\|\Lambda_{k+1}\|_2 \leqslant \|\Lambda_k\|_2$. So, the convergence analysis of E_k becomes the convergence analysis of Λ_k. That is, we now come to investigate $\Lambda_{k+1} = \Lambda_k - h(\Lambda_k^2 + \Lambda_k)$, where diagonal matrix Λ_k is denoted by $\Lambda_k = \mathrm{diag}(\sigma_{1,k}, \cdots, \sigma_{i,k}, \cdots, \sigma_{n,k})$. Thus, the ith decoupled subsystem of $\Lambda_{k+1} = \Lambda_k - h(\Lambda_k^2 + \Lambda_k)$ is described as

$$\sigma_{i,k+1} = \sigma_{i,k} - h(\sigma_{i,k}^2 + \sigma_{i,k}), \tag{8.16}$$

$\forall i = 1, 2, \cdots, n$ and $k = 0, 1, 2, \cdots$, with $-1 < \sigma_{i,0} < 1$ due to $\|AX_0 - I\|_2 < 1$.

To ensure that $\|E_k\|_2$ and $\|\Lambda_k\|_2$ decrease to zero, as k tends to positive infinity (i.e., $\lim_{k\to\infty} E_k = 0$ and $\lim_{k\to\infty} \Lambda_k = 0$), we keep $|\sigma_{i,k+1}| < |\sigma_{i,k}|$ (if $\sigma_{i,k} \neq 0$) for each $i = 1, 2, \cdots, n$. That is,

$$|\sigma_{i,k+1}| = |\sigma_{i,k} - h(\sigma_{i,k}^2 + \sigma_{i,k})| < |\sigma_{i,k}|, \; i = 1, 2, \cdots, n. \tag{8.17}$$

In view of $-1 < \sigma_{i,0} < 1$ and $h > 0$, the solution of inequality (8.17) is

$$0 < h < \frac{2}{1 + \sigma_{i,k}}, \; -1 < \sigma_{i,k} < 1, \; \forall i = 1, 2, \cdots, n.$$

Since $-1 < \sigma_{i,k} < 1$, we have $1 < 2/(1 + \sigma_{i,k})$ and $2/(1 + \sigma_{i,k}) \to 1$ when $\sigma_{i,k} \to 1$. Thus, we take $0 < h \leqslant 1$, satisfying the above inequalities as well as convergence condition (8.17), which guarantees $\lim_{k\to\infty} \Lambda_k = 0$ and $\lim_{k\to\infty} E_k = \lim_{k\to\infty} U\Lambda_k U^{\mathrm{T}} = 0$. From the definition of error matrix $E_k = AX_k - I$, it follows that $\lim_{k\to\infty} X_k = A^{-1}$. The proof is thus complete.

Following the above analysis procedure, now we come to investigate the bounds of step size h in the general DTZD model (8.4). Continuing from Equations (8.15) and (8.16), to simplify the analysis, let us consider the residual error matrix $E_k = AX_k - I$ in the diagonal form $E_k = \mathrm{diag}(\sigma_{1,k}, \cdots, \sigma_{i,k}, \cdots, \sigma_{n,k})$. Thus, the ith decoupled subsystem of $E_{k+1} = E_k - h(E_k + I)\Phi(E_k)$ is described as $\sigma_{i,k+1} = \sigma_{i,k} - h(\sigma_{i,k} + 1)\phi(\sigma_{i,k})$, $\forall i = 1, 2, \cdots, n$ and $k = 0, 1, 2, \cdots$, with $-1 < \sigma_{i,0} < 1$ due to $\|AX_0 - I\|_2 < 1$. Similarly, for the convergence of E_k, we require

$$|\sigma_{i,k+1}| = |\sigma_{i,k} - h(\sigma_{i,k} + 1)\phi(\sigma_{i,k})| < |\sigma_{i,k}|, \; i = 1, 2, \cdots, n. \tag{8.18}$$

Because activation function $\phi(\cdot)$ is a monotonically increasing odd function, we have $\phi(-\sigma_{i,k}) = -\phi(\sigma_{i,k})$, and

$$\phi(\sigma_{i,k}) \begin{cases} > 0, & \text{if } \sigma_{i,k} > 0, \\ = 0, & \text{if } \sigma_{i,k} = 0, \\ < 0, & \text{if } \sigma_{i,k} < 0. \end{cases}$$

In view of $-1 < \sigma_{i,0} < 1$ and $h > 0$, the solution of inequality (8.18) is

$$0 < h < \frac{2\sigma_{i,k}}{(1 + \sigma_{i,k})\phi(\sigma_{i,k})}, \; -1 < \sigma_{i,k} < 1, \; \forall i = 1, 2, \cdots, n.$$

Thus, to satisfy the above inequalities as well as convergence condition (8.18), we take

$$0 < h < \min_{|\sigma_{i,k}| < 1} w(\sigma_{i,k}), \; \text{with } w(\sigma_{i,k}) = \frac{2\sigma_{i,k}}{(1 + \sigma_{i,k})\phi(\sigma_{i,k})},$$

which ensures that $E_k \to 0$ and $X_k \to A^{-1}$ as $k \to +\infty$.

Remark 4 For monotonically increasing odd activation function $\phi(\cdot)$, the resultant function $w(\sigma_{i,k})$ is generally no longer monotonically increasing or decreasing. Therefore, $\min_{|\sigma_{i,k}| < 1} w(\sigma_{i,k})$ has to be evaluated corresponding to a specific type of $\phi(\cdot)$. With the aid of Figure 8.2(a), the bounds of step size h are listed below for the diagonal situation, within which the general DTZD model (8.4), starting from an initial state X_0 satisfying $\|AX_0 - I\|_2 < 1$, converges to theoretical inverse A^{-1}.

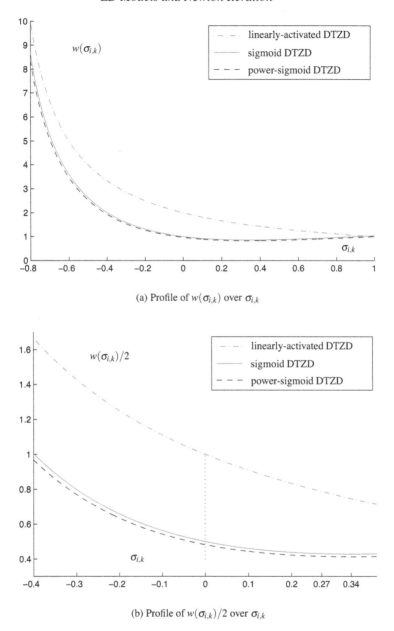

(a) Profile of $w(\sigma_{i,k})$ over $\sigma_{i,k}$

(b) Profile of $w(\sigma_{i,k})/2$ over $\sigma_{i,k}$

FIGURE 8.2 Profiles of $w(\sigma_{i,k})$ and $w(\sigma_{i,k})/2$ resulting from different activation functions.

1) When the array of linear activation function $\phi(e_{ij}) = e_{ij}$ is used, the bound is $0 < h \leqslant 1$, the same as in Theorem 15.

2) When the array of bipolar sigmoid activation function is used with $\xi = 4$, the minimum value of $w(\sigma)$, $\min_{|\sigma_{i,k}|<1} w(\sigma_{i,k}) = 0.8579$, is achieved with $\sigma_{i,k} = 0.3416$. Thus, the bound of step size h in this case is $0 < h < 0.8579$.

3) When the array of power activation function $\phi(e_{ij}) = e_{ij}^p$ is used with odd integer $p \geqslant 3$, the minimum value of $w(\sigma_{i,k})$, $\min_{|\sigma_{i,k}|<1} w(\sigma_{i,k}) = 1$, is achieved with $\sigma_{i,k} \to 1$. Thus, the bound of step size h in this case is $0 < h \leqslant 1$.

4) When the array of power-sigmoid activation function is used with $\xi = 4$ and $p = 3$, the minimum value of $w(\sigma_{i,k})$, $\min_{|\sigma_{i,k}|<1} w(\sigma_{i,k}) = 0.8270$, is achieved with $\sigma_{i,k} = 0.3416$. Thus, the bound of step size h in this case is $0 < h < 0.8270$.

5) When another type of activation function array is used, the bound of step size h can be obtained in a similar way.

Note that, for the general nonsingular situation, the upper bound of step size h for (8.4) is usually smaller than the above one (except for the case of using the array of linear activation functions). Besides, it is worth comparing the so-called "approximation and diagonalization" analysis approach in this chapter and the analysis approach of fixed point iteration in Chapters 2 and 6 (specifically, Proposition 4, Theorem 7, and Example 6.4), which is left to interested readers as a topic of exercise.

8.4.2 Optimal Value of Step Size h

For superior convergence of DTZD model (8.4), it is desired to find the optimal (or suboptimal) value of step size h in order to keep $\max_{1\leqslant i\leqslant n} |\sigma_{i,k+1}|$ of (8.18) as small as we can, i.e.,

$$h_{\mathrm{opt}} = \arg\min_h \max_{1\leqslant i\leqslant n} |\sigma_{i,k+1}| = \arg\min_h \max_{1\leqslant i\leqslant n} |\sigma_{i,k} - h(\sigma_{i,k}+1)\phi(\sigma_{i,k})|.$$

The ideal situation of h_{opt} is to make $|\sigma_{i,k+1}| = 0$, i.e., $|\sigma_{i,k} - h(\sigma_{i,k}+1)\phi(\sigma_{i,k})| = 0, \forall i = 1,2,\cdots,n$. However, it is difficult to determine h_{opt} from the above relation, since the values of $\sigma_{i,k}$, $i = 1,2,\cdots,n$, are unknown. Evidently, if step size h is too small, the convergence process may be very slow; whereas, if h is too large, DTZD model (8.4) may have zigzag trajectories or even diverge. Typical trajectories of state matrix X_k of DTZD model (8.4) are shown in Figure 8.3, where different values of step size h are tried, in addition to using the array of bipolar sigmoid activation functions. Specifically speaking, if $h = 0.3$, the convergence is slow, and 25 iterations are needed to

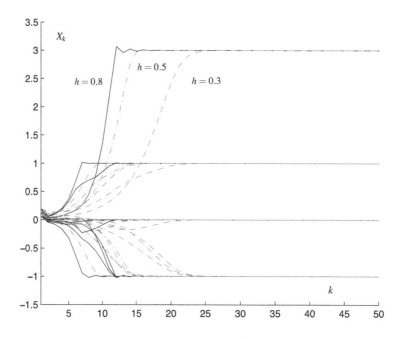

FIGURE 8.3 Neural states of DTZD model (8.4) using bipolar sigmoid activation functions with different h for online matrix inversion [118].

generate the solution. If $h = 0.5$, DTZD model (8.4) converges smoothly to the theoretical solution within 14 iterations. In comparison, if $h = 0.8$, the DTZD model's solution becomes zig-zag.

To approach the ideal situation of finding h_{opt}, we need to make $|\sigma_{i,k+1}| = |\sigma_{i,k} - h(\sigma_{i,k} + 1)\phi(\sigma_{i,k})| = 0$, i.e.,

$$h = h(\sigma_{i,k}) = \frac{\sigma_{i,k}}{(\sigma_{i,k} + 1)\phi(\sigma_{i,k})} = \frac{w(\sigma_{i,k})}{2}. \tag{8.19}$$

Note that the value of $\sigma_{i,k}$ for $i = 1, 2, \cdots, n$ is different from each other, and thus the value of h is theoretically different with respect to different $\sigma_{i,k}$. If $\|E_k\|_2 = \vartheta$, then $0 \leqslant |\sigma_{i,k}| \leqslant \vartheta$ holds for each $i = 1, 2, \cdots, n$. In view of Figure 8.2(b), if $\|E_k\|_2 = \vartheta$ is small enough (e.g., with $\vartheta < 0.3$), then $\forall \sigma_{i,k}$, with $i = 1, 2, \cdots, n$,

$$\frac{\vartheta}{(\vartheta + 1)\phi(\vartheta)} = \frac{w(\vartheta)}{2} \leqslant h(\sigma_{i,k}) \leqslant \frac{\vartheta}{(-\vartheta + 1)\phi(\vartheta)} = \frac{w(-\vartheta)}{2}, \tag{8.20}$$

in the context of using the array of linear, sigmoid (with $\xi = 4$), or power-sigmoid (with $\xi = 4$ and $p = 3$) activation functions. It follows from the convergence result (i.e., $\vartheta \to 0$) that bound (8.20) shrinks finally to $h = h(\sigma_{i,k}) = w(0)/2$, $\forall \sigma_{i,k}$ with $i = 1, 2, \cdots, n$, which is exploited as the practical substitute for optimal step size h_{opt}. In addition, $h_{opt} \approx h = w(0)/2$ agrees relatively well with Figure 8.3, which has already been mentioned in the preceding paragraph. With respect to different types of activation functions, the detailed discussion on the optimal (or nearly optimal) choices of h is given in the following.

1) When the array of linear activation functions is used in DTZD model (8.4) [i.e., the linear model (8.5)], we have $h_{opt} \approx h = w(0)/2 = 1$, which also satisfies the bound $0 < h \leqslant 1$ given in Theorem 15.

2) When the array of bipolar sigmoid activation functions is used in DTZD model (8.4) with $\xi = 4$, we have $h_{opt} \approx h = w(0)/2 = 0.5$, which also satisfies the rough bound $0 < h < 0.8579$ discussed in Remark 4.

3) When the array of power-sigmoid activation functions is used in DTZD model (8.4) with $\xi = 4$ and $p = 3$, we have $h_{opt} \approx h = w(0)/2 = 0.4820$, which also satisfies the rough bound $0 < h < 0.8270$ discussed in Remark 4.

8.5 Illustrative Examples

In the previous three sections, we have presented the theoretical results about DTZD model (8.4), in addition to showing that the general nonlinear form of Newton iteration for matrix inversion is given in (8.4). In this section, we present the following three illustrative examples, which substantiate the theoretical results.

Example 8.1 For illustration and comparison, we are now applying DTZD model (8.4) [including Newton iteration (8.6) as a special case] to compute the inverse of the following matrix $A \in \mathbb{R}^{3\times3}$ [30]:

$$A = \begin{bmatrix} 1 & 1 & 2 \\ 3 & 2 & 3 \\ 1 & 1 & 1 \end{bmatrix}.$$

Different activation function array $\Phi(\cdot)$ and step size h are tried in the numerical experiments of

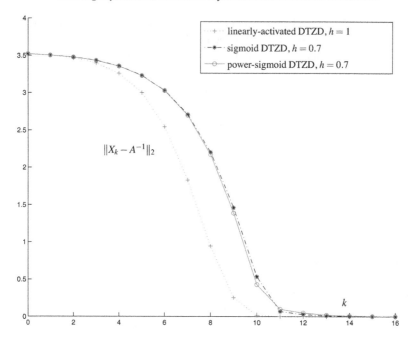

FIGURE 8.4 Solution–error trajectories of DTZD model (8.4) for online matrix inversion.

DTZD model (8.4). Starting from initial state $X_0 = 2A^T / \mathrm{tr}(AA^T)$, DTZD model (8.4) generates the inverse of A denoted by $\bar{X}$ within a few iterations:

$$\bar{X} = \begin{bmatrix} -1.0000 & 1.0000 & -1.0000 \\ 0.0000 & -1.0000 & 3.0000 \\ 1.0000 & 0.0000 & -1.0000 \end{bmatrix},$$

rounded to four decimal places. We have the following important facts.

1) Neural states X_k of DTZD model (8.4) using bipolar sigmoid activation functions with different step size h are depicted in Figure 8.3. It shows that, when using optimal step size $h = 0.5$, the DTZD model converges to $\bar{X}$ within 14 iterations.

2) The situations of using other types of activation functions to construct DTZD model (8.4) are depicted as well, i.e., in Figure 8.4. Note that the corresponding solution-error trajectories $\|X_k - A^{-1}\|_2$ are shown there. From the figure, we see that, when using the array of linear activation functions with step size $h = 1$, the array of bipolar sigmoid activation functions with step size $h = 0.7$, and the array of power-sigmoid activation functions with step size $h = 0.7$, the resultant DTZD models all converge to $\bar{X}$ within 14 iterations.

For comparison, to solve the same matrix-inverse problem (i.e., the above-presented one), two gradient-dynamics models were applied to compute such an inverse $\bar{X}$, for which, however, roughly 1000 iterations were needed [30].

Example 8.2 In this example, we would like to compare the performance of DTZD model (8.4) when using different types of activation functions or different values of step size. To do so, four groups of 1000 full-rank random matrices $A \in \mathbb{R}^{3 \times 3}$ are tested. Note that the matrix entries are uniformly distributed random numbers in $[-5, 5]$ generated by using MATLAB® [57], and the initial state of DTZD model (8.4) is chosen as $X_0 = 2A^T / \mathrm{tr}(AA^T)$. The corresponding numerical results are shown in Table 8.1, where the following notation is introduced.

TABLE 8.1 Performance Comparison on DTZD Model (8.4) for Inverting Online 4 Groups of 1000 Random Matrices

	Average number of iterations					
Precision	$\text{DTZD}_{\text{lin}}^{0.80}$	$\text{DTZD}_{\text{lin}}^{1.0}$	$\text{DTZD}_{\text{sig}}^{0.50}$	$\text{DTZD}_{\text{sig}}^{0.60}$	$\text{DTZD}_{\text{ps}}^{0.482}$	$\text{DTZD}_{\text{ps}}^{0.60}$
$< 10^{-12}$	24.4970	10.800	13.872	23.884	13.876	24.761
$< 10^{-8}$	18.9150	10.038	11.949	17.076	12.158	17.917
$< 10^{-4}$	13.2880	9.104	10.875	11.224	11.224	11.919
$< 10^{-2}$	10.2560	8.133	10.359	8.430	10.153	8.548

TABLE 8.2 Performance Comparison on DTZD Model (8.4) Starting from Different Initial States to Invert Online Positive-Definite Matrices

	Average number of iterations	
Precision	$X_0 = 2A^{\text{T}}/\text{tr}(AA^{\text{T}})$	$X_0 = 2I/\text{tr}(A)$
$\|AX_k - I\|_2 < 10^{-12}$	25.639	19.015
$\|AX_k - I\|_2 < 10^{-8}$	15.634	9.994
$\|AX_k - I\|_2 < 10^{-4}$	14.795	9.066
$\|AX_k - I\|_2 < 10^{-2}$	13.943	8.167

1) The result related to DTZD model (8.4) using the array of linear activation functions with step size $h = 1$ (namely, Newton iteration) is denoted by $\text{DTZD}_{\text{lin}}^{1.0}$.

2) The result related to DTZD model (8.4) using the array of bipolar sigmoid activation functions with step size $h = 0.50$ is denoted by $\text{DTZD}_{\text{sig}}^{0.50}$.

3) The result related to DTZD model (8.4) using the array of power-sigmoid activation functions with step size $h = 0.60$ is denoted by $\text{DTZD}_{\text{ps}}^{0.60}$.

As seen from Table 8.1, when the prescribed precision is very small, better performance is achieved by using the array of linear activation functions with optimal step size $h = 1$. On the other hand, when using the array of bipolar sigmoid activation functions, better performance is achieved with optimal step size $h = 0.5$; and, when using the array of power-sigmoid activation functions, better performance is achieved with optimal step size $h = 0.482$. Note that, if the prescribed precision is not very small, DTZD model (8.4) with a large value of h may perform better.

Example 8.3 This numerical-experiment example compares the performance of DTZD model (8.4) starting from different initial states X_0 for inverting the positive-definite matrices. To do so, four groups of 1000 positive-definite random matrices $A \in \mathbb{R}^{3 \times 3}$ are tested, with the matrix entries uniformly distributed in $[-5, 5]$ as well, and the initial state X_0 of DTZD model (8.4) is chosen as $2A^{\text{T}}/\text{tr}(AA^{\text{T}})$ or preferably $X_0 = 2I/\text{tr}(A)$. The numerical-experiment results are shown in Table 8.2, which summarize that $X_0 = 2I/\text{tr}(A)$ is a better initial state for DTZD model (8.4) to invert positive-definite matrix A within a fewer iterations.

8.6 New DTZD Models Aided with Line-Search Algorithm

In this section, more DTZD models, which are different from DTZD model (8.4) and Newton iteration (8.6), are presented by employing more equally spaced multiple-point backward difference formulas [105]. Furthermore, the line-search algorithm [55] is used to obtain the appropriate step size h for fast convergence.

8.6.1 More Generalized Models

Based on the polynomial-interpolation theory [57], Lagrange interpolating polynomial is constructed by using corresponding discrete-time values of a target function. Then, the approximate first-order numerical-differentiation formulas are derived in terms of multiple sampling nodes (or to say, points). In addition, the relatively high computational-precision of the numerical-differentiation approximation to $\dot{X}(t)$ is achieved via the equally spaced difference formulas involving multiple sampling nodes. The approximation order of the resultant numerical differentiation is thus $O(\tau^{\kappa-1})$, where τ is the sampling gap (or to say, period of time), and κ here denotes the number of sampling nodes. It is worth noting that the aforementioned Euler forward difference formula $\dot{X}(t = k\tau) \approx (X((k+1)\tau) - X(k\tau))/\tau$ only has the precision order $O(\tau)$ for approximating $\dot{X}(t)$. From CTZD model (8.2), to obtain higher approximation precision [e.g., order $O(\tau^2)$] and better convergence as well as to further investigate DTZD models, more new DTZD models are generalized and presented below via the equally spaced multiple-point backward difference formulas.

Specifically, according to [57], the equally spaced three-point backward difference formula is written as

$$\dot{X}((k+2)\tau) \approx \frac{3X((k+2)\tau) - 4X((k+1)\tau) + X(k\tau)}{2\tau}, \tag{8.21}$$

where $k = 0, 1, 2, \cdots$ denotes again the update index. Note that (8.21) has a precision order $O(\tau^2)$ for approximating the corresponding time derivative $\dot{X}(t)$. This means that formula (8.21) might be (much) better than Euler forward difference formula when we derive the DTZD models. Thus, we have the following proposition [105].

Proposition 6 *Given nonsingular matrix $A \in \mathbb{R}^{n \times n}$, if monotonically increasing odd function array $\Phi(\cdot)$ is used, a new iteration model (involving three sampling-nodes and with two initial states X_0 and X_1) generalized form CTZD model (8.2) for the inverse of matrix A is written as*

$$X_{k+2} = \frac{4}{3}X_{k+1} - \frac{1}{3}X_k - hX_{k+1}\Phi(AX_{k+1} - I), \tag{8.22}$$

where $k = 0, 1, 2, \cdots$, and $h > 0$ denotes a suitable step size.

Proof. From CTZD model (8.2), we have

$$A\dot{X}((k+2)\tau) = -\gamma\Phi\Big(AX((k+2)\tau) - I\Big). \tag{8.23}$$

Using the presented backward difference formula (8.21), we discretize the Equation (8.23) as

$$A(3X_{k+2} - 4X_{k+1} + X_k) = -2\gamma\tau\Phi(AX_{k+2} - I), \tag{8.24}$$

where we denote $X_{k+2} = X((k+2)\tau)$, with $k = 0, 1, 2, \cdots$.

As matrix A is nonsingular (i.e., invertible), DTZD model (8.24) is written in the following form:

$$3X_{k+2} = 4X_{k+1} - X_k - 2\gamma\tau A^{-1}\Phi(AX_{k+2} - I).$$

(8.25)

Similar to Subsection 8.2.2, in view of the convergence of $X(t)$ to A^{-1} [in other words, $X(t) \approx A^{-1}$ for time instant t being large enough], we simply replace A^{-1} in DTZD model (8.25) with X_{k+2}. Then we have

$$3X_{k+2} = 4X_{k+1} - X_k - 2\gamma\tau X_{k+2}\Phi(AX_{k+2} - I).$$

(8.26)

However, X_{k+2} in Equation (8.26) is the unknown matrix to be iteratively solved. In order to make Equation (8.26) more computable (or to say, explicitly computable), similar to the idea of Gauss-Seidel method [57], it is reasonable that X_{k+2} on the right-hand side of (8.26) be replaced by X_{k+1}. Thus, DTZD model (8.26) becomes

$$X_{k+2} = \frac{4}{3}X_{k+1} - \frac{1}{3}X_k - hX_{k+1}\Phi(AX_{k+1} - I),$$

(8.27)

where $h = 2\gamma\tau/3 > 0$ denotes the step size, which would be selected appropriately via the line-search algorithm [55] (presented in the ensuing subsection) for the convergence to the theoretical inverse A^{-1}. Thus, the proof is complete.

For presentation convenience, Equation (8.27) is called the three-point DTZD model. In addition, if the array of linear activation functions is used, the three-point DTZD model (8.27) reduces to

$$X_{k+2} = \frac{4}{3}X_{k+1} - \frac{1}{3}X_k - hX_{k+1}(AX_{k+1} - I).$$

(8.28)

Evidently, the three-point DTZD model (8.28) is different from Newton iteration (8.6), in the sense that DTZD model (8.28) is a three-point iteration, while Newton iteration (8.6) is a two-point iteration. It is worth mentioning here that the two initial states X_0 and X_1 of the presented three-point DTZD model (8.28) [or (8.27)] are chosen, respectively, as $X_0 = 2A^T/\text{tr}(AA^T)$ and $X_1 = 2X_0 - X_0AX_0$ [of which the latter, for simplicity, is generated directly from Newton iteration (8.6), with $k = 0$].

Furthermore, by employing the equally spaced multiple-point backward difference formulas, we have the following proposition [105] for matrix inversion.

Proposition 7 *Given nonsingular matrix $A \in \mathbb{R}^{n \times n}$, if monotonically increasing odd function array $\Phi(\cdot)$ is used, many more iteration models (involving multiple sampling points) generalized from CTZD model (8.2) for the inverse of matrix A are achieved, such as*

- *the four-point DTZD model (or termed the generalized four-point nonlinearly activated Newton iteration)*

$$X_{k+3} = \frac{18}{11}X_{k+2} - \frac{9}{11}X_{k+1} + \frac{2}{11}X_k - hX_{k+2}\Phi(AX_{k+2} - I),$$

- *the five-point DTZD model (or termed the generalized five-point nonlinearly activated Newton iteration)*

$$X_{k+4} = \frac{48}{25}X_{k+3} - \frac{36}{25}X_{k+2} + \frac{16}{25}X_{k+1} - \frac{3}{25}X_k - hX_{k+3}\Phi(AX_{k+3} - I),$$

- *the six-point DTZD model (or termed the generalized six-point nonlinearly activated Newton iteration)*

$$X_{k+5} = \frac{300}{137}X_{k+4} - \frac{300}{137}X_{k+3} + \frac{200}{137}X_{k+2}$$
$$- \frac{75}{137}X_{k+1} + \frac{12}{137}X_k - hX_{k+4}\Phi(AX_{k+4} - I),$$

- *and the seven-point DTZD model (or termed the generalized seven-point nonlinearly activated Newton iteration)*

$$X_{k+6} = \frac{360}{147}X_{k+5} - \frac{450}{147}X_{k+4} + \frac{400}{147}X_{k+3}$$
$$- \frac{225}{147}X_{k+2} + \frac{72}{147}X_{k+1} - \frac{10}{147}X_k - hX_{k+5}\Phi(AX_{k+5} - I).$$

Proof. Following the proof of Proposition 6, with the aid of the equally spaced multiple-point backward difference formulas shown as follows (which involve four through seven sampling points):

- the four-point backward-difference formula

$$\dot{X}((k+3)\tau) \approx \frac{1}{6\tau}\Big(11X((k+3)\tau) - 18X((k+2)\tau) + 9X((k+1)\tau) - 2X(k\tau)\Big),$$

- the five-point backward-difference formula

$$\dot{X}((k+4)\tau) \approx \frac{1}{12\tau}\Big(25X((k+4)\tau) - 48X((k+3)\tau)$$
$$+ 36X((k+2)\tau) - 16X((k+1)\tau) + 3X(k\tau)\Big),$$

- the six-point backward-difference formula

$$\dot{X}((k+5)\tau) \approx \frac{1}{60\tau}\Big(137X((k+5)\tau) - 300X((k+4)\tau) + 300X((k+3)\tau)$$
$$- 200X((k+2)\tau) + 75X((k+1)\tau) - 12X(k\tau)\Big),$$

- and the seven-point backward-difference formula

$$\dot{X}((k+6)\tau) \approx \frac{1}{60\tau}\Big(147X((k+6)\tau) - 360X((k+5)\tau) + 450X((k+4)\tau)$$
$$- 400X((k+3)\tau) + 225X((k+2)\tau) - 72X((k+1)\tau) + 10X(k\tau)\Big).$$

Based on the above multiple-point backward-difference formulas, the multiple-point DTZD models are thus obtained, which completes the proof.

The above multiple-point DTZD models [including the three-point DTZD model (8.28)], all generalized from CTZD model (8.2), are evidently different from the original Newton iteration (8.6). Before ending this subsection, it is worth mentioning that the initial values of the above multiple-point DTZD models are normally chosen via Newton iteration (8.6) starting from $X_0 = 2A^T/\text{tr}(AA^T)$.

8.6.2 Line-Search Algorithm

In this subsection, the line-search algorithm [55] is presented to obtain the appropriate step-size value of h for the DTZD models [e.g., model (8.28)]. The line-search algorithm is mainly about the determination of the search-direction and the appropriate step size. In other words, in each iteration, a search direction is obtained, and then different step sizes are tried along the direction for a better solution point [55].

The line-search algorithm employed in this chapter is thus divided into the following two main steps. First, according to ZD design formula (8.1), we obtain the search direction as the convergence direction of the matrix-valued error function $E = AX(t) - I$. This guarantees that the state matrix $X(t)$ of the ZD model converges to the theoretical inverse A^{-1}. Secondly, after the search direction is determined, the step size h is obtained by trying different values of step size via the multiplication or division technique. That is, when the error E decreases, the step size is multiplied by a real number $\eta > 1$; and, when E increases, the step size is divided by $\eta > 1$. In details, the following computation procedure is used to obtain the appropriate value of step size in the kth iteration for the three-point DTZD model (8.28), as an example.

1) If the kth iteration error $\|E_{k+1}\|_F$ decreases (i.e., $\|E_{k+1}\|_F < \|E_k\|_F$), the current values $\|E_{k+1}\|_F$, X_{k+1} and h are saved, and then step size h is multiplied by η. The computation procedure used to determine the step size h is described as follows.

```
While 1
    h = h × η
    Calculate X_{k+1} and E_{k+1}
    if E_{k+1} ⩽ bestERR
        Save to bestERR, bestX, and bestH
    else
        Restore bestERR, bestX, and bestH
        break
    end
end
```

2) If $\|E_{k+1}\|_F$ increases (i.e., $\|E_{k+1}\|_F \geqslant \|E_k\|_F$), the step size h is divided by $\eta > 1$. It is worth mentioning that, when step size h is divided by η for too many times, it would make $h \to 0$ and may yield wrong information. Thus, to avoid this undesired result, we restore the already saved best values "bestERR," "bestX," and "bestH" if $h \leqslant 10^{-20}$. The computation procedure is shown in the following box.

```
While 1
    h = h ÷ η
    Calculate X_{k+1} and E_{k+1}
    if E_{k+1} ⩽ bestERR
        Save to bestERR, bestX, and bestH
        break
    end
    if h ⩽ 10^{-20}
        Restore bestERR, bestX, and bestH
        break
    end
end
```

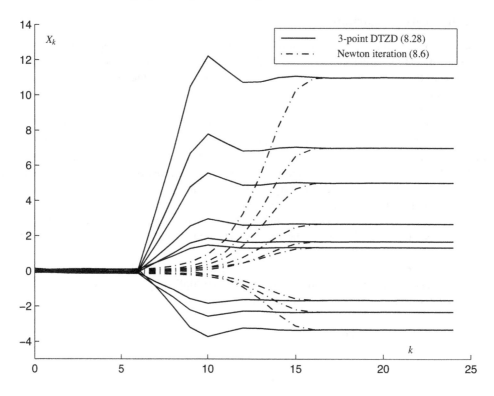

FIGURE 8.5 Neural states of three-point DTZD model (8.28) and Newton iteration (8.6) for matrix inversion.

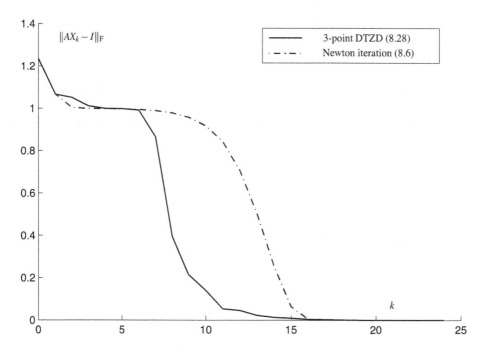

FIGURE 8.6 Comparison on residual errors $\|AX_k - I\|_F$ synthesized by three-point DTZD model (8.28) and Newton iteration (8.6) for matrix inversion.

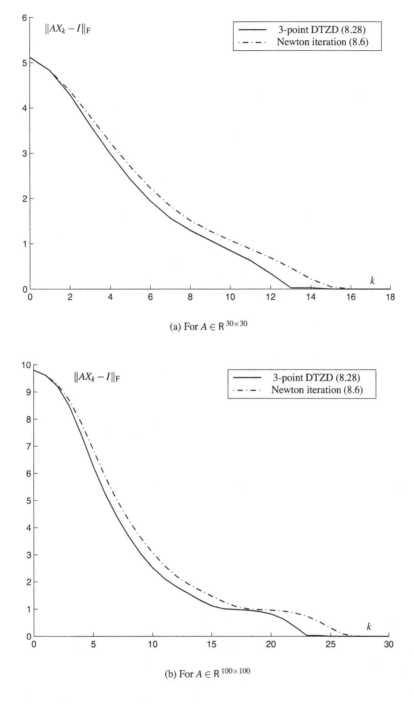

FIGURE 8.7 Comparison on residual errors synthesized by DTZD model (8.28) and Newton iteration (8.6) for matrix inversion with different dimensions.

Following the above discussion and analysis, the appropriate step size h is obtained in each iteration for three-point DTZD model (8.28) (as well as other multiple-point DTZD models). In the ensuing section, four matrices of different dimensions are considered and tested to illustrate the performance of the presented DTZD model (8.28) aided with the line-search algorithm.

8.6.3 Illustrative Examples

For illustrative purposes, numerical-experiment results are presented and discussed to show the convergence of three-point DTZD model (8.28) aided with the line-search algorithm for online matrix inversion. For comparison, Newton iteration (8.6) is applied as well.

Now, let us consider the matrix inversion problem $AX = I$ with the following matrix $A \in \mathbb{R}^{3 \times 3}$, as well as its theoretical inverse A^{-1} (which is given for checking the solution correctness of the DTZD model):

$$A = \begin{bmatrix} 7 & 0 & -5 \\ -2 & 1 & 0 \\ -1 & 3 & -4 \end{bmatrix} \text{ and } A^{-1} = \begin{bmatrix} 1.333 & 5.000 & -1.667 \\ 2.667 & 11.000 & -3.333 \\ 1.667 & 7.000 & -2.333 \end{bmatrix}.$$

For the numerical experiment of DTZD model (8.28), two successive initial states (i.e., X_0 and X_1) and initial step size h are chosen respectively as $X_0 = 2A^{\mathrm{T}} / \mathrm{tr}(AA^{\mathrm{T}})$, $X_1 = 2X_0 - X_0 A X_0$ and $h = 1$. In addition, the real number $\eta = 2n$ is set for adjusting the value of h. Specifically, Figure 8.5 illustrates the state matrix X_k of DTZD model (8.28) aided with the line-search algorithm for the inversion. It is seen that neural state X_k of three-point DTZD model (8.28) converges to theoretical inverse A^{-1}. For comparison, Newton iteration (8.6) is employed as well to solve the same matrix-inversion problem with initial state $X_0 = 2A^{\mathrm{T}} / \mathrm{tr}(AA^{\mathrm{T}})$. Its convergence performance is also seen in Figure 8.5, which shows that state X_k of Newton iteration (8.6) converges to theoretical inverse A^{-1} as well. It is worth noting here that, at a beginning stage [e.g., $k \in (0, 15)$], DTZD model (8.28) has (much) superior convergence, as compared with Newton iteration (8.6). This is also seen in Figure 8.6, which illustrates the convergence of residual errors $\|AX_k - I\|_{\mathrm{F}}$ of the above two models [i.e., DTZD model (8.28) and Newton iteration (8.6)].

To further investigate DTZD model (8.28) and its performance for matrix inversion, illustrative examples based on three randomly generated matrices with different dimensions (i.e., $\mathbb{R}^{30 \times 30}$ or $\mathbb{R}^{100 \times 100}$) are conducted. The numerical results are shown in Figure 8.7, which further illustrates the superior convergence of DTZD model (8.28) aided with the line-search algorithm.

Before ending this section, it is worth mentioning that the numerical verification (i.e., via numerical experiments) of other multiple-point DTZD models for matrix inversion are left again for interested readers to complete as a topic of exercise.

8.7 Summary

In this chapter, we have developed and investigated the discrete-time zeroing dynamics (DTZD) for online matrix inversion. Such a DTZD model has been regarded as a generalized form of Newton iteration for matrix inversion. Corresponding to the use of different types of activation functions, the initial state and step size have been analyzed for such a DTZD model, so as to ensure superior convergence and better stability [118]. Moreover, by employing more equally spaced multiple-point backward-difference formulas, the new multiple-point DTZD models have been developed and investigated for online matrix inversion. Besides, to obtain the appropriate step size, the line-search algorithm has been employed. Numerical-experiment results have illustrated the effectiveness of the presented DTZD models (including Newton iteration as a special case) for online matrix inversion.

Chapter 9

Moore–Penrose Inversion

Abstract

In this chapter, we generalize, investigate, and analyze the zeroing dynamics (ZD) models for time-varying full-rank matrix Moore–Penrose inversion. Theoretical analysis results show the superior convergence and better performance of the ZD models as compared with the gradient dynamics (GD) models. In addition, the computer simulation results and the application to inverse kinematic control of a redundant robot arm illustrate the feasibility and effectiveness of the ZD models for the time-varying full-rank matrix Moore–Penrose inversion.

9.1 Introduction

Moore–Penrose inverse (i.e., pseudoinverse) is considered to be one of the fundamental problems widely encountered in various science and engineering fields, such as ocean data assimilation [26], kinematic control [64], acoustic field control [34], and fault tolerant control [79]. Owing to its fundamental roles, much effort has been directed toward computational aspects of fast matrix Moore–Penrose inversion, and many algorithms have thus been proposed, such as Newton iteration [5,89], successive matrix squaring algorithms [12], Greville recursive method [137], and singular value decomposition [86]. Note that these methods usually aim at constant matrices. However, real-time matrix Moore–Penrose inversion problem generally exists; for example, the real-time solution of time-varying Moore–Penrose inverse can be utilized to formulate and solve the inverse kinematic problem of a redundant robot arm. In this chapter, the ZD models are generalized, investigated, and analyzed for time-varying full-rank matrix Moore–Penrose inversion [131], which is based on matrix-valued indefinite error functions. Furthermore, through computer simulations (including the robotic application), we substantiate the excellent convergence of the ZD models solving in real time for time-varying Moore–Penrose inverse.

9.2   Preliminaries

To lay a basis for further discussion, the following preliminaries of matrix Moore–Penrose inverse is given.

Definition 3   Consider a matrix $A \in \mathbb{R}^{m \times n}$. If X satisfies one or several of the following Penrose equations:

$$AXA = A,$$

$$XAX = X,$$

$$(AX)^{\mathrm{T}} = AX,$$

$$(XA)^{\mathrm{T}} = XA,$$

then X is called the generalized inverse of A. In addition, if X satisfies all of the above four equations, X is called the Moore–Penrose inverse (also referred to as pseudoinverse) of A, denoted by A^{+}. It is worth pointing out that Moore–Penrose inverse is unique and has minimal Frobenius norm among all the generalized inverses.

Given matrix $A \in \mathbb{R}^{m \times n}$, if $\mathrm{rank}(A) = \min\{m, n\}$ (i.e., A is of full rank), then AA^{T} (or $A^{\mathrm{T}}A$) is nonsingular. The unique Moore–Penrose inverse $A^{+} \in \mathbb{R}^{n \times m}$ for matrix A is given as

$$A^{+} = \begin{cases} A^{\mathrm{T}}(AA^{\mathrm{T}})^{-1}, & \text{if } m < n, \\ A^{-1}, & \text{if } m = n, \\ (A^{\mathrm{T}}A)^{-1}A^{\mathrm{T}}, & \text{if } m > n, \end{cases} \tag{9.1}$$

where the upper part, middle part, and lower part of Equation (9.1) correspond to the right Moore–Penrose inverse, the inverse and the left Moore–Penrose inverse, respectively. So, if full-rank matrix A is rectangular (i.e., $m \neq n$), the Moore–Penrose inverse A^{+} satisfies one of the following two matrix equations:

$$A^{+}AA^{\mathrm{T}} = A^{\mathrm{T}}, \text{ if } m < n, \tag{9.2}$$

$$A^{\mathrm{T}}AA^{+} = A^{\mathrm{T}}, \text{ if } m > n. \tag{9.3}$$

Note that, if $\mathrm{rank}(A) < \min\{m, n\}$ (i.e., A is rank-deficient), AA^{T} (or $A^{\mathrm{T}}A$) is singular. Thus, A^{+} can not be obtained via Equation (9.1). In this chapter, we only consider the time-varying full-rank matrix Moore–Penrose inverse, and A hereafter is defined as a full-rank matrix. The corresponding research on the time-varying rank-deficient matrix Moore–Penrose inverse is left for interested readers as a topic of extension or exercise.

9.3 ZD Models for Moore–Penrose Inverse

In this section, we present the ZD models for the Moore–Penrose inverse computation. In addition, two design methods of the ZD models for time-varying Moore–Penrose inversion are presented and investigated as follows.

9.3.1 Intrinsically Nonlinear Method of ZD Design

In this subsection, the matrix-valued indefinite error functions based on Equations (9.2) and (9.3) are employed to construct the ZD models solving for the right and left Moore–Penrose inverses.

9.3.1.1 ZD Model for Right Moore–Penrose Inverse

Firstly, as the right Moore–Penrose inverse $A^+(t)$ satisfies $A^+(t)A(t)A^T(t) = A^T(t)$, we define the following matrix-valued error function:

$$E(t) = X(t)A(t)A^T(t) - A^T(t), \tag{9.4}$$

where $X(t)$ corresponds to the theoretical Moore–Penrose inverse that we are to solve for. To make $E(t)$ converge to zero, its time derivative $\dot{E}(t)$ is described in the following form (i.e., the so-called ZD design formula):

$$\dot{E}(t) = -\gamma\Phi\Big(E(t)\Big) = -\gamma\Phi\Big(X(t)A(t)A^T(t) - A^T(t)\Big), \tag{9.5}$$

where design parameter $\gamma > 0$ is used to scale the convergence rate. In addition, $\Phi(\cdot) : \mathbb{R}^{n \times m} \longrightarrow \mathbb{R}^{n \times m}$ denotes a matrix-to-matrix activation function array. Being the ijth processing element of $\Phi(\cdot)$, any monotonically increasing odd activation function $\phi(\cdot)$ can be used. In this chapter, the array of linear activation functions and the array of power-sigmoid activation functions are used.

Secondly, it follows from Equation (9.4) that

$$\dot{E}(t) = \dot{X}(t)A(t)A^T(t) + X(t)\Big(\dot{A}(t)A^T(t) + A(t)\dot{A}^T(t)\Big) - \dot{A}^T(t). \tag{9.6}$$

Substituting Equations (9.4) and (9.6) into Equation (9.5) leads to the following ZD model solving for the right Moore–Penrose inverse $A^+(t)$:

$$\dot{X}(t)A(t)A^T(t) = -\gamma\Phi\Big(X(t)A(t)A^T(t) - A^T(t)\Big)$$
$$- X(t)\Big(\dot{A}(t)A^T(t) + A(t)\dot{A}^T(t)\Big) + \dot{A}^T(t). \tag{9.7}$$

For the purpose of modeling such a ZD model, we may need to transform Equation (9.7) into the following explicit form (with its block diagram depicted in Figure 9.1):

$$\dot{X}(t) = \dot{X}(t)\Big(I - A(t)A^T(t)\Big) - \gamma\Phi\Big(X(t)A(t)A^T(t) - A^T(t)\Big)$$
$$- X(t)\Big(\dot{A}(t)A^T(t) + A(t)\dot{A}^T(t)\Big) + \dot{A}^T(t). \tag{9.8}$$

Besides, it follows from ZD model (9.8) that its ijth neuron's dynamic equation is written as the following explicit dynamics:

$$\dot{x}_{ij} = \sum_{l=1}^{m}\sum_{p=1}^{n}\dot{x}_{il}(\delta_{lj} - a_{lp}\alpha_{pj}) - \gamma\phi\Big(\sum_{l=1}^{m}\sum_{p=1}^{n}x_{il}a_{lp}\alpha_{pj} - \alpha_{ij}\Big)$$
$$- \sum_{l=1}^{m}\sum_{p=1}^{n}x_{il}\dot{a}_{lp}\alpha_{pj} - \sum_{l=1}^{m}\sum_{p=1}^{n}x_{il}a_{lp}\dot{\alpha}_{pj} + \dot{\alpha}_{ij}, \tag{9.9}$$

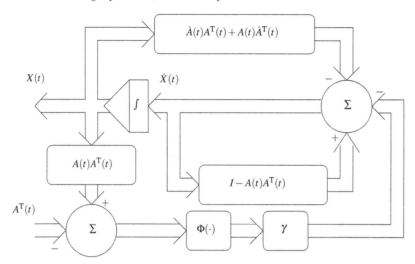

FIGURE 9.1 Block diagram of ZD model (9.7) for time-varying Moore–Penrose inversion.

where x_{ij} is the ijth entry of $X(t)$, a_{ij} is the ijth entry of $A(t)$, α_{ij} is the ijth entry of $A^\mathrm{T}(t)$, and δ_{ij} is the ijth entry of identity matrix I. We set $AA^\mathrm{T} = B$, $\dot{A}A^\mathrm{T} = C$ and $A\dot{A}^\mathrm{T} = D$, and b_{ij}, c_{ij} and d_{ij} denote the ijth entries of matrices B, C, and D, respectively. Thus, ZD model (9.9) is written as follows (with its ith row circuit schematic depicted in Figure 9.2):

$$\dot{x}_{ij} = \sum_{l=1}^{m} \dot{x}_{il}(\delta_{lj} - b_{lj}) - \gamma\phi\left(\sum_{l=1}^{m} x_{il}b_{lj} - \alpha_{ij}\right) - \sum_{l=1}^{m} x_{il}c_{lj} - \sum_{l=1}^{m} x_{il}d_{lj} + \dot{\alpha}_{ij}. \tag{9.10}$$

Thirdly, we have the following theoretical results on global exponential convergence of such a ZD model.

Theorem 16 *Given a smoothly time-varying matrix $A(t) \in \mathbb{R}^{m\times n}$ of full rank, if monotonically increasing odd activation function array $\Phi(\cdot)$ is used, then state matrix $X(t)$ of ZD model (9.7), starting from any initial state $X(0)$, converges to the time-varying theoretical right Moore–Penrose inverse of $A(t)$, i.e., $A^+(t)$.*

Proof Let $\tilde{X}(t) = X(t) - A^+(t)$ denote the difference between the solution $X(t)$ generated by ZD model (9.7) and the theoretical solution $A^+(t)$ of $X(t)A(t)A^\mathrm{T}(t) - A^\mathrm{T}(t) = 0$. In view of $A^+(t)A(t)A^\mathrm{T}(t) - A^\mathrm{T}(t) = 0$, its time derivative is depicted as

$$\dot{A^+}(t)A(t)A^\mathrm{T}(t) + A^+(t)\left(\dot{A}(t)A^\mathrm{T}(t) + A(t)\dot{A}^\mathrm{T}(t)\right) - \dot{A}^\mathrm{T}(t) = 0.$$

By using the above identity and ZD model (9.7), with $X(t) = \tilde{X}(t) + A^+(t)$, it follows that $\tilde{X}(t)$ is the solution to the ensuing dynamics, with the initial state $\tilde{X}(0) = X(0) - A^+(0)$:

$$\dot{\tilde{X}}(t)A(t)A^\mathrm{T}(t) + \tilde{X}(t)\left(\dot{A}(t)A^\mathrm{T}(t) + A(t)\dot{A}^\mathrm{T}(t)\right) = -\gamma\Phi\left(\tilde{X}(t)A(t)A^\mathrm{T}(t)\right). \tag{9.11}$$

From Equation (9.4) and $X(t) = \tilde{X}(t) + A^+(t)$, we have $E(t) = \tilde{X}(t)A(t)A^\mathrm{T}(t)$. Equation (9.11) is thus rewritten as $\dot{E}(t) = -\gamma\Phi(E(t))$, which is a compact matrix form of the following set of $n \times m$ equations:

$$\dot{e}_{ij}(t) = -\gamma\phi(e_{ij}(t)), \quad \forall i \in \{1, 2, \cdots, n\} \text{ and } j \in \{1, 2, \cdots, m\}. \tag{9.12}$$

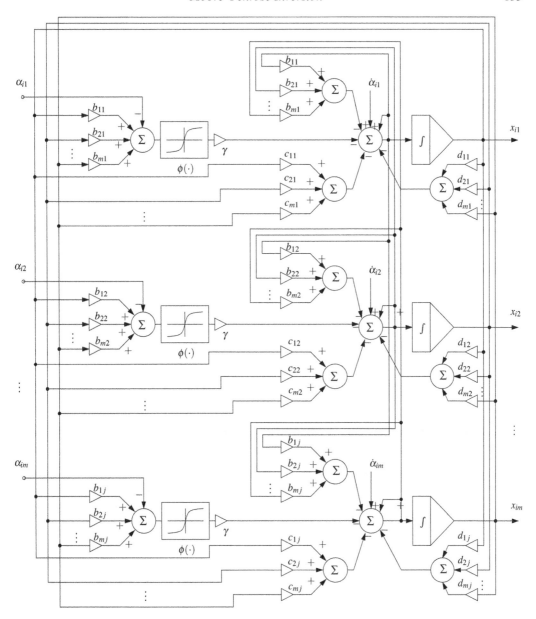

FIGURE 9.2 Circuit schematic of the ith neuron row in ZD model (9.8) as derived from Equation (9.10).

Evidently, we define a Lyapunov function candidate $v_{ij} = e_{ij}^2/2 \geqslant 0$ for the ijth subsystem (9.12) with its time derivative

$$\frac{\mathrm{d}v_{ij}}{\mathrm{d}t} = e_{ij}\dot{e}_{ij} = -\gamma e_{ij}\phi(e_{ij}). \tag{9.13}$$

Because $\phi(\cdot)$ is a monotonically increasing odd function, we have $\phi(-e_{ij}) = -\phi(e_{ij})$ and

$$\phi(e_{ij}) \begin{cases} > 0, & \text{if } e_{ij} > 0 \\ = 0, & \text{if } e_{ij} = 0 \\ < 0, & \text{if } e_{ij} < 0, \end{cases}$$

which guarantee the negative-definiteness of $\dot{v}_{ij}$; i.e., $\dot{v}_{ij} < 0$ for $e_{ij} \neq 0$ and $\dot{v}_{ij} = 0$ for $e_{ij} = 0$. By Lyapunov theory [78, 100, 123, 131, 132], $e_{ij}(t)$ globally converges to zero for any $i \in \{1, 2, \cdots, n\}$ and $j \in \{1, 2, \cdots, m\}$. Thus, in view of $E(t) = \tilde{X}(t)A(t)A^{\mathrm{T}}(t)$ and the uniqueness of the right Moore–Penrose inverse, we have $\tilde{X}(t) \to 0 \in \mathbb{R}^{n \times m}$ as $t \to \infty$, i.e., state matrix $X(t)$ globally converges to the theoretical time-varying Moore–Penrose inverse $A^+(t)$. The proof on global convergence of ZD model (9.7) is thus complete.

9.3.1.2 ZD Model for Left Moore–Penrose Inverse

Similarly, we get the ZD model for left time-varying Moore–Penrose inverse by constructing the matrix-valued error function using Equation (9.3) as

$$E(t) = A^{\mathrm{T}}(t)A(t)X(t) - A^{\mathrm{T}}(t). \tag{9.14}$$

The corresponding ZD model is obtained as

$$A^{\mathrm{T}}(t)A(t)\dot{X}(t) = -\gamma \Phi\Big(A^{\mathrm{T}}(t)A(t)X(t) - A^{\mathrm{T}}(t)\Big)$$
$$- \Big(\dot{A}^{\mathrm{T}}(t)A(t) + A^{\mathrm{T}}(t)\dot{A}(t)\Big)X(t) + \dot{A}^{\mathrm{T}}(t). \tag{9.15}$$

Corollary 4 *Given a smoothly time-varying matrix $A(t) \in \mathbb{R}^{m \times n}$ of full rank, if monotonically increasing odd activation function array $\Phi(\cdot)$ is used, then state matrix $X(t)$ of ZD model (9.15), starting from any initial state $X(0)$, converges to the time-varying theoretical left Moore–Penrose inverse of matrix $A(t)$, i.e., $A^+(t)$.*

9.3.2 Extended Nonlinearization Method of ZD Design

By using the error functions (9.4) and (9.14), ZD models (9.7) and (9.15) seem relatively complicated and lengthy. In this subsection, other error functions are employed to construct the ZD models for Moore–Penrose inverse so as to obtain simplified neural-dynamic models.

For the right Moore–Penrose inverse, the simplified matrix-valued error function is set as

$$E(t) = X(t)A(t) - I. \tag{9.16}$$

Then, we restrict the time derivative of error function E to be

$$\dot{E}(t) = -\gamma E(t) = -\gamma\big(X(t)A(t) - I\big). \tag{9.17}$$

In view of Equation (9.16), we obtain

$$\dot{E}(t) = \dot{X}(t)A(t) + X(t)\dot{A}(t). \tag{9.18}$$

Expanding design formula (9.17) via Equation (9.18) leads to

$$\dot{X}(t)A(t) = -\gamma\big(X(t)A(t) - I\big) - X(t)\dot{A}(t). \tag{9.19}$$

Furthermore, in order to make ZD model (9.19) more computable and make $X(t)$ converge to the unique solution, multiplying $A^{\mathrm{T}}(t)$ in both sides of Equation (9.19) yields

$$\dot{X}(t)A(t)A^{\mathrm{T}}(t) = -\gamma\Big(X(t)A(t)A^{\mathrm{T}}(t) - A^{\mathrm{T}}(t)\Big) - X(t)\dot{A}(t)A^{\mathrm{T}}(t). \tag{9.20}$$

Inspired by Equation (9.5), Equation (9.20) can be improved as

$$\dot{X}(t)A(t)A^{\mathrm{T}}(t) = -\gamma\Phi\Big(X(t)A(t)A^{\mathrm{T}}(t) - A^{\mathrm{T}}(t)\Big) - X(t)\dot{A}(t)A^{\mathrm{T}}(t), \tag{9.21}$$

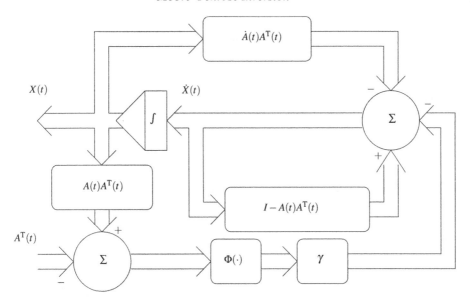

FIGURE 9.3 Block diagram of ZD model (9.21) for time-varying Moore–Penrose inversion.

which is the simplified ZD model for right Moore–Penrose inverse, with the block diagram depicted in Figure 9.3.

Similarly, to get the simplified ZD model for left Moore–Penrose inverse, we define the error function as

$$E(t) = A(t)X(t) - I. \tag{9.22}$$

Following the same design procedure of ZD model (9.21), we have the following simplified ZD model for left Moore–Penrose inverse computation:

$$A^T(t)A(t)\dot{X}(t) = -\gamma\Phi\left(A^T(t)A(t)X(t) - A^T(t)\right) - A^T(t)\dot{A}(t)X(t). \tag{9.23}$$

9.4 Comparison between ZD and GD Models

In order to keep the completeness of the chapter, the ZD models are compared with the GD (i.e., gradient dynamics) models for time-varying Moore–Penrose inversion in this section. By following the design methods in literature [99], and defining the scalar-valued error functions as

$$\mathcal{E}(t) = \|X(t)A(t)A^T(t) - A^T(t)\|_F^2/2, \quad \text{if } m < n, \tag{9.24}$$

$$\mathcal{E}(t) = \|A^T(t)A(t)X(t) - A^T(t)\|_F^2/2, \quad \text{if } m > n, \tag{9.25}$$

we have the following GD models for time-varying Moore–Penrose inverse:

$$\dot{X}(t) = -\gamma\left(X(t)A(t)A^T(t) - A^T(t)\right)A(t)A^T(t), \quad \text{if } m < n, \tag{9.26}$$

$$\dot{X}(t) = -\gamma A^T(t)A(t)\left(A^T(t)A(t)X(t) - A^T(t)\right), \quad \text{if } m > n. \tag{9.27}$$

Evidently, GD models (9.26) and (9.27) are designed to solve the right and left time-varying Moore–Penrose inverses, respectively. Moreover, we compare the ZD and GD models for Moore–Penrose inversion via the following facts.

1) The design of ZD models is based on the elimination of every entry of the matrix-valued error function [e.g., Equations (9.4) and (9.14)]. In contrast, the design of GD models is based on the elimination of the norm-based scalar-valued error function [e.g., Equations (9.24) and (9.25)]. The matrix-valued error function can make the resultant ZD models monitor and force every entry of the error to zero. Thus, more information is used for network learning, and better performance can be achieved for the ZD models, as compared with the GD models.

2) The ZD models exploit the time-derivative information of coefficient matrices $A(t)$ and $A^{\mathrm{T}}(t)$ during their real-time solving processes. This is the reason why the ZD models globally exponentially converge to the exact solution of the time-varying Moore–Penrose inverse problem. In contrast, the GD models do not exploit such important information, and thus may not be effective enough in solving such a time-varying problem. In essence, the ZD method is based on a prediction thought, while the GD method belongs to a tracking approach.

3) Different types of linear and nonlinear functions can be used as the activation functions of the ZD models, while the classic GD models only exploit the linear activation functions. Non-linearity always exists. Even if the linear activation functions are used, nonlinear phenomena may appear in their hardware implementation. Thus, different types of nonlinear activation functions can make the ZD models perform better in practice.

4) The ZD models are generally depicted in implicit dynamics [e.g., $\dot{X}(t)A(t)A^{\mathrm{T}}(t) = \cdots$, and $A^{\mathrm{T}}(t)A(t)\dot{X}(t) = \cdots$], which frequently arise in analog electronic circuits and systems due to Kirchhoff's rules. In contrast, the GD models are depicted in explicit dynamics, i.e., $\dot{X}(t) = \cdots$, which are usually associated with classic Hopfield-type recurrent neural networks. Furthermore, the implicit dynamic equations can preserve physical parameters in the coefficient matrices. They can describe the usual and unusual parts of a dynamic system in the same form. Thus, implicit systems have higher abilities in representing dynamic systems, as compared with explicit systems. Besides, the implicit dynamic systems can be mathematically transformed to explicit dynamic systems, if needed.

5) As shown in previous chapters (especially, Chapters 1, 2, 4, 6 and 8), and also this chapter, the connections between Newton iteration and the ZD models are established. That is, Newton iteration for solving constant problems is a special case of the ZD models (by considering only the use of linear activation functions and fixing the step size to be 1). On the other hand, the ZD models can be viewed as the general forms of Newton iteration.

6) The ZD approach converts time-varying problems to linear or nonlinear equations to solve. In contrast, the GD approach converts time-varying problems to minimization problems to solve.

7) The derivation of the ZD models is less complicated and only requires the mathematical knowledge of BS (bachelor of science) level. In contrast, the derivation of GD models is much more complex and needs intricate mathematical knowledge of MS (master of science) or PhD (doctor of philosophy) level.

8) As shown in the last part of the book (i.e., Part VII), new fractals can be yielded by using the ZD models to solve nonlinear equations in complex domain. Besides, it is found that new fractals incorporate the well-known Newton fractals (generated by Newton iteration) as special cases, though these two kinds of fractals are different. In contrast, the GD-related fractal has not been reported.

9.5 Simulation and Verification

In order to verify the efficacy of the presented ZD models and the correctness of the neural-dynamic solution, illustrative examples based on the following three time-varying matrices are presented below.

Example 9.1 A time-varying matrix and its right Moore–Penrose inverse are, respectively,

$$A(t) = \begin{bmatrix} \sin 1.5t & \cos 1.5t & -\sin 1.5t \\ -\cos 1.5t & \sin 1.5t & \cos 1.5t \end{bmatrix}$$

and

$$A^+(t) = \begin{bmatrix} 0.5\sin 1.5t & -0.5\cos 1.5t \\ \cos 1.5t & \sin 1.5t \\ -0.5\sin 1.5t & 0.5\cos 1.5t \end{bmatrix}.$$

Example 9.2 A time-varying matrix and its left Moore–Penrose inverse are, respectively,

$$A(t) = \begin{bmatrix} \sin 1.5t & \cos 1.5t \\ -\cos 1.5t & \sin 1.5t \\ \sin 1.5t & \cos 1.5t \end{bmatrix} \text{ and } A^+(t) = \begin{bmatrix} 0.5\sin 1.5t & -\cos 1.5t & 0.5\sin 1.5t \\ 0.5\cos 1.5t & \sin 1.5t & 0.5\cos 1.5t \end{bmatrix}.$$

Example 9.3 A time-varying matrix and its right Moore–Penrose inverse are, respectively,

$$A(t) = \begin{bmatrix} \sin t & \cos t & -\sin t \\ -\cos t & \sin t & \cos t \end{bmatrix} \text{ and } A^+(t) = \begin{bmatrix} 0.5\sin t & -0.5\cos t \\ \cos t & \sin t \\ -0.5\sin t & 0.5\cos t \end{bmatrix}.$$

Note that the variation frequency of Example 9.1 is one and a half times that of Example 9.3.

9.5.1 Simulation Techniques

While Section 9.3 presents the main theoretical results of the presented ZD models, the following MATLAB® simulation techniques [57] are investigated to show the characteristics of such ZD models.

9.5.1.1 Kronecker Product and Vectorization

Review the ZD and GD models. Their dynamic equations are all described in matrix form, which cannot be directly simulated. Thus, the techniques of Kronecker product and vectorization [104,107] are needed and used to transform such matrix-form differential equations to vector-form differential equations.

Take ZD model (9.7), for example. Based on the Kronecker product and vectorization techniques, for simulation proposes, we transform ZD model (9.7) to the following vector-form differential equation:

$$((AA^T)^T \otimes I)\,\mathrm{vec}(\dot{X}) = \mathrm{vec}(\dot{A}^T) - \left((\dot{A}A^T)^T \otimes I + (A\dot{A}^T)^T \otimes I\right)\mathrm{vec}(X)$$
$$- \gamma\Phi\left(((AA^T)^T \otimes I)\,\mathrm{vec}(X) - \mathrm{vec}(A^T)\right). \tag{9.28}$$

The Kronecker product can be generated by using MATLAB routine "kron" readily; in other words, $A \otimes B$ can be generated by code `kron(A,B)`. To generate vec(X), the routine "reshape" can be used. Specifically, if the matrix X has n rows and m columns, then the code of vectorizing X is `reshape(X,m*n,1)`, which generates a column vector, $\mathrm{vec}(X) = [x_{11}, \cdots, x_{n1}, x_{12}, \cdots, x_{n2}, \cdots, x_{1m}, \cdots, x_{nm}]^T$.

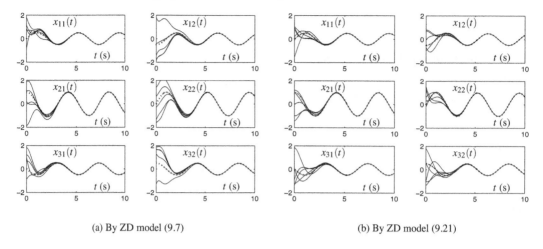

(a) By ZD model (9.7)   (b) By ZD model (9.21)

FIGURE 9.4 Neural states of ZD models for Example 9.1 with linear activation functions and $\gamma = 1$ used, where dotted curves denote theoretical solution $A^+(t)$, and subplot ranges are all $[0,10] \times [-2,2]$.

9.5.1.2 ODE with Mass Matrix

In the simulation of ZD model (9.7), MATLAB routine "ode45" is preferred because "ode45" can solve the initial-value ODE problem with a mass matrix, e.g., $M(t,\mathbf{x})\dot{\mathbf{x}} = \mathbf{g}(t,\mathbf{x})$, where nonsingular matrix $M(t,\mathbf{x})$ on the left-hand side of such an equation is termed the mass matrix, and $\mathbf{g}(t,\mathbf{x})$ is the right-hand side of vector-form ZD model (9.28).

To solve the ODE with a mass matrix, routine "odeset" should also be used. Its "Mass" property should be assigned to be function handle @MatrixM, which returns the value of mass matrix $M(t,\mathbf{x})$. Note that, if $M(t,\mathbf{x})$ does not depend on state vector $\mathbf{x}$ and if function "MatrixM" is to be invoked with only one input argument t, then the value of the "MStateDep" property of routine "odeset" can be set as "none."

9.5.2 Verification of Examples

In this subsection, computer simulations are conducted by applying the presented ZD and GD models to compute Moore–Penrose inverses of time-varying matrices in Examples 9.1 through 9.3, and the corresponding simulative and comparative results are illustrated in Figures 9.4 through 9.11.

As seen from the neural states of the presented ZD models for different matrices' Moore–Penrose inverses in Figures 9.4 through 9.6, as well as Figure 9.10(a), the solutions of the ZD models always converge to the theoretical time-varying Moore–Penrose inverse exactly. In addition, Figures 9.6 and 9.7 show that the convergence speed becomes much faster if design parameter γ increases. We can also observe from Figure 9.7 that, using the same value of γ, the solution errors (i.e., $\|X(t) - A^+(t)\|_F$) synthesized by using power-sigmoid activation functions tend to zero nearly twice as fast as those synthesized by using linear activation functions.

As shown in Figures 9.8 and 9.10(b), with $\gamma = 1$, the neural states of the GD models always lag behind the theoretical solution $A^+(t)$. It can be seen from Figure 9.9 that the solution errors of the GD models are considerably large. Comparing the simulation results in Figures 9.9 and 9.11(b), we see that the GD models perform worse in handling the time-varying Moore–Penrose inversion, when the variation frequency of matrix $A(t)$ increases. In contrast, the ZD models converge to the theoretical solution $A^+(t)$ exactly, as shown in Figures 9.4 through 9.7, 9.10(a) and 9.11(a). These

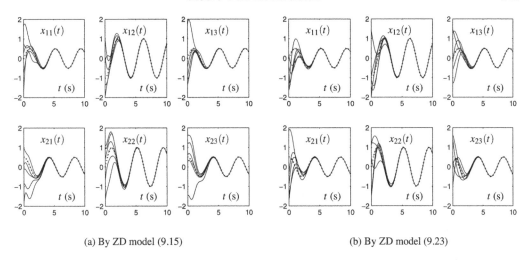

(a) By ZD model (9.15)   (b) By ZD model (9.23)

FIGURE 9.5 Neural states of ZD models for Example 9.2 with linear activation functions and $\gamma = 1$ used, where subplot ranges are all $[0, 10] \times [-2, 2]$.

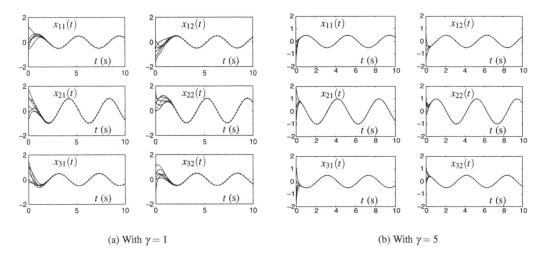

(a) With $\gamma = 1$ (b) With $\gamma = 5$

FIGURE 9.6 Neural states of ZD model (9.21) for Example 9.1, with power-sigmoid activation functions used, where subplot ranges are all $[0, 10] \times [-2, 2]$.

results substantiate the superior performance of the ZD models for time-varying Moore–Penrose inversion, as compared with the GD models.

9.6 Application to Robot Arm

As mentioned in Section 9.1, the online solution of Moore–Penrose inverse is essential for the inverse kinematic control of the redundant robot arm. Therefore, this section presents the application

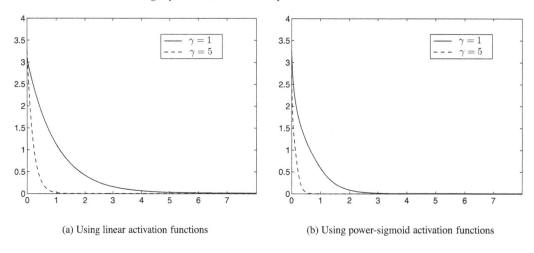

(a) Using linear activation functions   (b) Using power-sigmoid activation functions

FIGURE 9.7 Solution errors $\|X(t) - A^+(t)\|_F$ of ZD model (9.21) for Example 9.1 over time t (s).

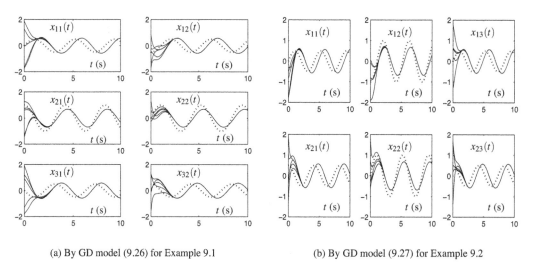

(a) By GD model (9.26) for Example 9.1   (b) By GD model (9.27) for Example 9.2

FIGURE 9.8 Neural states of GD models, with $\gamma = 1$ and subplot ranges all $[0,10] \times [-2,2]$.

of the presented neural-dynamic models to the kinematic control of a general (planar) redundant robot arm via real-time time-varying Moore–Penrose inversion.

9.6.1   Inverse Kinematics

Consider a redundant robot arm of which the end-effector position vector $\mathbf{r}(t) \in \mathbb{R}^m$ in Cartesian space is related to the joint-space vector $\theta(t) \in \mathbb{R}^n$ through the following forward kinematic equation:

$$\mathbf{r}(t) = \mathbf{f}(\theta(t)), \tag{9.29}$$

where $\mathbf{f}(\cdot)$ is a continuous nonlinear mapping function with known structure and parameters for a given robot arm. The inverse kinematic problem is to find the joint variable $\theta(t)$ for any given $\mathbf{r}(t)$ through the inverse mapping of Equation (9.29), i.e., $\theta(t) = \mathbf{f}^{-1}(\mathbf{r}(t))$ if it exists.

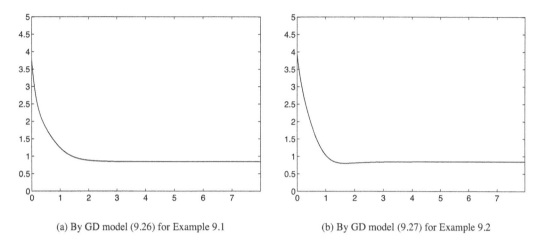

(a) By GD model (9.26) for Example 9.1   (b) By GD model (9.27) for Example 9.2

FIGURE 9.9 Solution errors $\|X(t) - A^+(t)\|_F$ of GD models, with $\gamma = 1$ over time t (s).

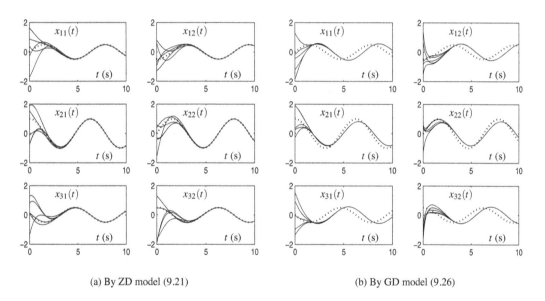

(a) By ZD model (9.21)   (b) By GD model (9.26)

FIGURE 9.10 Comparison on ZD model (9.21) and GD model (9.26), with $\gamma = 1$ for Example 9.3, where subplot ranges are all $[0, 10] \times [-2, 2]$.

Unfortunately, it is usually difficult to find an analytical solution of $\mathbf{f}^{-1}$ due to the redundancy and nonlinearity of function $\mathbf{f}(\cdot)$. The inverse kinematic problem is thus usually solved at the velocity level. Differentiating Equation (9.29) with respect to time t yields a linear relation between velocities $\dot{\mathbf{r}}$ and $\dot{\theta}$:

$$J(\theta(t))\dot{\theta}(t) = \dot{\mathbf{r}}(t), \tag{9.30}$$

where $J(\theta) \in \mathbb{R}^{m \times n}$ is the Jacobian matrix defined as $J(\theta) = \partial\mathbf{f}(\theta)/\partial\theta$. Equation (9.30) is under-determined and may admit an infinite number of solutions, because, in a redundant robot arm, $m < n$.

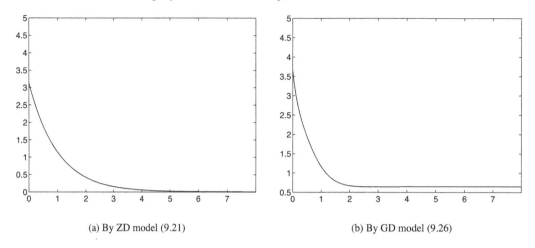

(a) By ZD model (9.21) (b) By GD model (9.26)

FIGURE 9.11 Solution errors $\|X(t) - A^+(t)\|_F$ of ZD model (9.21) and GD model (9.26), with $\gamma = 1$ for Example 9.3 over time t (s).

The pseudoinverse-type solution to Equation (9.30), widely used by robotics researchers, is generally formulated as a minimum-norm particular solution plus a homogeneous solution:

$$\dot{\theta}(t) = J^+(t)\dot{\mathbf{r}}(t) + (I - J^+(t)J(t))\mathbf{z}(t), \tag{9.31}$$

where $J^+(t) \in \mathbb{R}^{n \times m}$ denotes the Moore–Penrose inverse of $J(t)$. The vector $\mathbf{z}(t) \in \mathbb{R}^n$ is arbitrary, which can be selected as the negative gradient of a performance index to be minimized.

Since J is of full row rank and $m < n$, the Moore–Penrose inverse of J, i.e., J^+, is defined as $J^T(JJ^T)^{-1}$. The usual numerical solutions for $J^+(t)$ are computationally intensive and may have relatively large computational errors. Moreover, with multiple tasks and constraints included within Equation (9.31) via $\mathbf{z}(t)$, heavy-burden computation process may hinder real-time applications, especially in high-DOF (degrees of freedom) robotic systems.

Defining $A(t) = J(t)$, we can re-exploit ZD model (9.7) to solve for $J^+(t)$ in a theoretically parallel and error-free manner so as to expedite the computational process and achieve better solution precision. That is, $J^+(t) = X(t)$, where $X(t)$ is the state matrix generated in real time by the presented ZD model (9.7).

9.6.2 Simulations Based on Five-Link Robot Arm

In this subsection, we apply the two neural-dynamic models [i.e., ZD model (9.7) and GD model (9.26)] to the path-tracking redundancy resolution of a five-link planar robot arm (which has five degrees of freedom). The dimension of Jacobian matrix of the robot arm is 2×5 and the degrees of redundancy is 3, since only the end-effector positioning is considered here. In the computer simulations, the end-effector of the five-link robot arm is expected to track a square path.

In the first simulation, ZD model (9.7) with a power-sigmoid activation function array and design parameter $\gamma = 10^4$ is employed for the redundancy resolution of the five-link robot arm. The desired motion of the end-effector is a square with the side-length being 0.46 m. The motion duration is 40 s, and the initial joint states $\theta(0) = [-\pi/12, \pi/12, \pi/6, -\pi/4, \pi/3]^T$ rad. Figure 9.12 illustrates the simulated motion trajectories of the robot arm in the planar operational-space, of which the end-effector trajectory is sufficiently close to the desired square path. The corresponding joint variables and joint velocities are depicted in Figure 9.13. As further shown in Figure 9.14, the

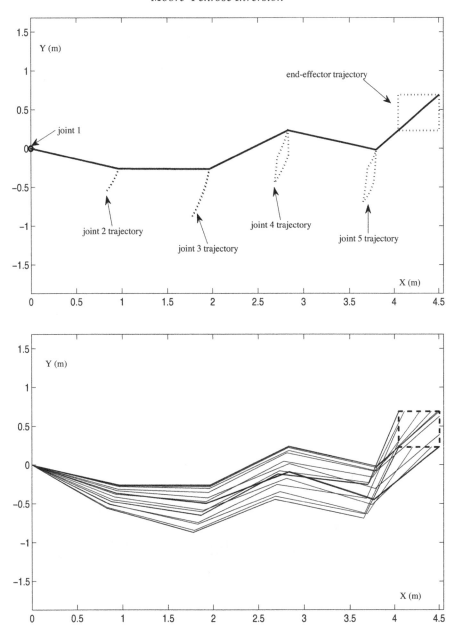

FIGURE 9.12 Motion trajectories of five-link planar robot arm synthesized by ZD model (9.7).

maximal position and velocity errors of the end-effector are less than 3×10^{-6} m and 4×10^{-9} m/s, respectively.

For comparison, as the second simulation, GD model (9.26) is also simulated to solve for $J^{+}(t)$ in real time. Using the same design parameter $\gamma = 10^4$, the end-effector position and velocity errors synthesized by the GD model (i.e., less than 1.1×10^{-5} m and 1.2×10^{-6} m/s, respectively) are relatively larger than the ones by the ZD model. In addition, to achieve the precision similar to that of the ZD model, the design parameter γ of the GD model should be larger than 10^8, which is a very stringent restriction on the potential analog/digital system design. Evidently, the ZD model is supe-

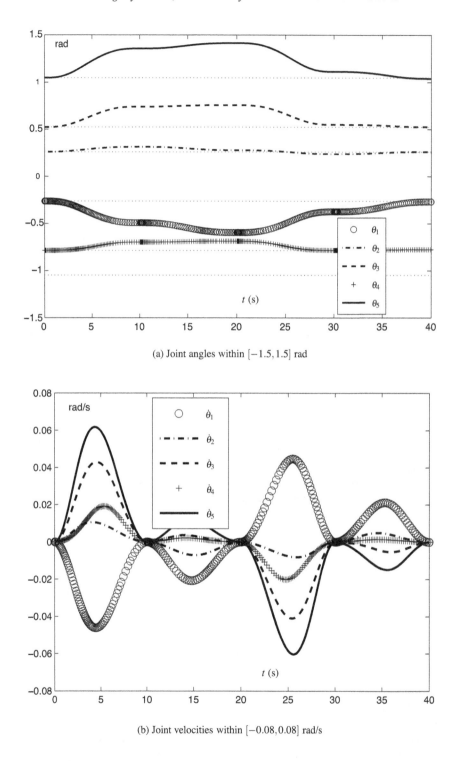

(a) Joint angles within $[-1.5, 1.5]$ rad

(b) Joint velocities within $[-0.08, 0.08]$ rad/s

FIGURE 9.13 Joint variables of five-link planar robot arm synthesized by ZD model (9.7).

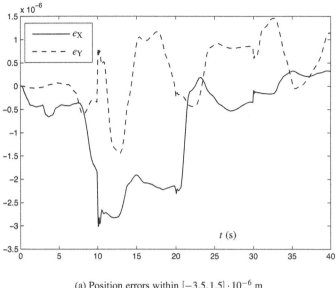

(a) Position errors within $[-3.5, 1.5] \cdot 10^{-6}$ m

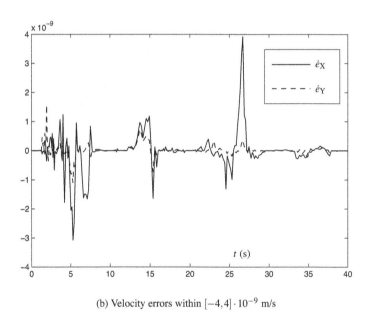

(b) Velocity errors within $[-4, 4] \cdot 10^{-9}$ m/s

FIGURE 9.14 End-effector errors of five-link planar robot arm tracking a square path synthesized by ZD model (9.7).

rior in the inverse kinematic control of the redundant robot arm, as compared with the GD model. The main reason is that the ZD model methodologically exploits the time-derivative information of matrix $J(t)$ [i.e., $\dot{J}(t)$], and thus it is more effective on the real-time Moore–Penrose inversion. In contrast, GD model (9.26) does not utilize such important information, and thus may not be as effective as ZD model (9.7) in the inverse kinematic control of the redundant robot arm(s).

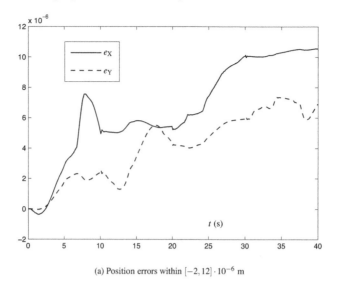

(a) Position errors within $[-2, 12] \cdot 10^{-6}$ m

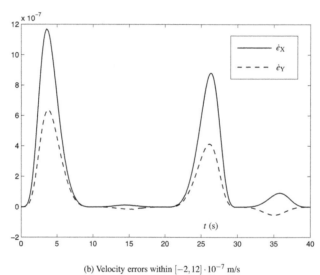

(b) Velocity errors within $[-2, 12] \cdot 10^{-7}$ m/s

FIGURE 9.15 End-effector errors of five-link planar robot arm tracking a square path synthesized by GD model (9.26).

9.7 Summary

For time-varying Moore–Penrose inversion, ZD has been generalized and developed in this chapter, which provides an effective online-computing approach for the time-varying full-rank matrices. The simulation of such ZD models have been investigated. Theoretical analysis and simulation results have both substantiated better performance (i.e., superior convergence and higher precision) of the ZD models, as compared with the GD models. In addition, the application of the ZD model to the kinematic control of the redundant robot arm via real-time time-varying Moore–Penrose inversion has been presented with effectiveness well shown.

Part IV

Matrix Square Root Finding

Chapter 10

ZD Models and Newton Iteration

Abstract

In this chapter, we investigate the zeroing dynamics (ZD) models for time-varying matrix square root finding, which is evidently different and extended from the time-varying scalar square root finding investigated in Chapter 1. For the purpose of potential hardware implementation, the discrete-time zeroing dynamics (DTZD) model is constructed and developed, which incorporates Newton iteration as a special case. Besides, to obtain an appropriate step-size value (in each iteration), the line-search algorithm is employed for the DTZD model. Numerical-experiment results substantiate the effectiveness of the proposed DTZD model aided with the line-search algorithm, in addition to the link and explanation to Newton iteration for matrix square root finding.

10.1 Introduction

The problem of matrix square root is widely encountered in many scientific areas [10, 20, 31, 59, 60, 110]. Due to its fundamental roles, much effort has been directed toward the matrix square root finding. Being one of the most useful methods, Newton iteration [31] has been investigated for matrix square root finding, owing to its good properties of convergence and stability. In addition, many variants of Newton iteration have been developed and analyzed by many researchers. For example, the simple form of Newton iteration is well known and computationally less expensive [31]. Similar iterations have been designed differently, such as Denman and Beavers (DB) method [20] based on the matrix sign function iteration, and Meini iteration [59] based on a cyclic reduction (CR) algorithm. In addition, numerically stable Schur decomposition method [31] has been presented for matrix square root finding. Generally speaking, the methods reported in [10, 20, 31, 53, 59, 60] (including the ones mentioned above) are related to the numerical algorithms (which may be less efficient enough for large-scale or online problems solving, due to their serial-processing nature). Being another important type of solution, many parallel-processing computational schemes, including various neural-dynamic solvers, have been developed, analyzed, and implemented on specific architectures. The neural-dynamic approach is now regarded as a powerful alternative to online computation because of its parallel distributed nature, and more importantly, the suitability of hardware implementation.

For the purpose of potential hardware implementation, in this chapter, the DTZD model is generalized and investigated for matrix square root finding [130]. Different activation functions and different step sizes (which can be obtained via the line-search algorithm [53]) are employed and investigated for such a model. When the linear activation functions and step size $h = 1$ are used, the DTZD model reduces exactly to Newton iteration for matrix square root finding.

10.2   Problem Formulation and ZD Models

Now, let us consider the following time-varying matrix square root equation:

$$X^2(t) - A(t) = 0, \tag{10.1}$$

where, both being smoothly time-varying matrices, positive-definite $A(t) \in \mathbb{R}^{n \times n}$ and its time derivative $\dot{A}(t) \in \mathbb{R}^{n \times n}$ are assumed to be known numerically (or at least measurable accurately). In addition, let $X(t)$, which is to be solved for, denote the time-varying square root of $A(t)$. For simplicity, the following square-root existence condition [108] is assumed.

Square-root existence condition: If smoothly time-varying matrix $A(t) \in \mathbb{R}^{n \times n}$ is positive-definite (in general sense [98]) at any time instant $t \in [0, +\infty)$, then there exists a time-varying matrix square root $X(t) \in \mathbb{R}^{n \times n}$ for $A(t)$.

10.2.1   CTZD Model

To solve for time-varying matrix square root by Zhang *et al*'s method, the following matrix-valued error function is defined first (instead of using any scalar-valued norm-based energy function associated with the GD method and model):

$$E(t) = X^2(t) - A(t) \in \mathbb{R}^{n \times n}.$$

Secondly, the time derivative $\dot{E}(t) \in \mathbb{R}^{n \times n}$ is designed such that every entry $e_{ij}(t) \in \mathbb{R}$ of $E(t) \in \mathbb{R}^{n \times n}$ converges to zero, with $i, j = 1, 2, \cdots, n$; in mathematics, to choose $\dot{e}_{ij}(t)$ such that $\lim_{t \to \infty} e_{ij}(t) = 0, \forall i, j \in \{1, 2, \cdots, n\}$. A general form of $\dot{E}(t)$ is

$$\dot{E}(t) = -\Gamma \Phi\big(E(t)\big), \tag{10.2}$$

where design parameter $\Gamma \in \mathbb{R}^{n \times n}$ is a positive-definite matrix used to scale the convergence rate. Note that the matrix-valued design parameter Γ can be set as large as the hardware would permit or selected appropriately for simulative/experimental purposes. For simplicity, in this chapter, we assumed $\Gamma = \gamma I$ with the scalar-valued design parameter $\gamma > 0$. In addition, the activation function array $\Phi(\cdot) : \mathbb{R}^{n \times n} \to \mathbb{R}^{n \times n}$ is a matrix-valued entry-to-entry mapping, in which each scalar-valued processing-unit $\phi(\cdot)$ is a monotonically increasing odd activation function as detailed in Chapter 2 (specifically, in Subsection 2.3.2).

Thirdly, expanding the ZD design formula (10.2) leads to the following CTZD model:

$$X(t)\dot{X}(t) + \dot{X}(t)X(t) = -\gamma \Phi\big(X^2(t) - A(t)\big) + \dot{A}(t), \tag{10.3}$$

where $X(t)$, starting from initial condition $X(0) = X_0 \in \mathbb{R}^{n \times n}$, denotes the state matrix corresponding to the time-varying theoretical matrix square root of $A(t)$. Besides, $\dot{X}(t)$ and $\dot{A}(t)$ denote the

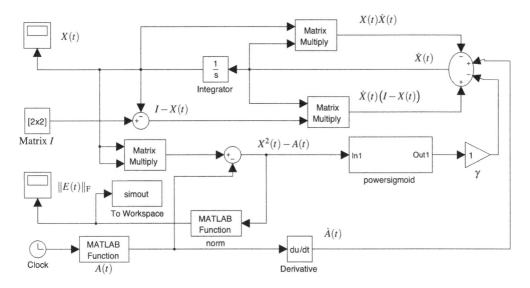

FIGURE 10.1 Simulink model of CTZD (10.3) applied to time-varying matrix square root finding.

time-derivatives of $X(t)$ and $A(t)$, respectively. Note that CTZD model (10.3) methodologically and systematically exploits the time-derivative information of the problem-matrix [i.e., $\dot{A}(t)$], and thus it can be more effective in solving the time-varying matrix root problem.

For example, we use CTZD model (10.3) to find the square root of time-varying matrix

$$A(t) = \begin{bmatrix} 36 + 3\sin 2t \cos 2t & 21\sin 2t \\ 28\cos 2t & 64 + 3\sin 2t \cos 2t \end{bmatrix},$$

with the following theoretical solution given to check the CTZD solution correctness:

$$X^*(t) = \begin{bmatrix} 6 & 1.5\sin 2t \\ 2\cos 2t & 8 \end{bmatrix}.$$

Such a square-root finding example is constructed and simulated by using MATLAB® Simulink® [103,124], with the overall Simulink model depicted in Figure 10.1 [130]. The simulation results are shown in Figures 10.2 and 10.3. From Figure 10.2, we see that the CTZD model solution trajectories (denoted by solid curves) fit well with those of the theoretical matrix square root of $A(t)$ (denoted by dotted curves). In addition, the convergence speed is expedited by increasing the value of γ, which is shown in Figure 10.3.

Furthermore, by considering the constant matrix square root problem only [i.e., with the time derivative $\dot{A}(t) = 0$], CTZD model (10.3) reduces to

$$X(t)\dot{X}(t) + \dot{X}(t)X(t) = -\gamma\Phi\big(X^2(t) - A\big). \tag{10.4}$$

Evidently, the above two CTZD models, i.e., models (10.3) and (10.4), are depicted in implicit dynamics, instead of explicit dynamics, as discussed in details in Chapter 9 (specifically, in Section 9.4).

From the analysis results of [130] and references therein, we have the following lemma for the principal square root, which focuses on the constant situation of (10.1) (i.e., $X^2 - A = 0$).

Lemma 4 *Let $A \in \mathbb{R}^{n \times n}$ have no nonpositive real eigenvalues, then there is an unique real square root X of A, which is denoted by $A^{1/2}$ and called the principal square root of A.*

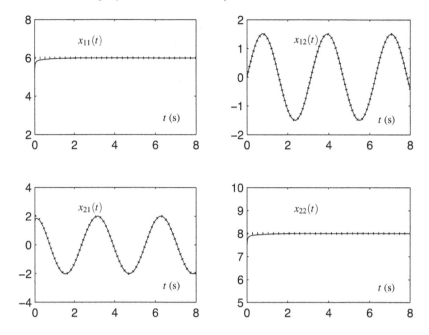

FIGURE 10.2 Time-varying matrix square root finding via CTZD model (10.3), with theoretical solution denoted by dotted curves and CTZD solution denoted by solid curves.

10.2.2 DTZD Model

For the purpose of potential hardware implementation via digital circuits, it is more preferable to discretize CTZD model (10.4) by using Euler forward-difference rule, $\dot{X}(t = k\tau) \approx \left(X((k+1)\tau) - X(k\tau)\right)/\tau$, where $\tau > 0$ denotes the sampling gap. In general, we have $X_k = X(t = k\tau)$ for presentation convenience. Thus, the following theorem about the DTZD model solving for the constant matrix square root is obtained.

Theorem 17 *Given matrix $A \in \mathbb{R}^{n \times n}$, which satisfies Lemma 4, if a monotonically increasing odd function array $\Phi(\cdot)$ is used, the following DTZD model (10.4) solving for matrix square root $A^{1/2}$ is established:*

$$\begin{cases} X_k D_k + D_k X_k = -\Phi(X_k^2 - A), \\ X_{k+1} = X_k + h D_k, \end{cases} \tag{10.5}$$

where $k = 0, 1, 2, \cdots$, $h = \tau\gamma > 0$ and $D_k = (X_{k+1} - X_k)/h$. In addition, step size h should be selected appropriately to guarantee the convergence of X_k to theoretical matrix square root $A^{1/2}$.

Proof From the presented CTZD model (10.4), we have

$$X(k\tau)\dot{X}(k\tau) + \dot{X}(k\tau)X(k\tau) = -\gamma\Phi\left(X^2(k\tau) - A\right). \tag{10.6}$$

Then, using Euler forward-difference rule [i.e., $\dot{X}(k\tau) \approx \left(X((k+1)\tau) - X(k\tau)\right)/\tau$], with $h = \tau\gamma > 0$ and $D_k = (X_{k+1} - X_k)/h$, we discretize equation (10.6) as

$$\begin{cases} X_k D_k + D_k X_k = -\Phi(X_k^2 - A), \\ X_{k+1} = X_k + h D_k, \end{cases}$$

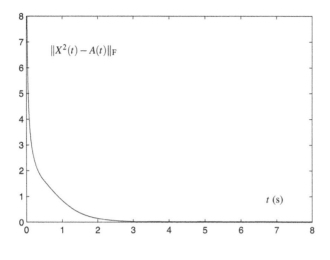

(a) With $\gamma = 1$

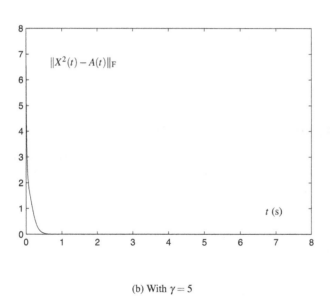

(b) With $\gamma = 5$

FIGURE 10.3 Residual errors $\|X^2(t) - A(t)\|_{\mathrm{F}}$ synthesized by CTZD model (10.3).

where step size h would be selected appropriately via the line-search algorithm [53] (presented in the ensuing section as well as Chapter 8, Subsection 8.6.2). The proof is thus complete.

For modeling and implementation purposes, the presented DTZD model (10.5) is reformulated as follows:

$$\begin{cases} D_k = X_k D_k + D_k(X_k + I) + \Phi(X_k^2 - A), \\ X_{k+1} = X_k + hD_k. \end{cases} \tag{10.7}$$

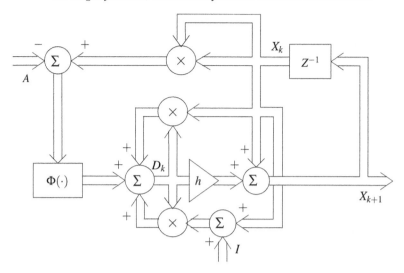

FIGURE 10.4 Block diagram of DTZD model (10.5) for constant matrix square root finding.

The corresponding block diagram of DTZD model (10.5) is thus shown in Figure 10.4, which is actually built based on Equation (10.7).

10.3   Link and Explanation to Newton Iteration

As we know, much effort has been directed towards computational aspects of matrix square root finding, and many algorithms, such as Newton iteration and its variants, have been investigated and analyzed [20, 31, 53, 59]. To lay a basis for further discussion, the following lemma about Newton iteration for matrix square root is given [31, 53].

Lemma 5 *Given matrix $A \in \mathbb{R}^{n \times n}$, which satisfies Lemma 4, Newton iteration for matrix square root finding is formulated as*

$$\begin{cases} X_k H_k + H_k X_k = A - X_k^2, \\ X_{k+1} = X_k + H_k, \end{cases} \tag{10.8}$$

which, starting from a suitable initial value X_0, converges to the principal square root $A^{1/2}$.

It is worth mentioning that Newton iteration (10.8) has good property of convergence (i.e., with a quadratic rate of convergence) and stability [31, 53]. To keep the completeness of this chapter, the relationship between Newton iteration (10.8) and DTZD model (10.5) is presented in this section as follows.

Consider DTZD model (10.5). When we use the array of linear activation function $\phi(e_{ij}) = e_{ij}$, DTZD model (10.5) reduces to

$$\begin{cases} X_k D_k + D_k X_k = -(X_k^2 - A), \\ X_{k+1} = X_k + h D_k. \end{cases} \tag{10.9}$$

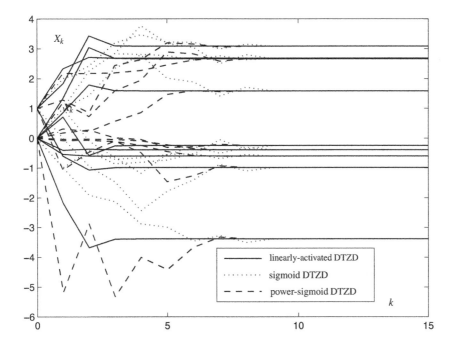

FIGURE 10.5 Neural states of DTZD model (10.5) using three different types of activation function array for square root finding of matrix $A \in \mathbb{R}^{3 \times 3}$.

For $h = 1$ and $H_k = D_k$, equation (10.9) becomes

$$\begin{cases} X_k H_k + H_k X_k = A - X_k^2, \\ X_{k+1} = X_k + H_k, \end{cases}$$

which is exactly Newton iteration (10.8) for matrix square root finding [31,53]. In other words, we discover that a general form of Newton iteration for constant matrix square root finding is given by DTZD model (10.5).

10.4 Line-Search Algorithm

As detailed in Subsection 8.6.2, the line-search algorithm [53] can be used to obtain the appropriate step-size value h of DTZD models. In this section, step size h of DTZD model (10.5) is also selected via the line-search algorithm for the convergence of X_k to the theoretical square root $A^{1/2}$.

Specifically, the design of line-search algorithm is mainly based on the determination of a search-direction and the appropriate step-size value along that direction. Such an algorithm employed in this chapter can be divided into the following two main steps and can also be seen in Subsection 8.6.2.

First, a search direction is obtained in each iteration. According to ZD design formula (10.2), we can have the search-direction as the convergence-direction of the matrix-valued error $E = X^2 - A$. This guarantees that the state matrix X_k of DTZD model (10.5) can converge to the theoretical square root $A^{1/2}$. Secondly, along the above search-direction, the appropriate step size can be obtained by trying different step-size values, via multiplication or division techniques. That is, when the error-value decreases, the step size is multiplied by a real number $\eta > 1$; and, when the error-value

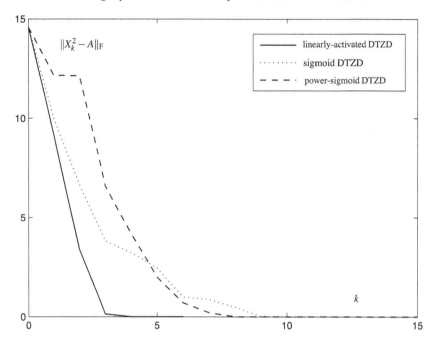

FIGURE 10.6 Residual errors $\|X_k^2 - A\|_{\mathrm{F}}$ synthesized by DTZD model (10.5) using three different types of activation function array for square root finding of matrix $A \in \mathbb{R}^{3 \times 3}$.

increases, the step size is divided by η. For example, η can be 2 or $2n$, where n is the dimension of matrix A. Note that, in this chapter, we prefer the former (i.e., $\eta = 2$). Please see and compare Subsection 8.6.2 for details.

10.5   Illustrative Examples

In this section, numerical-experiment results are presented to illustrate the convergence of DTZD model (10.5) aided with the line-search algorithm for matrix square root finding. Besides, different activation function arrays $\Phi(\cdot)$ are comparatively employed in DTZD model (10.5) starting from the initial value $X_0 = I$.

Now, let us consider the following randomly generated positive-definite matrix $A \in \mathbb{R}^{3 \times 3}$:

$$A = \begin{bmatrix} -5.4266 & 3.4240 & -1.6190 \\ -8.7163 & 4.3043 & -2.1848 \\ 5.3466 & 2.8412 & 6.3228 \end{bmatrix},$$

of which the theoretical square root matrix is

$$A^{1/2} = \begin{bmatrix} -0.9874 & 1.5867 & -0.3875 \\ -3.3827 & 3.0864 & -0.6057 \\ 2.6683 & -0.2413 & 2.6853 \end{bmatrix},$$

given for checking the solution correctness of DTZD model (10.5).

For the numerical experiment of the proposed DTZD model (10.5), the initial step size h is chosen as 1. Figure 10.5 illustrates the state matrix X_k (i.e., the kth state value, where $k = 0, 1, 2, \cdots$)

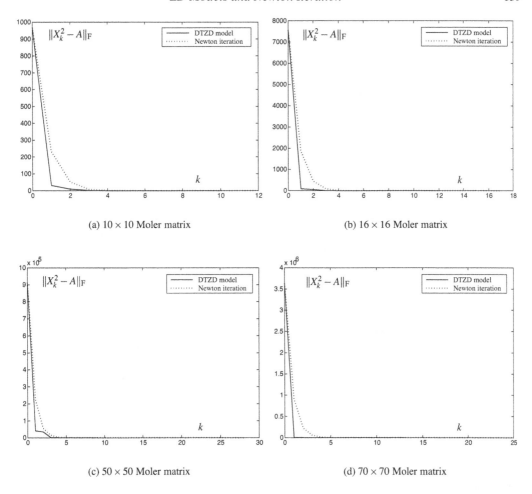

FIGURE 10.7 Comparison on residual errors $\|X_k^2 - A\|_F$ synthesized by DTZD model (10.9) and Newton iteration (10.8) for square root finding of Moler matrices with different dimensions, where subfigure ranges are $[0, 12] \times [0, 1] \cdot 10^3$, $[0, 18] \times [0, 8] \cdot 10^3$, $[0, 30] \times [0, 1] \cdot 10^6$, and $[0, 25] \times [0, 4] \cdot 10^6$.

of DTZD model (10.5) aided with the line-search algorithm. As seen from Figure 10.5, state matrix X_k of DTZD model (10.5) converges to the theoretical square root $A^{1/2}$, by using the three different types of activation functions (i.e., linear, sigmoid, and power-sigmoid activation functions). In addition, the residual errors $\|X_k^2 - A\|_F$ are seen in Figure 10.6, which illustrate the convergence performance of DTZD model (10.5) aided with the line-search algorithm.

To further investigate the performance of DTZD model (10.5) for matrix square root finding, as well as to compare it with Newton iteration [note that DTZD model (10.5) incorporates Newton iteration (10.8) as a special case], illustrative examples and results are presented based on the Moler matrix [31] (which is symmetric and positive-definite) with different dimensions. Specifically, Figure 10.7 illustrates the residual errors $\|X_k^2 - A\|_F$ synthesized by DTZD model (10.9) [i.e., the linear activation functions are used in DTZD model (10.5)] aided with the line-search algorithm and Newton iteration (10.8) for the square root finding of the Moler matrix. From Figure 10.7, we see that DTZD model (10.9) aided with the line-search algorithm has superior convergence, as compared with the standard Newton iteration (10.8).

TABLE 10.1   Performance Comparison between DTZD Model (10.9) and Other Methods for Square Root Finding of Moler Matrices with Different Dimensions

	Number of iterations									
	$A \in \mathbb{R}^{10 \times 10}$					$A \in \mathbb{R}^{30 \times 30}$				
Precision	DTZD	NI*	DB	CR	IN	DTZD	NI	DB	CR	IN
10^{-12}	10	12	–	11	12	9	27	–	–	–
10^{-10}	10	12	12	11	12	9	24	–	16	17
10^{-8}	9	11	11	10	11	9	20	–	13	14
10^{-4}	6	7	7	6	7	7	14	–	7	8

*NI = Newton Iteration

Furthermore, the numerical experiments based on two Moler matrices with different dimensions (i.e., $\mathbb{R}^{10 \times 10}$ and $\mathbb{R}^{30 \times 30}$) are conducted to show the comparison between DTZD model (10.9) and other methods (i.e., Newton iteration [31], DB iteration [20], CR iteration [59], and improved Newton (IN) iteration [31]). The corresponding numerical results are shown in Table 10.1, in which the symbol "–" means that the prescribed precision cannot be achieved by the corresponding method. From the table, we see that the prescribed precision can be achieved in a few iterations by using DTZD model (10.9) aided with the line-search algorithm. In contrast, the prescribed precision $\|X_k^2 - A\|_F < 10^{-12}$ cannot be achieved by some other methods, such as DB, CR, and IN iterations, for square root of the 30×30 Moler matrix root finding. Moreover, as seen from Table 10.1, compared with other methods, when the prescribed precision and the dimension of Moler matrix become higher, better performance can be achieved by DTZD models. These examples also illustrate the effectiveness of DTZD model (10.9).

10.6   Summary

The DTZD model (10.5) has been generalized, developed, and investigated in this chapter for constant matrix square root finding, which incorporates Newton iteration (10.8) as a special case. In addition, to obtain an appropriate step-size value, the line-search algorithm has been re-employed for the DTZD model. Numerical-experiment results have substantiated the effectiveness of the proposed DTZD model aided with the line-search algorithm (as well as Newton iteration).

Chapter 11

ZD Model Using Hyperbolic Sine Activation Functions

Abstract

In this chapter, to pursue the superior convergence and robustness properties, a special type of activation function (i.e., the hyperbolic sine activation function) is applied to the zeroing dynamics model for real-time solution of time-varying matrix square root. Theoretical analysis and simulative results illustrate the superior performance of the ZD model using hyperbolic sine activation functions in the context of model-implementation errors, in comparison with that using linear activation functions.

11.1 Model and Activation Functions

In this section, for readers' convenience as well as for completeness, the design procedure of ZD model (10.3) for time-varying matrix square root finding is repeated and compared below.

To solve for time-vary matrix square root $A^{1/2}(t)$, by Zhang et al.'s method [99–102, 104, 107, 114, 117, 118, 122, 128, 129, 134], the following matrix-valued error function can be defined first:

$$E(t) = X^2(t) - A(t) \in \mathbb{R}^{n \times n}.$$

Secondly, the error function's time-derivative $\dot{E}(t) \in \mathbb{R}^{n \times n}$ can be made, such that every entry $e_{ij}(t) \in \mathbb{R}$ of $E(t) \in \mathbb{R}^{n \times n}$ converges to zero, with $i, j = 1, 2, \cdots, n$; in mathematics, to choose $\dot{e}_{ij}(t)$, such that $\lim_{t \to \infty} e_{ij}(t) = 0$, $\forall i, j \in \{1, 2, \cdots, n\}$. A general form of $\dot{E}(t)$ is given by Zhang et al. as follows [99–102, 104, 107, 114, 117, 118, 122, 128, 129, 134]:

$$\frac{\mathrm{d}E(t)}{\mathrm{d}t} = -\Gamma \Phi(E(t)), \tag{11.1}$$

where design parameter Γ and activation function array $\Phi(\cdot)$ are described as follows.

1) $\Gamma \in \mathbb{R}^{n \times n}$ is a positive-definite (diagonal) matrix used to scale the convergence rate. For simplicity, Γ can be γI, with scalar $\gamma > 0 \in \mathbb{R}$ and I denoting the identity matrix.

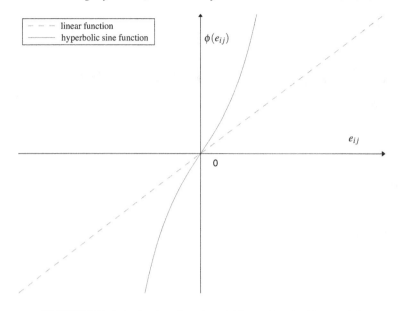

FIGURE 11.1 Activation function $\phi(\cdot)$ used in the ZD model.

2) $\Phi(\cdot) : \mathbb{R}^{n \times n} \to \mathbb{R}^{n \times n}$ denotes an activation function matrix array. In general, any monotonically increasing odd activation function $\phi(\cdot)$, being the ijth element of $\Phi(\cdot)$, can be used
for the construction of the neural-dynamic model. In this chapter, the following two types of
activation functions are discussed and compared (which are shown in Figure 11.1):

 i) linear activation function $\phi(e_{ij}) = e_{ij}$, and

 ii) hyperbolic sine activation function $\phi(e_{ij}) = \exp(\xi e_{ij})/2 - \exp(-\xi e_{ij})/2$, with design
 parameter $\xi \geqslant 1$.

Thirdly, expanding ZD design formula (11.1), we obtain the following implicit dynamic equation of the ZD model for real-time matrix square root finding:

$$X(t)\dot{X}(t) + \dot{X}(t)X(t) = -\gamma\Phi\big(X^2(t) - A(t)\big) + \dot{A}(t), \qquad (11.2)$$

where $X(t)$, starting from an initial condition $X(0) \in \mathbb{R}^{n \times n}$, denotes the state matrix corresponding
to theoretical time-varying matrix square root $X^*(t)$ of $A(t)$. It is worth mentioning that, when using
a linear activation function array $\Phi(\cdot)$, the general nonlinearly activated ZD model (11.2) reduces
to the following linearly activated one:

$$X(t)\dot{X}(t) + \dot{X}(t)X(t) = -\gamma X^2(t) + \gamma A(t) + \dot{A}(t). \qquad (11.3)$$

11.2   Convergence Analysis

It follows from Kronecker-product and vectorization techniques [104, 107] that time-varying matrix
square root Equation (10.1) can be written as [110]:

$$\big(I \otimes X(t)\big)\text{vec}\big(X(t)\big) - \text{vec}\big(A(t)\big) = 0, \qquad (11.4)$$

where, as presented also in Chapter 9 (specifically, Subsection 9.5.1), symbol $\otimes$ denotes the Kronecker product, and operator $\mathrm{vec}(\cdot) : \mathbb{R}^{n \times n} \to \mathbb{R}^{n^2 \times 1}$ [e.g., $\mathrm{vec}(A(t))$] generates a column vector obtained by stacking all column vectors of the input matrix argument [e.g., $A(t)$] together. For simulative purposes, based on the techniques of Kronecker product and vectorization, the matrix differential Equation (11.2) can be transformed to a vector differential equation, and thus we have the following theorem [110].

Theorem 18 *The matrix-form differential Equation (11.2) [i.e., ZD model (11.2)] is reformulated as the following vector-form differential equation:*

$$(I \otimes X + X^{\mathrm{T}} \otimes I)\mathrm{vec}(\dot{X}) = -\gamma \Phi\big((I \otimes X)\mathrm{vec}(X) - \mathrm{vec}(A)\big) + \mathrm{vec}(\dot{A}), \qquad (11.5)$$

where argument t is dropped for presentation convenience, and activation function array $\Phi(\cdot)$ in (11.5) is defined as before except that its dimensions are changed to be $\mathbb{R}^{n^2 \times 1} \to \mathbb{R}^{n^2 \times 1}$. Besides, $M(t,x) = I \otimes X + X^{\mathrm{T}} \otimes I$ denotes the nonsingular mass matrix of a standard ordinary-differential-equation (ODE) problem.

Proof By vectorizing ZD model (11.2) based on Kronecker product, the left-hand side of Equation (11.2) is

$$\mathrm{vec}(X\dot{X} + \dot{X}X) = \mathrm{vec}(X\dot{X}) + \mathrm{vec}(\dot{X}X)$$
$$= (I \otimes X)\mathrm{vec}(\dot{X}) + (X^{\mathrm{T}} \otimes I)\mathrm{vec}(\dot{X})$$
$$= (I \otimes X + X^{\mathrm{T}} \otimes I)\mathrm{vec}(\dot{X}).$$

At the same time, the right-hand side of Equation (11.2) is

$$\mathrm{vec}(-\gamma \Phi(X^2 - A) + \dot{A}) = -\gamma \mathrm{vec}\big(\Phi(X^2 - A) + \dot{A}\big)$$
$$= -\gamma \mathrm{vec}\big(\Phi(X^2 - A)\big) + \mathrm{vec}(\dot{A}). \qquad (11.6)$$

Note that the aforementioned activation function array $\Phi(\cdot)$ is also vectorized, i.e., from $\mathbb{R}^{n^2 \times 1}$ to $\mathbb{R}^{n^2 \times 1}$. Thus, we have

$$\mathrm{vec}\big(\Phi(X^2 - A)\big) = \Phi\big(\mathrm{vec}(X^2 - A)\big) = \Phi\big(\mathrm{vec}(X^2) - \mathrm{vec}(A)\big)$$
$$= \Phi\big((I \otimes X)\mathrm{vec}(X) - \mathrm{vec}(A)\big). \qquad (11.7)$$

Combining Equations (11.6) and (11.7) yields the following vectorization of the right-hand side of Equation (11.2):

$$\mathrm{vec}(-\gamma \Phi(X^2 - A) + \dot{A}) = -\gamma \Phi\big((I \otimes X)\mathrm{vec}(X) - \mathrm{vec}(A)\big) + \mathrm{vec}(\dot{A}).$$

Evidently, the vectorization of both sides of matrix-form differential Equation (11.2) should be equal, which generates the vector-form differential Equation (11.5). The proof is thus complete.

In addition, for ZD model (11.2) which solves for the time-varying matrix square root of $A(t)$, we have the following theorems on its convergence.

Theorem 19 *Consider smoothly time-varying matrix $A(t) \in \mathbb{R}^{n \times n}$ in nonlinear Equation (10.1), which satisfies the square-root existence condition. If a monotonically increasing odd activation function array $\Phi(\cdot)$ is used, then error function $E(t) = X^2(t) - A(t) \in \mathbb{R}^{n \times n}$ of ZD model (11.2), starting from randomly generated positive-definite (or negative-definite) diagonal initial state-matrix $X(0) \in \mathbb{R}^{n \times n}$, can converge to zero [which implies that state matrix $X(t) \in \mathbb{R}^{n \times n}$ of ZD model (11.2) can converge to theoretical positive-definite (or negative-definite) time-varying matrix square root $X^*(t)$ of $A(t)$].*

Proof From the compact form of ZD design formula $\dot{E}(t) = -\Gamma\Phi(E(t))$, a set of n^2 decoupled differential equations can be written equivalently as follows:

$$\dot{e}_{ij}(t) = -\gamma\phi\big(e_{ij}(t)\big), \tag{11.8}$$

for any $i \in \{1,2,3,\cdots,n\}$ and $j \in \{1,2,3,\cdots,n\}$. Thus, to analyze the equivalent $i j$th subsystem (11.8), we define a Lyapunov function candidate $v_{ij}(t) = e_{ij}^2(t)/2 \geqslant 0$ with its time-derivative

$$\frac{\mathrm{d}v_{ij}(t)}{\mathrm{d}t} = e_{ij}(t)\dot{e}_{ij}(t) = -\gamma e_{ij}(t)\phi\big(e_{ij}(t)\big).$$

Note that $\phi(\cdot)$ is a monotonically increasing odd activation function. Therefore, we have $\phi(-e_{ij}(t)) = -\phi(e_{ij}(t))$ and the following result:

$$e_{ij}(t)\phi(e_{ij}(t)) \begin{cases} > 0, & \text{if } e_{ij}(t) \neq 0, \\ = 0, & \text{if } e_{ij}(t) = 0, \end{cases}$$

which guarantees the final negative-definiteness of $\dot{v}_{ij}$ (i.e., $\dot{v}_{ij} < 0$ for $e_{ij} \neq 0$, while $\dot{v}_{ij} = 0$ for $e_{ij} = 0$ only). By Lyapunov theory [78, 100, 123, 131, 132], equilibrium point $e_{ij} = 0$ of Equation (11.8) is asymptotically stable; i.e., $e_{ij}(t)$ converges to zero, for any $i \in \{1,2,3,\cdots,n\}$ and $j \in \{1,2,3,\cdots,n\}$. In other words, the matrix-valued error function $E(t)$ is convergent to zero. In addition, we have $E(t) = X^2(t) - A(t)$; or, equivalently, $X^2(t) = A(t) + E(t)$. In view of $E(t) \to 0$ as $t \to +\infty$, we have $X^2(t) \to A(t)$ [i.e., $X(t) \to X^*(t)$] as $t \to +\infty$. That is, state matrix $X(t)$ of ZD model (11.2) can converge to the theoretical time-varying matrix square root $X^*(t)$ of $A(t)$.

Furthermore, when $X(t)$ of ZD model (11.2) starting from a randomly generated positive-definite diagonal initial state-matrix $X(0)$, it can converge to the positive-definite time-varying matrix square root $A^{1/2}(t)$ [i.e., a form of $X^*(t)$]. This can be proved by contradiction. Suppose that state matrix $X(t)$ starting from a positive-definite diagonal initial state-matrix $X(0)$ converges to the negative-definite time-varying matrix square root $-A^{1/2}(t)$ [i.e., the other form of $X^*(t)$], then state-matrix $X(t)$ must pass through at least one 0-eigenvalue state, which leads to the contradiction that the left- and right-hand sides of ZD model (11.2) can not hold. So, starting from a randomly generated positive-definite diagonal initial state-matrix $X(0)$, state matrix $X(t)$ of ZD model (11.2) converges to the positive-definite time-varying matrix square root $A^{1/2}(t)$. Similarly, it can be proved that, starting from a randomly generated negative-definite diagonal initial state-matrix $X(0)$, state matrix $X(t)$ of ZD model (11.2) converges to the negative-definite time-varying matrix square root $-A^{1/2}(t)$. The proof is thus complete.

Theorem 20 *In addition to Theorem 19, if a linear activation function array $\Phi(\cdot)$ is used, then the matrix-valued error function $E(t) = X^2(t) - A(t) \in \mathbb{R}^{n \times n}$ of ZD model (11.2), starting from a randomly generated positive-definite (or negative-definite) diagonal initial-state-matrix $X(0) \in \mathbb{R}^{n \times n}$, can exponentially converge to zero with convergence rate γ, which corresponds to the convergence of state matrix $X(t) \in \mathbb{R}^{n \times n}$ of ZD model (11.2) to $X^*(t)$. Moreover, if the hyperbolic sine activation function array is used, then the superior convergence can be achieved for ZD model (11.2), as compared with that using the linear activation function array.*

Proof For linear activation function $\phi(e_{ij}) = e_{ij}$, we have the following two parts.

1) From the $i j$th subsystem (11.8), we have $\dot{e}_{ij} = -\gamma e_{ij}$, and thus $e_{ij}(t) = \exp(-\gamma t)\, e_{ij}(0)$. In other words, the matrix-valued error function $E(t) \in \mathbb{R}^{n \times n}$ can be expressed explicitly as

$$E(t) = \begin{bmatrix} e_{11}(0) & e_{12}(0) & \cdots & e_{1n}(0) \\ e_{21}(0) & e_{22}(0) & \cdots & e_{2n}(0) \\ \vdots & \vdots & \ddots & \vdots \\ e_{n1}(0) & e_{n2}(0) & \cdots & e_{nn}(0) \end{bmatrix} \exp(-\gamma t) = E(0)\exp(-\gamma t).$$

This evidently shows that error function $E(t)$ exponentially converges to zero, with convergence rate γ for ZD model (11.2) activated by the linear activation function array. In addition, we have $E(t) = X^2(t) - A(t)$ and then $X^2(t) = A(t) + E(t)$, i.e., $X^2(t) = A(t) + E(0)\exp(-\gamma t)$. In view of $E(0)\exp(-\gamma t) \to 0$ exponentially as $t \to +\infty$, we have again $X^2(t) \to A(t)$ and $X(t) \to X^*(t)$ as $t \to +\infty$. That is, state matrix $X(t)$ of ZD model (11.2) can converge to the theoretical time-varying matrix square root $X^*(t)$ of $A(t)$.

2) Define a Lyapunov function candidate $v = \|E(t)\|_{\mathrm{F}}^2/2 = \mathrm{trace}(E^{\mathrm{T}}(t)E(t))/2 = \mathrm{vec}^{\mathrm{T}}(E(t))\,\mathrm{vec}\big(E(t)\big)/2 \geqslant 0$ for ZD model (11.2). Since

$$\mathrm{vec}\big(E(t)\big) = [e_{11}, \cdots, e_{n1}, e_{12}, \cdots, e_{n2}, \cdots, e_{1n}, \cdots, e_{nn}]^{\mathrm{T}} \to 0$$

is equivalent to $E(t) \to 0$, then we use $\mathrm{vec}\big(E(t)\big)$ instead of $E(t)$ to analyze the Lyapunov function candidate and its time derivative. Thus, defining e_i as the ith element of $\mathrm{vec}\big(E(t)\big)$, we have

$$v(t) = \sum_{i=1}^{n^2} e_i^2(t)/2, \quad \text{and} \quad \dot{v}(t) = \sum_{i=1}^{n^2} e_i(t)\dot{e}_i(t) = -\gamma \sum_{i=1}^{n^2} e_i(t)\phi(e_i(t)).$$

So, if the linear (lin) activation function array is used, we have

$$v(t)_{\mathrm{lin}} = \sum_{i=1}^{n^2} e_i^2(t)/2 \geqslant 0, \quad \text{and} \quad \dot{v}(t)_{\mathrm{lin}} = -\gamma \sum_{i=1}^{n^2} e_i^2(t) \leqslant 0.$$

This implies the matrix-valued error function $E(t)$ of linearly activated ZD model (11.3) can converge to zero according to the aforementioned discussion and Lyapunov theory.

On the other hand, if the hyperbolic sine function array is used, the corresponding Lyapunov function candidate is still $v(t)_{\mathrm{hs}} = v(t) \geqslant 0$. By Taylor series expansion [25, 51, 57], the aforementioned hyperbolic sine function is formulated as

$$\phi(e_i) = (\exp(\xi e_i) - \exp(-\xi e_i))/2$$
$$= \left(2(\xi e_i) + 2 \cdot \frac{(\xi e_i)^3}{3!} + 2 \cdot \frac{(\xi e_i)^5}{5!} + \cdots\right)/2$$
$$= \xi e_i + \frac{(\xi e_i)^3}{3!} + \frac{(\xi e_i)^5}{5!} + \cdots$$
$$= \sum_{r=1}^{+\infty} \frac{\xi^{2r-1}(e_i)^{2r-1}}{(2r-1)!}.$$

Thus, we have the following derivation for the hyperbolic sine (hs) situation:

$$\dot{v}(t)_{\mathrm{hs}} = -\gamma \sum_{i=1}^{n^2} e_i\phi(e_i) = -\gamma \sum_{i=1}^{n^2} e_i \sum_{r=1}^{+\infty} \frac{\xi^{2r-1}(e_i)^{2r-1}}{(2r-1)!}$$
$$= -\gamma \sum_{i=1}^{n^2} \sum_{r=1}^{+\infty} \frac{\xi^{2r-1}(e_i)^{2r}}{(2r-1)!}$$
$$\ll -\gamma \sum_{i=1}^{n^2} e_i^2 = \dot{v}(t)_{\mathrm{lin}} \leqslant 0.$$

From the above analysis, we see that Lyapunov function candidate $v(t)_{\mathrm{hs}}$ diminishes to zero when the hs activation function array is used, with much faster convergence rate than that using the linear activation function array. This implies that, when the hs function array is exploited, nonlinearly activated ZD model (11.2) possesses superior convergence in comparison with linearly activated ZD model (11.3). The proof is thus complete.

Before ending this section, it is worth pointing out here that one more advantage of using the hs activation functions over the linear activation functions lies in the extra design parameter ξ, which is an effective factor of the convergence rate. When there is an upper bound on γ due to the hardware implementation, the parameter ξ will be another effective factor to expedite the convergence of ZD model (11.2). The convergence for the array of hs activation function can be much faster than that for the array of linear activation function, when using the same level of design parameters ξ and γ. This is because the error function $E(t)$ in (11.2) is amplified by the hyperbolic sine activation function for the whole error range $(-\infty, +\infty)$ (i.e., the larger slope and absolute value of the hs activation function in the whole error range, as shown in Figure 11.1).

11.3   Robustness Analysis

In the analog implementation or simulation of the neural dynamics, we usually assume that it is under ideal conditions. However, there always exist some implementation errors in hardware implementation [132]. The differentiation errors of $A(t)$, as well as the model-implementation error, appear most frequently in the hardware implementation. For these implementation errors possibly appearing in model (11.2), we investigate the robustness of the ZD model by considering the following matrix-valued differential equation as perturbed with a large model-implementation error:

$$X(t)\dot{X}(t) + \dot{X}(t)X(t) = -\gamma\Phi(X^2(t) - A(t)) + \dot{A}(t) + \Delta\omega, \tag{11.9}$$

where $\Delta\omega \in \mathbb{R}^{n \times n}$ denotes the combination of the general model-implementation errors (including the differentiation errors of matrix $A(t)$ as a part). The following lemmas on the robustness of largely perturbed ZD model (11.9) can correspondingly be achieved [110, 132].

Lemma 6 *Consider the above perturbed ZD model with a large model implementation error $\Delta\omega \in \mathbb{R}^{n \times n}$ finally depicted in Equation (11.9). If $0 \leqslant \|\Delta\omega\|_F \leqslant \varepsilon < \infty$ for any $t \in [0, +\infty]$, then the maximal steady-state residual error $\max_{t \to \infty} \|E(t)\|_F$ is upper bounded by some positive scalar, provided that design parameter $\gamma > 0$ is large enough (the so-called design parameter requirement). Furthermore, the maximal steady-state residual error $\max_{t \to \infty} \|E(t)\|_F$ decreases to zero as γ tends to positive infinity.*

Lemma 7 *In addition to the general robustness results given in Lemma 6, the largely perturbed ZD model (11.9) possesses the following properties.*

   1) With the linear activation function array used, the maximal steady-state residual error $\max_{t \to \infty} \|E(t)\|_F$ can be written as a positive scalar under the design parameter requirement.

   2) With the hs activation function array used, the superior convergence and robustness properties exist for the whole error range $(-\infty, +\infty)$; i.e., the design parameter requirement can be removed in this situation, and the value of the maximal steady-state residual error $\max_{t \to \infty} \|E(t)\|_F$ can be made (much) smaller [which can be further done by increasing γ or ξ], as compared with the situation of using the linear activation function array.

11.4 Illustrative Examples

The above two sections present the convergence and robustness results of ZD model (11.2) [together with a largely perturbed ZD model (11.9)] for real-time solution of time-varying matrix square root problem (10.1). In this section, computer-simulation results are provided to verify the superior characteristics of using the hs activation function array to those of using linear activation function array.

Example 11.1 For illustration and comparison, let us consider Equation (10.1) with the following time-varying matrix $A(t)$:

$$A(t) = \begin{bmatrix} 16 + \sin(4t)\cos(4t) & 7\sin(4t) \\ 7\cos(4t) & 9 + \sin(4t)\cos(4t) \end{bmatrix}. \tag{11.10}$$

Note that the theoretical time-varying matrix square root $X^*(t)$ of $A(t)$ is given as

$$X^*(t) = \begin{bmatrix} 4 & \sin(4t) \\ \cos(4t) & 3 \end{bmatrix},$$

which is used to check the correctness of the ZD solution $X(t)$.

Based on the aforementioned vector-form ZD model (11.5) proposed in Theorem 18, we can obtain the simulated ZD state matrix $X(t)$. As illustrated in Figure 11.2, starting from randomly generated positive-definite diagonal initial state $X(0) \in [0,2]^{2\times 2}$, state matrix $X(t)$ of ZD model (11.2) using hs activation functions converges to the theoretical time-varying solution much faster, as compared with that using linear activation functions. Besides, in order to further investigate the convergence performance, we monitor and show the residual errors $\|E(t)\|_F$ during the problem solving processes of ZD model (11.2). As seen from Figure 11.3, the residual errors $\|E(t)\|_F$ of ZD model (11.2) all decrease rapidly to zero, where the convergence rate of ZD model (11.2) using hs activation functions appears to be 5 times faster than that using linear activation functions. From these figures, we can confirm well the theoretical results given in Theorems 19 and 20.

To show the robustness characteristics of the largely perturbed ZD model (11.9), the following large model implementation error $\Delta\omega$ is specially added:

$$\Delta\omega = \begin{bmatrix} 10^2 & 10^2 \\ 10^2 & 10^2 \end{bmatrix}.$$

As we can see from Figure 11.4, with the large model-implementation error added, the maximal steady-state residual errors $\max_{t\to\infty}\|E(t)\|_F$ of the perturbed ZD model (11.9) are still bounded. Besides, with $\gamma = 1$, ZD model (11.9) using linear activation functions results in a very large residual error, which is shown evidently in Figure 11.4(a). In contrast, as shown comparatively in Figure 11.4(a) and (b), when hs activation functions are used, the convergence time of the perturbed ZD model (11.9) is faster than that using linear activation functions, and the maximal steady-state residual error of the perturbed ZD model (11.9) is around 50 times smaller than that using linear activation functions. Furthermore, Figure 11.4(b) is about using hs activation functions with design parameter $\xi = 3$, while Figure 11.5(a) and (b) are about $\xi = 5$ and $\xi = 7$, respectively. It is observed that, by increasing ξ, the maximal steady-state residual error of ZD model (11.9) is decreased very effectively (e.g., for $\xi = 7$, which is around 133 times smaller than that using linear activation functions). In addition, comparing Figure 11.4(b) and Figure 11.6, we see that, as design parameter γ increases form 1 to 10 and even to 100, the maximal steady-state residual error is decreased effectively from around 3.53 to 1.99 and then to 0.59. Note that, when γ is much larger (e.g., 10^6 or 10^9), the maximal steady-state residual error would become tiny. Besides, if the very

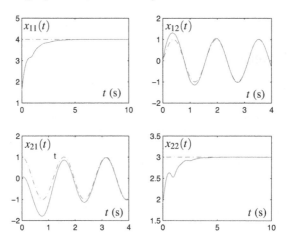

(a) Using linear activation functions

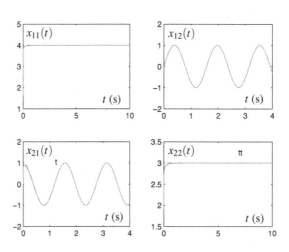

(b) Using hyperbolic sine activation functions with $\xi = 3$

FIGURE 11.2 Time-varying matrix square root finding of (11.10) via ZD model (11.2), with $\gamma = 1$, where ZD solutions $X(t)$ are denoted by solid curves, and theoretical time-varying square root $A^{1/2}(t)$ is denoted by dash-dotted curves.

large model-implementation error (e.g., with $\Delta\omega_{ij} = 10000$, $\forall i, j = 1, 2$) is added to the ZD model, the robustness results are shown in Figure 11.7, from which the same conclusion is drawn. That is, using hs activation functions results in a much smaller maximal steady-state residual error than using linear activation functions [e.g., around 5000 times smaller, as seen comparatively from Figure 11.7(a) and (b)].

Example 11.2 In order to further investigate the efficacy of the ZD models [including (11.2) and (11.9)] using hs activation functions, let us consider Equation (10.1) with the following symmetric

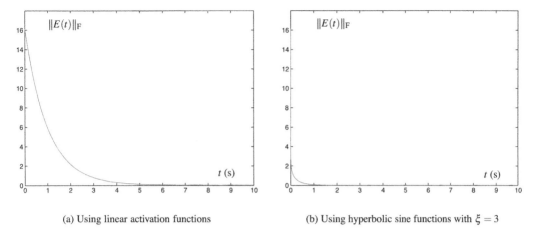

(a) Using linear activation functions

(b) Using hyperbolic sine functions with $\xi = 3$

FIGURE 11.3 Residual errors of ZD model (11.2), with $\gamma = 1$ for time-varying matrix square root finding of (11.10), where labeled subfigure ranges are both $[0, 10] \times [0, 16]$.

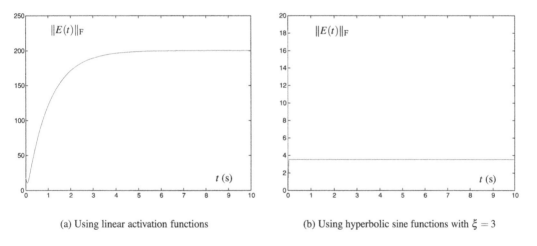

(a) Using linear activation functions

(b) Using hyperbolic sine functions with $\xi = 3$

FIGURE 11.4 Residual errors of largely perturbed ZD model (11.9) for time-varying matrix square root finding of (11.10), where subfigure ranges are $[0, 10] \times [0, 250]$ and $[0, 10] \times [0, 20]$.

positive-definite time-varying matrix $A(t)$, with its theoretical time-varying square root $X^*(t)$ given as well for comparison purposes:

$$A(t) = \begin{bmatrix} 5 + 0.25\sin^2(6t) & 2\sin(6t) + 0.5\cos(6t) & 4 + 0.25\sin(6t)\cos(6t) \\ 2\sin(6t) + 0.5\cos(6t) & 4.25 & 2\cos(6t) + 0.5\sin(6t) \\ 4 + 0.25\sin(6t)\cos(6t) & 2\cos(6t) + 0.5\sin(6t) & 5 + 0.25\cos^2(6t) \end{bmatrix}$$

$$X^*(t) = \begin{bmatrix} 2 & 0.5\sin(6t) & 1 \\ 0.5\sin(6t) & 2 & 0.5\cos(6t) \\ 1 & 0.5\cos(6t) & 2 \end{bmatrix}. \tag{11.11}$$

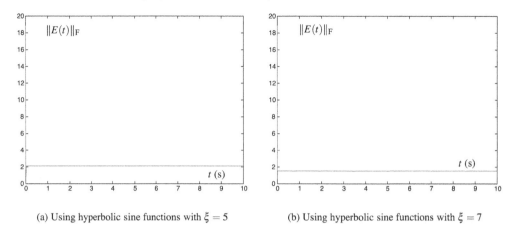

(a) Using hyperbolic sine functions with $\xi = 5$ (b) Using hyperbolic sine functions with $\xi = 7$

FIGURE 11.5 Residual errors of largely perturbed ZD model (11.9) using hyperbolic sine activation functions with different values of ξ, where subfigure ranges are both $[0, 10] \times [0, 20]$.

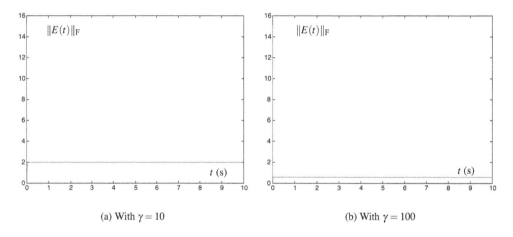

(a) With $\gamma = 10$ (b) With $\gamma = 100$

FIGURE 11.6 Residual errors of largely perturbed ZD model (11.9) using hyperbolic sine activation functions with different values of γ, where subfigure ranges are both $[0, 10] \times [0, 16]$.

As illustrated in Figure 11.8, starting from randomly generated positive-definite diagonal initial state $X(0) \in [0, 2]^{3 \times 3}$, state matrix $X(t)$ of ZD model (11.2) using hs activation functions converges faster than that using linear activation functions. In order to further investigate the convergence performance, we monitor and show the residual errors $\|E(t)\|_F$ during the problem solving processes of ZD model (11.2). Specifically, from Figure 11.9, we observe that the residual errors $\|E(t)\|_F$ of ZD model (11.2) decrease rapidly to zero, where the convergence rate of ZD model (11.2) using hs activation functions is also 5 times faster than that using linear activation functions. These results substantiate again the theoretical results given in Theorems 19 and 20.

To comparatively show the robustness characteristics of the largely perturbed ZD model, we exploit once more the large model-implementation error, with $\Delta \omega_{ij} = 10^2$, $\forall i, j \in \{1, 2, 3\}$, which is added in ZD model (11.9). With design parameter $\gamma = 1$ and two types of activation functions used,

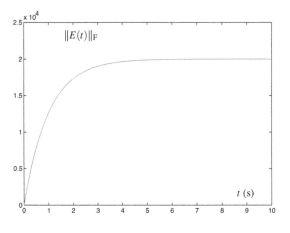

(a) Using linear activation functions

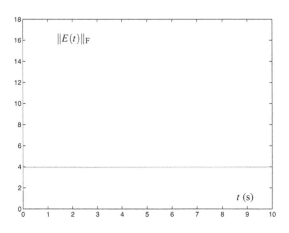

(b) Using hyperbolic sine activation functions with $\xi = 5$

FIGURE 11.7 Residual errors of very largely perturbed ZD model (11.9) for time-varying matrix square root finding of (11.10), where subfigure ranges are $[0, 10] \times [0, 2.5] \cdot 10^4$ and $[0, 10] \times [0, 18]$.

the robustness performance of the largely perturbed ZD model (11.9) is shown in Figure 11.10, where the maximal steady-state residual errors are still bounded. In addition, as shown in Figure 11.10(a) and (b), when hs activation functions with $\xi = 5$ are used, the convergence time of the largely perturbed ZD model (11.9) is faster than that using linear activation functions, and the maximal steady-state residual error of the largely perturbed ZD model (11.9) is around 100 times smaller than that using linear activation functions. In summary, compared with the situation of using linear activation functions, superior robustness performance is achieved for the ZD models by using hs activation functions.

Example 11.3 In order to further investigate the efficacy of ZD model (11.2) using hs activation functions for larger-dimension matrices, let us consider Equation (10.1) with the following time-

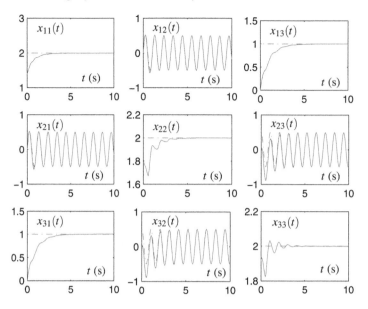

(a) Using linear activation functions

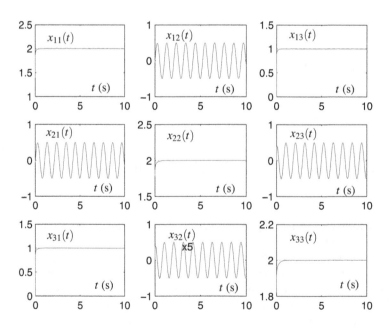

(b) Using hyperbolic sine activation functions with $\xi = 5$

FIGURE 11.8 Finding time-varying matrix square root (11.11) via ZD model (11.2) with $\gamma = 1$.

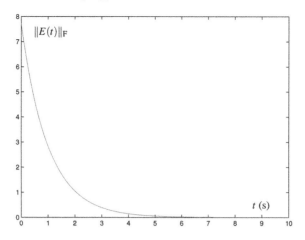

(a) Using linear activation functions

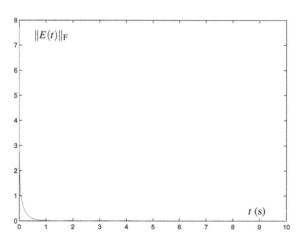

(b) Using hyperbolic sine activation functions with $\xi = 5$

FIGURE 11.9 Residual errors of ZD model (11.2), with $\gamma = 1$ solving for time-varying matrix square root (11.11), where subfigure ranges are both $[0,10] \times [0,8]$.

varying Toeplitz matrix $A(t)$:

$$
A(t) = \begin{bmatrix}
a_1(t) & a_2(t) & a_3(t) & \cdots & a_n(t) \\
a_2(t) & a_1(t) & a_2(t) & \cdots & a_{n-1}(t) \\
a_3(t) & a_2(t) & a_1(t) & \cdots & a_{n-2}(t) \\
\vdots & \vdots & \vdots & \ddots & \vdots \\
a_n(t) & a_{n-1}(t) & a_{n-2}(t) & \cdots & a_1(t)
\end{bmatrix} \in \mathbb{R}^{n \times n}.
\tag{11.12}
$$

Let $a_1(t) = n + \sin(5t)$, and $a_k(t) = \cos(5t)/(k-1)$ with $k = 2,3,\cdots,n$. Specifically, Figure 11.11 shows the simulation results of ZD model (11.2) using hs activation functions for the time-varying square root finding of the above Toeplitz matrix $A(t)$ in the situations of $n = 4$ and $n = 10$. As seen from Figure 11.11(a) and (b), the residual errors $\|E(t)\|_\text{F}$ of ZD model (11.2) for factorizing Toeplitz matrices with different dimensions (i.e., $\mathbb{R}^{4 \times 4}$ and $\mathbb{R}^{10 \times 10}$) both diminish to zero, which implies that

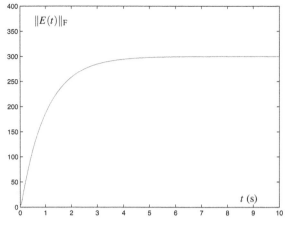

(a) Using linear activation functions

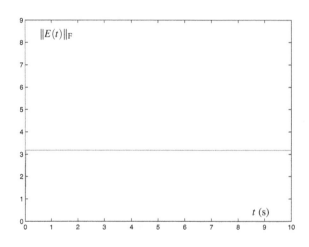

(b) Using hyperbolic sine activation functions with $\xi = 5$

FIGURE 11.10 Residual errors of largely perturbed ZD model (11.9) solving for (11.11).

the state matrices correspondingly converge to the time-varying square roots of the two Toeplitz matrices, respectively. These further substantiate the efficacy of ZD model (11.2) using hyperbolic sine activation functions on solving for the time-varying square roots of larger-dimension matrices.

Before ending this section, it is worth noting that, because of the similarity of the results and figures to the above ones, the corresponding simulations of the ZD models starting from negative-definite initial states are not presented (though they have been conducted successfully as well as consistently with the theoretical results given in this chapter) and left to interested readers to complete as a topic of exercise.

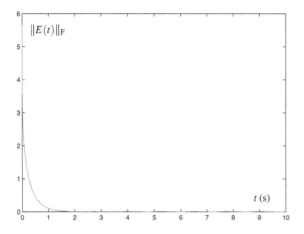

(a) Factorizing a Toeplitz matrix with $n = 4$

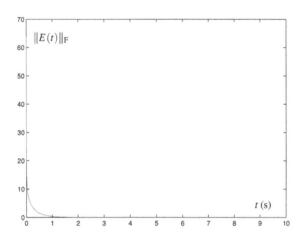

(b) Factorizing a Toeplitz matrix with $n = 10$

FIGURE 11.11 Residual errors of ZD model (11.2) using hyperbolic sine activation functions for time-varying square root finding of Toeplitz matrices $A(t)$ with different dimensions.

11.5 Summary

In this chapter, the convergence and robustness properties of the ZD models using two types of activation functions have been analyzed and substantiated for the real-time solution of time-varying matrix square root. Specifically, theoretical analysis has illustrated that superior convergence and robustness can be achieved readily for the ZD models even in the context of (very) large model-implementation errors by using hs activation functions, as compared with those using linear activation functions. Computer-simulation results have further substantiated the efficacy and superiority of the ZD models using hs activation functions for the time-varying matrix square root finding.

Part V

Time-Varying Quadratic Optimization

Chapter 12

ZD Models for Quadratic Minimization

Abstract

For potential digital hardware implementation, the DTZD models are proposed and investigated in this chapter for time-varying quadratic minimization (QM). The proposed DTZD models can utilize the time-derivative information of time-varying coefficients. For comparison, a discrete-time GD model is also developed and presented to solve the same time-varying QM problem. Numerical-experiment results illustrate the efficacy and superiority of the proposed DTZD models for time-varying QM, in comparison with the discrete-time GD model.

12.1 Introduction

QM is a class of mathematical unconstrained optimization, which is widely encountered in numerous scientific and engineering fields, such as signal processing [6,45,61], electromagnetic field [35], and plasma science [91]. Many algorithms/methods have been analyzed in the past decades for the QM problem solving, e.g., gradient method [6], descent method [73], Newton method [6,47,50,73], and quasi-Newton method [6,18,91]. Due to its important role, most of these algorithms/methods have been numerically performed on today's digital computers. As a powerful alternative to solve the QM problem, neural-dynamic models, algorithms, and methods have emerged owing to their features of distributed parallel computation and convenient hardware implementation. Note that the convergence and robustness properties of the CTZD model for time-varying quadratic optimization have been investigated and analyzed in the authors' previous work [117, 132]. Instead of using a scalar-valued norm-based nonnegative energy function associated with the GD model, the design of the CTZD model is based on a vector-valued indefinite lower-unbounded error function. Moreover, the CTZD model takes advantage of the time-derivative information of time-varying coefficients associated with the given time-varying problem to be solved, and thus obtains much better convergence performance as compared with the GD model.

In this chapter, three DTZD models for time-varying QM problem solving are proposed and investigated for the purposes of potential hardware (e.g., digital circuits) implementation and numerical-algorithm development [120]. The corresponding discrete-time models are DTZDK model, DTZDU model, and S-DTZD model, which denote the DTZD model with known time-derivative information of coefficients, the DTZD model with unknown time-derivative information of coefficients, and the DTZD model with a simplified form, respectively. In addition, as the authors think and propose, if a computational scheme originating from neural dynamics is simulated for the purpose of hardware (e.g., circuits) implementation or implemented on hardware as a prototype, it is viewed as a model of neural dynamics; and, if such a computational scheme is coded as a program and performed on digital computers, it is viewed as a numerical algorithm of the neural dynamics.

12.2   Problem Formulation and CTZD Model

In this section, let us consider the following time-varying QM problem [120]:

$$\text{minimize } f(\mathbf{x}(t),t) = \mathbf{x}^{\mathrm{T}}(t)P(t)\mathbf{x}(t)/2 + \mathbf{q}^{\mathrm{T}}(t)\mathbf{x}(t), \tag{12.1}$$

where the smoothly time-varying coefficient matrix $P(t) \in \mathbb{R}^{n \times n}$ is positive-definite and symmetric at any time instant $t \in [0, +\infty)$, and coefficient vector $\mathbf{q}(t) \in \mathbb{R}^n$ is also smoothly time-varying. The vector $\mathbf{x}(t) \in \mathbb{R}^n$ is unknown, and, as our solution objective, is to be calculated at any time instant $t \in [0, +\infty)$. In fact, to solve (12.1), we only need to solve for time-varying vector $\mathbf{x}(t) \in \mathbb{R}^n$, such that time-varying linear equation system (or to say, time-varying linear system) $P(t)\mathbf{x}(t) + \mathbf{q}(t) = 0$ of time-varying QM problem (12.1) holds true. In this chapter, we assume that $\mathbf{x}^*(t)$ is the time-varying theoretical solution of time-varying QM (12.1), satisfying

$$\frac{\partial f(\mathbf{x}(t),t)}{\partial \mathbf{x}(t)}\bigg|_{\mathbf{x}(t)=\mathbf{x}^*(t)} = P(t)\mathbf{x}^*(t) + \mathbf{q}(t) = 0. \tag{12.2}$$

With $\mathbf{x}^*(t)$, the time-varying theoretical minimum trajectory of QM (12.1) can be denoted as $\mathbf{x}^{*\mathrm{T}}(t)P(t)\mathbf{x}^*(t)/2 + \mathbf{q}^{\mathrm{T}}(t)\mathbf{x}^*(t)$.

To monitor and control the solving process of the above time-varying linear Equation (12.2), as well as QM (12.1), let us define the following error function $\mathbf{e}(t)$ [which, in this chapter, is vector-valued, indefinite, and lower-unbounded, in addition to having a zero trajectory $\mathbf{e}(t) = 0$ corresponding to the theoretical time-varying solution of linear Equation (12.2) and QM (12.1)]: $\mathbf{e}(t) = P(t)\mathbf{x}(t) + \mathbf{q}(t) \in \mathbb{R}^n$. Then, by following Zhang *et al.*'s neural-dynamic design procedure (i.e., zeroing dynamics (ZD) design procedure) [99–102, 104, 107, 114, 117, 118, 122, 128, 129, 134], to make each element $e_i(t)$ of error vector $\mathbf{e}(t) \in \mathbb{R}^n$ converge to zero, the following ZD design formula is adopted:

$$\frac{\mathrm{d}\mathbf{e}(t)}{\mathrm{d}t} = -\gamma\mathbf{e}(t), \tag{12.3}$$

where design parameter $\gamma > 0$ is the reciprocal of a capacitance parameter used to scale the convergence rate, and should be set as large as the hardware would permit (e.g., in analog circuits or VLSI [8, 119]). Expanding ZD design formula (12.3) leads to the following implicit dynamic equation of the CTZD model:

$$P(t)\dot{\mathbf{x}}(t) = -\dot{P}(t)\mathbf{x}(t) - \gamma(P(t)\mathbf{x}(t) + \mathbf{x}(t)) - \dot{\mathbf{x}}(t), \tag{12.4}$$

where $\mathbf{x}(t)$, starting from initial condition $\mathbf{x}(0) \in \mathbb{R}^n$, denotes the neural state corresponding to the theoretical time-varying solution $\mathbf{x}^*(t) = -P^{-1}(t)\mathbf{q}(t)$ of (12.2).

12.3 DTZD Models

For potential hardware implementation (e.g., on digital circuits), the DTZD models are proposed and investigated, where three different situations about the use of time-derivative information of time-varying coefficients [i.e., about $\dot{P}(t)$ and $\dot{\mathbf{q}}(t)$] are discussed as follows.

12.3.1 With $\dot{P}(t)$ and $\dot{\mathbf{q}}(t)$ Known

To obtain the discrete-time model of CTZD model (12.4), we use the following Euler forward-difference rule [1, 3, 22, 25, 32, 40, 51, 57, 68]:

$$\dot{\mathbf{x}}(t = k\tau) \approx (\mathbf{x}((k+1)\tau) - \mathbf{x}(k\tau))/\tau,$$

where $\tau > 0$ and k (e.g., $k = 0, 1, 2, \ldots$) denote the sampling gap and the update index, respectively. For presentation convenience, we use $\mathbf{x}_k = \mathbf{x}(t = k\tau)$, $P_k = P(t = k\tau)$, $\mathbf{q}_k = \mathbf{q}(t = k\tau)$, $\dot{P}_k = \dot{P}(t = k\tau)$, $\dot{\mathbf{q}}_k = \dot{\mathbf{q}}(t = k\tau)$, and $P_k^{-1} = P^{-1}(t = k\tau)$. In view of the time derivatives being known (i.e., with $\dot{P}_k$ and $\dot{\mathbf{q}}_k$ known), the DTZDK model is thus obtained as

$$\frac{\mathbf{x}_{k+1} - \mathbf{x}_k}{\tau} = -P_k^{-1}\dot{P}_k\mathbf{x}_k - \gamma P_k^{-1}(P_k\mathbf{x}_k + \mathbf{q}_k) - P_k^{-1}\dot{\mathbf{q}}_k,$$

which is rewritten equivalently as

$$\begin{aligned}
\mathbf{x}_{k+1} &= \mathbf{x}_k - \tau P_k^{-1}\dot{P}_k\mathbf{x}_k - hP_k^{-1}(P_k\mathbf{x}_k + \mathbf{q}_k) - \tau P_k^{-1}\dot{\mathbf{q}}_k \\
&= \left((1-h)I - \tau P_k^{-1}\dot{P}_k\right)\mathbf{x}_k - hP_k^{-1}\mathbf{q}_k - \tau P_k^{-1}\dot{\mathbf{q}}_k,
\end{aligned} \tag{12.5}$$

where design parameter $h = \tau\gamma > 0$ denotes the step size, and $\mathbf{x}_k$ denotes the kth update or sample of $\mathbf{x}(t = k\tau)$, which starts from randomly generated initial state $\mathbf{x}_0$ close enough to theoretical initial $\mathbf{x}_0^* = -P_0^{-1}\mathbf{q}_0$. Note that different values of h lead to different convergence performance of DTZDK model (12.5), which can be seen later (i.e., in the ensuing simulative and numerical results presented in Section 12.5). Additionally, $\tau > 0$ should be set appropriately small for high solution precision.

12.3.2 With $\dot{P}(t)$ and $\dot{\mathbf{q}}(t)$ Unknown

In many engineering applications, it is difficult or impossible to obtain the analytical form or calculate the numerical value of $\dot{P}(t)$ and $\dot{\mathbf{q}}(t)$; i.e., $\dot{P}_k$ and $\dot{\mathbf{q}}_k$ are unknown. In view of this difficulty, we investigate the DTZD model with unknown time-derivative information of time-varying coefficients (thus abbreviated as DTZDU model) for time-varying QM problem solving. That is, simply utilizing Euler backward-difference rule to estimate $\dot{P}_k$ and $\dot{\mathbf{q}}_k$, we have $\dot{P}_k \approx (P_k - P_{k-1})/\tau$ and $\dot{\mathbf{q}}_k \approx (\mathbf{q}_k - \mathbf{q}_{k-1})/\tau$, of which τ and k are defined the same as in the above subsection. Thus, DTZDK model (12.5) becomes

$$\mathbf{x}_{k+1} = \mathbf{x}_k - P_k^{-1}\Delta P_k\mathbf{x}_k - hP_k^{-1}(P_k\mathbf{x}_k + \mathbf{q}_k) - P_k^{-1}\Delta\mathbf{q}_k, \tag{12.6}$$

with $\Delta P_k = P_k - P_{k-1}$ and $\Delta\mathbf{q}_k = \mathbf{q}_k - \mathbf{q}_{k-1}$. Note that $\Delta P_0 = P_0 - P_{-1}$ and $\Delta\mathbf{q}_0 = \mathbf{q}_0 - \mathbf{q}_{-1}$, which can be set as random values, P_0 and $\mathbf{q}_0$, or zeros in practice, because P_{-1} and $\mathbf{q}_{-1}$ are undefined (or to say, meaningless), and this will not affect much the performance of the DTZDU model. Further simplifying DTZDU model (12.6), we obtain the DTZDU model:

$$\begin{aligned}
\mathbf{x}_{k+1} &= P_k^{-1}P_{k-1}\mathbf{x}_k - h(\mathbf{x}_k + P_k^{-1}\mathbf{q}_k) - P_k^{-1}\mathbf{q}_k + P_k^{-1}\mathbf{q}_{k-1} \\
&= (P_k^{-1}P_{k-1} - hI)\mathbf{x}_k - (h+1)P_k^{-1}\mathbf{q}_k + P_k^{-1}\mathbf{q}_{k-1}.
\end{aligned} \tag{12.7}$$

12.3.3 With a Simplified Form

For comparison with DTZDK model (12.5) and DTZDU model (12.7) handling time-varying QM (12.1), we can simplify them by omitting the calculation part of time-derivative information of time-varying coefficients, and thus obtain the following simplified DTZD model (i.e., the S-DTZD model):

$$\mathbf{x}_{k+1} = \mathbf{x}_k - hP_k^{-1}(P_k\mathbf{x}_k + \mathbf{q}_k) = (1-h)\mathbf{x}_k - hP_k^{-1}\mathbf{q}_k. \tag{12.8}$$

12.4 GD Models

To show the comparison and superiority of the proposed DTZD models, a discrete-time GD model is investigated to solve the same time-varying QM problem (12.1). For better understanding and presentation convenience, in the first part of the section, we show the continuous-time GD model. That is, by following the GD design method [99,100,117,122,129,134], a scalar-valued norm-based nonnegative energy function is defined: $\mathscr{E}(t) = \|P(t)\mathbf{x}(t) + \mathbf{q}(t)\|_2^2/2$. The minimum point of $\mathscr{E}(t)$ is achieved with $\mathscr{E}(t) = 0$.

Then we need to obtain the derivative of $\mathscr{E}(t)$ with respect to $\mathbf{x}(t) \in \mathbb{R}^n$, i.e., the gradient, for designing a computational scheme along a descent direction of function $\mathscr{E}(t)$. So, $\partial\mathscr{E}/\partial\mathbf{x} = \partial(\|P(t)\mathbf{x}(t) + \mathbf{q}(t)\|_2^2/2)/\partial\mathbf{x}(t) = P^{\mathrm{T}}(t)(P(t)\mathbf{x}(t) + \mathbf{q}(t))$.

Finally, according to the GD design formula $\dot{\mathbf{x}}(t) = -\gamma(\partial\mathscr{E}/\partial\mathbf{x})$, we obtain the following GD model applied to the continuous-time time-varying QM solving:

$$\dot{\mathbf{x}}(t) = -\gamma P^{\mathrm{T}}(t)(P(t)\mathbf{x}(t) + \mathbf{q}(t)). \tag{12.9}$$

Now, in the second part of the section, to obtain a discrete-time GD model solving the time-varying QM problem (12.1), we re-employ Euler forward-difference rule in continuous-time GD model (12.9) as below: $(\mathbf{x}_{k+1} - \mathbf{x}_k)/\tau = -\gamma P_k^{\mathrm{T}}(P_k\mathbf{x}_k + \mathbf{q}_k)$. So, we obtain the following discrete-time GD model for the purpose of online time-varying QM (12.1) solving:

$$\mathbf{x}_{k+1} = \mathbf{x}_k - hP_k^{\mathrm{T}}(P_k\mathbf{x}_k + \mathbf{q}_k), \tag{12.10}$$

where $h = \tau\gamma > 0$ denotes the step size again and is selected as $h = 1/\mathrm{tr}(P_k^{\mathrm{T}}P_k)$.

12.5 Illustrative Example

Two classes of neural-dynamic models, i.e., the ZD and GD models, have been derived in the previous two sections for the time-varying QM problem solving. In this section, an illustrative example is fully simulated, investigated, and verified for substantiating the convergence and efficacy of the above-derived neural-dynamic models. Specifically, let us consider the following time-varying co-efficients of time-varying QM (12.1):

$$P(t) = \begin{bmatrix} 10+\sin(t) & \cos(t) \\ \cos(t) & 10+\sin(t) \end{bmatrix} \text{ and } \mathbf{q}(t) = \begin{bmatrix} \sin(t) \\ \cos(t) \end{bmatrix}.$$

The numerical-experiment results are shown in Figures 12.1 through 12.7 and Table 12.1. Specifically, Figure 12.1 illustrates the trajectories of state vector $\mathbf{x}_k = \mathbf{x}(k\tau) = [x_1(k\tau), x_2(k\tau)]^{\mathrm{T}}$

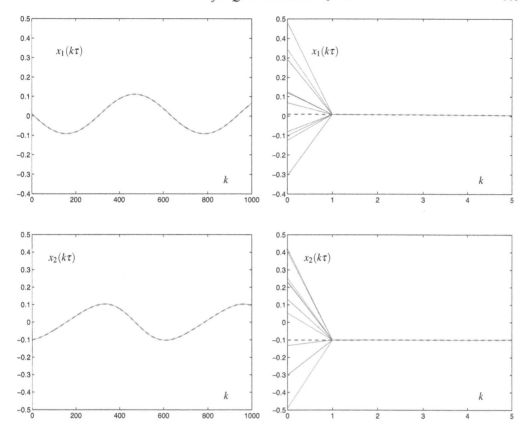

FIGURE 12.1 Neural states $\mathbf{x}_k = \mathbf{x}(k\tau) = [x_1(k\tau), x_2(k\tau)]^{\mathrm{T}}$ (denoted in solid lines) of DTZDK model (12.5) solving QM (12.1), with $h = 1$ and $\tau = 0.01$ starting from 10 random initial states, where dashed lines denote the theoretical solution, the ranges of the left two subfigures are $[0, 1000] \times [-0.4, 0.5]$ and $[0, 1000] \times [-0.5, 0.5]$, and the right two subfigures show the first five updates of the left two subfigures.

of DTZDK model (12.5). As shown in the figure, starting from 10 randomly generated initial state $\mathbf{x}_0$ [i.e., $\mathbf{x}(0)$], the two elements of neural state $\mathbf{x}(k\tau)$ both converge to the element trajectories of theoretical solution $\mathbf{x}^*(k\tau)$ of time-varying QM problem (12.1) with just one update, which can be seen clearly from the right two subfigures of Figure 12.1. For comparison, the trajectories of state vector $\mathbf{x}_k = \mathbf{x}(k\tau) = [x_1(k\tau), x_2(k\tau)]^{\mathrm{T}}$ of discrete-time GD model (12.10) are shown in Figure 12.2, where the convergence toward the theoretical trajectories requires at least five updates. In addition, we give Figures 12.3 and 12.4 to show the residual errors $\|\mathbf{e}_k\|_2 = \|P_k\mathbf{x}_k - \mathbf{q}_k\|_2$ of DTZDK model (12.5) and discrete-time GD model (12.10), respectively. As seen from Figure 12.3, the maximal steady-state residual errors (MSSREs) of DTZDK model (12.5) are very small (i.e., of order 10^{-5}, below the scale of order 10^{-4} in the lower subfigure). In contrast, the MSSREs of discrete-time GD model (12.10) are relatively larger (i.e., of order 10^{-2}), as depicted clearly in Figure 12.4. Note that the figures and other numerical results related to DTZDU model (12.7) and S-DTZD model (12.8) are omitted here for avoiding the redundancy of presentation in view of their similarities to those of DTZDK model (12.5) and left for interested readers to complete as a topic of exercise. But the values of MSSREs of DTZDU model (12.7) and S-DTZD model (12.8) are certainly different [e.g., the MSSRE of DTZDU model (12.7) is of order 10^{-4} in the situation of $h = 1$ and $\tau = 0.01$,

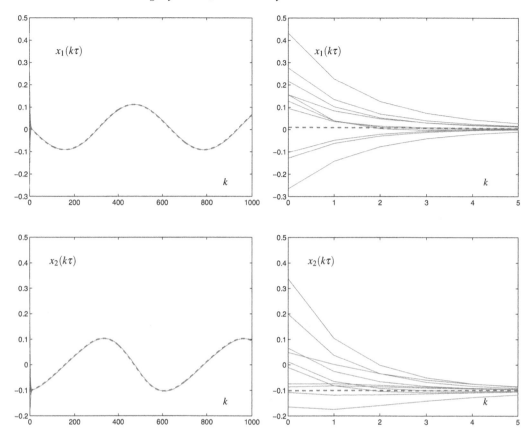

FIGURE 12.2 Neural states $\mathbf{x}_k = [x_1(k\tau), x_2(k\tau)]^{\mathrm{T}}$ of discrete-time GD model (12.10) solving QM (12.1), with $\tau = 0.01$ starting from 10 random initial states, where the ranges of the left two subfigures are $[0, 1000] \times [-0.3, 0.5]$ and $[0, 1000] \times [-0.2, 0.5]$, and the right two subfigures show the first five updates of the left two subfigures.

as shown in Figure 12.5]. Table 12.1 further displays the MSSREs of $\|\mathbf{e}_k\|_2$ of the four presented discrete-time models with different values of h and τ, of which 10 initial states $\mathbf{x}_0$ are randomly generated within region $[-0.5, 0.5]^2$.

More specifically, the following facts are illustrated via the numerical results given in Table 12.1, in which the symbol "−" is used to mean $h = 1/\operatorname{tr}(P_k^{\mathrm{T}} P_k)$.

1) By utilizing the time-derivative information of time-varying coefficients (including the estimation of the time derivatives), the MSSREs of DTZDK model (12.5) and DTZDU model (12.7) show a roughly $O(\tau^2)$ manner. For example, when $h = 1$ and $\tau = 0.001$, the MSSRE of DTZDK model (12.5) is 6.0658×10^{-7}; and, when τ decreases to 0.0001 (i.e., 10 times smaller), the MSSRE decreases to 6.0657×10^{-9} (i.e., correspondingly 100 times smaller). As another example, when $h = 1$ and $\tau = 0.001$, the MSSRE of DTZDU model (12.7) is 1.1797×10^{-6}; and, when $\tau = 0.0001$ (i.e., 10 times smaller), the MSSRE decreases to 1.1796×10^{-8} (i.e., correspondingly 100 times smaller again).

2) Without using the time-derivative information of time-varying coefficients, the MSSREs of S-DTZD model (12.8) and discrete-time GD model (12.10) show a roughly $O(\tau)$ manner. For example, when $h = 0.75$ and $\tau = 0.001$, the MSSRE of S-DTZD model (12.8) is 1.4815×10^{-3}; and, when τ decreases to 0.0001 (i.e., 10 times smaller), the MSSRE decreases to 1.4815×10^{-4} (i.e.,

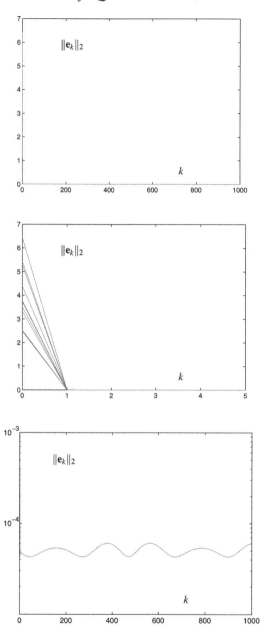

FIGURE 12.3 Residual errors $\|\mathbf{e}_k\|_2$ of DTZDK model (12.5) solving QM (12.1), with $h = 1$ and $\tau = 0.01$, where the middle subfigure shows the first five updates of the upper subfigure, and the lower subfigure shows steady-state error order 10^{-5} of the upper subfigure.

correspondingly 10 times smaller). Besides, the similar $O(\tau)$ pattern of the MSSREs of discrete-time GD model (12.10) can be seen clearly from the last row of Table 12.1.

3) Different values of step size h (which is suggested to satisfy $0 < h < 2$) also have an important effect on the MSSRE, no matter whether the time-derivative information of time-varying coefficients is used or not. For example, when $h = 1$ and $\tau = 0.0001$, the MSSRE of DTZDU model (12.7) shown in Table 12.1 is 1.1796×10^{-8}; when h decreases to 0.50, the MSSRE changes (or to say, increases)

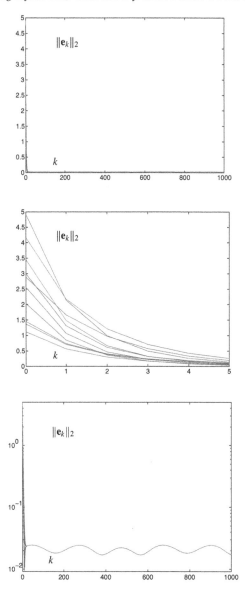

FIGURE 12.4 Residual errors $\|\mathbf{e}_k\|_2$ of discrete-time GD model (12.10) solving QM (12.1), with $\tau = 0.01$, where the middle subfigure shows the first five updates of the upper subfigure, and the lower subfigure shows steady-state error order 10^{-2} of the upper subfigure.

to 2.3593×10^{-8}; and, when h increases to 1.90, the MSSRE changes (or to say, decreases) to 6.2086×10^{-9}. In general, $h = 1$ is the default choice for DTZD models.

Moreover, in order to better understand the relationship between the MSSRE and the deign parameters (i.e., step size h and sampling gap τ) in both quantitative and qualitative manners, Figures 12.6 and 12.7 are given. Specifically, Figure 12.6 illustrates the relationship between the MSSRE and the sampling gap τ with $h = 1$ fixed. From this figure, we observe that the MSS-REs of the DTZD models using the time-derivative information have a roughly $O(\tau^2)$ manner, and the MSSREs of the DTZD models without using the time-derivative information and the GD model have a roughly $O(\tau)$ manner. The slopes of the former two lines of MSSREs (i.e., of the DTZDK and DTZDU models) are steeper than those of the latter two lines of MSSREs (i.e., of the

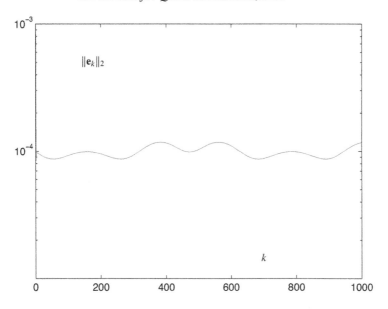

FIGURE 12.5 Order of residual error of DTZDU model (12.7) solving QM (12.1) with $h = 1$ and $\tau = 0.01$, which is 10^{-4}.

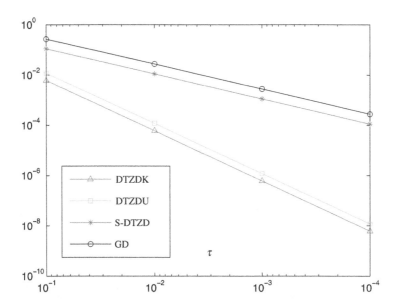

FIGURE 12.6 MSSREs of different discrete-time models solving time-varying QM (12.1) with $h = 1$ and with respect to different τ.

S-DTZD and GD models). In addition, it is evident that DTZDK model (12.5) is the best one, because it uses the accurate time-derivative information of time-varying coefficients; and that DTZDU model (12.7) performs slightly less favorably than DTZDK model (12.5), because only estimated time-derivative information of time-varying coefficients is used in DTZDU model (12.7). Among these discrete-time models, the proposed DTZD models all perform better than the discrete-time GD model (12.10) for the same time-varying QM problem solving. The relationship between the MSSRE and step size h is shown in Figure 12.7 with sampling gap $\tau = 0.0001$ fixed, which verifies again that the MSSRE is affected by h.

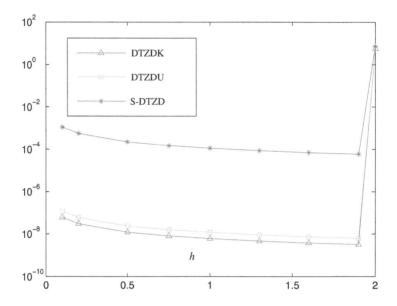

FIGURE 12.7 MSSREs of different DTZD models solving time-varying QM (12.1) with $\tau = 0.0001$ and with respect to different h.

TABLE 12.1   Maximal Steady-State Residual Errors of Different Discrete-Time Models Solving Time-Varying QM (12.1) Starting with Randomly Generated Initial States $\mathbf{x}_0$

Model	Step size	$\tau = 0.1$	$\tau = 0.01$	$\tau = 0.001$	$\tau = 0.0001$
DTZDK	$h = 0.10$	1.7352×10^{-1}	6.0191×10^{-4}	6.0653×10^{-6}	6.0657×10^{-8}
	$h = 0.20$	3.0588×10^{-2}	3.0278×10^{-4}	3.0328×10^{-6}	3.0328×10^{-8}
	$h = 0.50$	1.1935×10^{-2}	1.2131×10^{-4}	1.2131×10^{-6}	1.2131×10^{-8}
	$h = 0.75$	8.0696×10^{-3}	8.0882×10^{-5}	8.0877×10^{-7}	8.0876×10^{-9}
	$h = 1.00$	6.0673×10^{-3}	6.0664×10^{-5}	6.0658×10^{-7}	6.0657×10^{-9}
	$h = 1.30$	4.6765×10^{-3}	4.6665×10^{-5}	4.6660×10^{-7}	4.6659×10^{-9}
	$h = 1.60$	3.8031×10^{-3}	3.7915×10^{-5}	3.7911×10^{-7}	3.7911×10^{-9}
	$h = 1.90$	2.6495×10^{-1}	3.1929×10^{-5}	3.1925×10^{-7}	3.1925×10^{-9}
	$h = 2.00$	4.2978	5.1021	5.2605	5.6603
DTZDU	$h = 0.10$	3.1919×10^{-1}	1.1722×10^{-3}	1.1796×10^{-5}	1.1796×10^{-7}
	$h = 0.20$	5.3474×10^{-2}	5.8910×10^{-4}	5.8982×10^{-6}	5.8982×10^{-8}
	$h = 0.50$	2.3353×10^{-2}	2.3595×10^{-4}	2.3593×10^{-6}	2.3593×10^{-8}
	$h = 0.75$	1.5723×10^{-2}	1.5732×10^{-4}	1.5729×10^{-6}	1.5729×10^{-8}
	$h = 1.00$	1.1831×10^{-2}	1.1799×10^{-4}	1.1797×10^{-6}	1.1796×10^{-8}
	$h = 1.30$	9.1070×10^{-3}	9.0763×10^{-5}	9.0744×10^{-7}	9.0742×10^{-9}
	$h = 1.60$	7.3977×10^{-3}	7.3746×10^{-5}	7.3729×10^{-7}	7.3728×10^{-9}
	$h = 1.90$	5.2657×10^{-1}	6.2102×10^{-5}	6.2088×10^{-7}	6.2086×10^{-9}
	$h = 2.00$	4.4507	2.4771	2.5253	5.8078
S-DTZD	$h = 0.10$	1.058	1.1047×10^{-1}	1.1110×10^{-2}	1.1111×10^{-3}
	$h = 0.20$	4.9806×10^{-1}	5.5483×10^{-2}	5.5555×10^{-3}	5.5556×10^{-4}
	$h = 0.50$	2.1925×10^{-1}	2.2219×10^{-2}	2.2222×10^{-3}	2.2222×10^{-4}
	$h = 0.75$	1.4761×10^{-1}	1.4814×10^{-2}	1.4815×10^{-3}	1.4815×10^{-4}
	$h = 1.00$	1.1108×10^{-1}	1.1111×10^{-2}	1.1111×10^{-3}	1.1111×10^{-4}
	$h = 1.30$	8.5498×10^{-2}	8.5471×10^{-3}	8.5470×10^{-4}	8.5470×10^{-5}
	$h = 1.60$	6.9511×10^{-2}	6.9445×10^{-3}	6.9444×10^{-4}	6.9444×10^{-5}
	$h = 1.90$	2.8924×10^{-1}	5.8480×10^{-3}	5.8480×10^{-4}	5.8480×10^{-5}
	$h = 2.00$	2.3265	5.9784×10^{-1}	6.1511	7.2684
GD	$-$	2.6824×10^{-1}	2.7700×10^{-2}	2.7709×10^{-3}	2.7709×10^{-4}

In summary, the above numerical results have all verified the efficacy and superiority of the proposed DTZD models [especially, DTZDK model (12.5) and DTZDU model (12.7)] for time-varying quadratic minimization (12.1), as compared with discrete-time GD model (12.10).

12.6 Summary

In this chapter, by following Zhang *et al.*'s design method, three types of the DTZD models have been proposed to solve online the time-varying quadratic minimization (QM) problem. Different values of sampling gap τ and step size h have been shown to affect very much the MSSRE of the DTZD solution. Some important facts have been observed as follows.

1) The MSSREs of DTZDK model (12.5) and DTZDU model (12.7) using the time-derivative information of time-varying coefficients show a roughly $O(\tau^2)$ manner.

2) The MSSREs of S-DTZD model (12.8) without using the time-derivative information of time-varying coefficients show a roughly $O(\tau)$ manner.

3) Because step size h affects the convergence performance of the proposed DTZD models, no matter whether the time-derivative information of time-varying coefficients is used or not, a suitable value of h (usually satisfying $0 < h < 2$ and being chosen as 1) should be selected.

4) Compared with discrete-time GD model (12.10), the proposed DTZD models have much better convergence performance, especially when utilizing the time-derivative information of time-varying coefficients.

Chapter 13

ZD Models for Quadratic Programming

Abstract

In this chapter, we generalize and investigate the CTZD model for online solution of the time-varying convex quadratic programming (QP) problem subject to a time-varying linear equality constraint. For the purpose of potential hardware (e.g., digital circuits) implementation, the DTZD models are constructed and developed by using Euler forward-difference rule, which can also be effective numerical algorithms if implemented on digital computers. Computer-simulation and numerical-experiment results illustrate the efficacy (especially precision) of the presented CTZD and DTZD models for solving the time-varying QP problem.

13.1   Introduction

As an important branch of mathematical optimization, the QP subject to linear equality constraints has been theoretically analyzed, numerically handled, and widely applied in various scientific and engineering areas [4, 15, 28, 44, 48, 49, 62]. In view of its fundamental role, many algorithms have been proposed and investigated to solve the QP problem [44, 49]. In general, numerical algorithms performed on digital computers are considered to be well-accepted approaches to solve linear-equality constrained QP problems. As another general approach of solution, parallel-processing computational schemes, including neural dynamics, have played an important role in real-time and online computation, and can be regarded as one of the potential promising alternatives. Due to their parallel-distributed nature and hardware-implementation convenience, neural dynamics can perform excellently in many application fields [4, 15, 28]. Chua and Lin [15] developed a canonical nonlinear programming circuit (NPC) for simulating general nonlinear programs. Forti *et al.* [28] introduced a generalized circuit for nonsmooth programming, which derived from a natural extension of NPC. Bian and Xue [4] proposed a subgradient-based neural network to solve a nonsmooth, nonconvex optimization problem, of which the objective function was nonsmooth and nonconvex.

To date, most reported computational schemes are theoretically/intrinsically designed for constant problems solving or related to gradient methods. A few studies of other researchers have been published on dynamical methods for solving such time-varying problems in the literature at this stage. For example, Myung and Kim [62] proposed a time-varying two-phase (TVTP) algorithm, which, with a finite penalty parameter, can offer exact feasible solutions to a constrained time-varying nonlinear optimization. In this chapter, a novel CTZD model is introduced and developed for real-time optimal solution of the time-varying convex QP problem subject to a time-varying linear equality. Note that the authors' work is to find the time-varying optimal solution at any time instant, while Myung and Kim's work [62] considers the time-varying optimization problem when time t goes to infinity (or is large enough). Besides, the constraints in Myung and Kim's work [62] are constant. In contrast, the coefficients of the QP problem investigated in this work are time-varying, i.e., with a time-varying matrix-vector constraint. Furthermore, for the purpose of potential hardware implementation, the DTZD models are generalized and investigated for the online solution of the time-varying convex QP problem subject to a time-varying linear equality [39].

13.2   CTZD Model

Let us consider the following time-varying convex quadratic programming problem that is subject to a time-varying linear equality constraint [39, 117]:

$$\text{minimize}\quad \mathbf{x}^{\mathrm{T}}(t)P(t)\mathbf{x}(t)/2 + \mathbf{q}^{\mathrm{T}}(t)\mathbf{x}(t), \tag{13.1}$$

$$\text{subject to}\quad A(t)\mathbf{x}(t) = \mathbf{b}(t), \tag{13.2}$$

where Hessian matrix $P(t) \in \mathbb{R}^{n \times n}$ is smoothly time-varying, positive-definite, and symmetric for any time instant $t \in [0, +\infty) \subset \mathbb{R}$. The coefficient vector $\mathbf{q}(t) \in \mathbb{R}^n$ is assumed smoothly time-varying as well. In Equations (13.1) and (13.2), the time-varying decision vector $\mathbf{x}(t) \in \mathbb{R}^n$ is unknown and to be solved at any time instant $t \in [0, +\infty)$. In equality constraint (13.2), the coefficient matrix $A(t) \in \mathbb{R}^{m \times n}$, being of full row rank and vector $\mathbf{b}(t) \in \mathbb{R}^m$, are both assumed smoothly time-varying. According to the mathematical optimization method using Lagrange multipliers [25, 32, 39, 51, 57, 117], time-varying QP problem (13.1)–(13.2) can be transformed into the following time-varying linear equation system (or to say, time-varying linear system):

$$W(t)\mathbf{y}(t) = \mathbf{u}(t), \tag{13.3}$$

where

$$W(t) = \begin{bmatrix} P(t) & A^{\mathrm{T}}(t) \\ A(t) & 0 \end{bmatrix} \in \mathbb{R}^{(n+m) \times (n+m)},$$

$$\mathbf{y}(t) = \begin{bmatrix} \mathbf{x}(t) \\ v(t) \end{bmatrix} \in \mathbb{R}^{n+m}, \quad \mathbf{u}(t) = \begin{bmatrix} -\mathbf{q}(t) \\ \mathbf{b}(t) \end{bmatrix} \in \mathbb{R}^{n+m},$$

with $v(t) \in \mathbb{R}^m$ denoting Lagrange multiplier vector defined for (13.2).

Following Zhang *et al.*'s neural-dynamic design method (i.e., zeroing dynamics (ZD) design method) [102, 104, 107, 117, 118, 122, 129, 134], first, we can define a vector-valued error function $\mathbf{e}(t) = W(t)\mathbf{y}(t) - \mathbf{u}(t)$. Then, the time derivative $\dot{\mathbf{e}}(t)$ of error function $\mathbf{e}(t)$ is constructed as follows:

$$\dot{\mathbf{e}}(t) = -\gamma \Phi\big(\mathbf{e}(t)\big). \tag{13.4}$$

By expanding the above ZD design formula (13.4), the following implicit dynamic equation of a

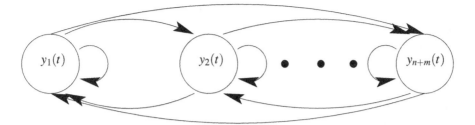

FIGURE 13.1 Neurons' connections and network architecture of CTZD model (13.5).

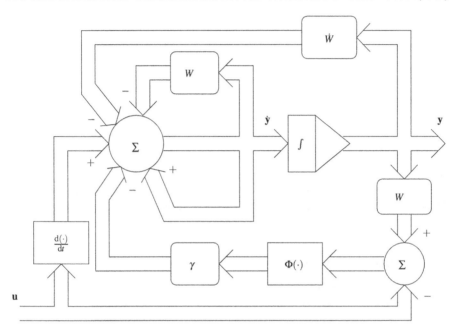

FIGURE 13.2 Block diagram realizing CTZD model (13.5).

CTZD model can readily be derived for solving in real time the time-varying QP problem (13.1)–(13.2):

$$W(t)\dot{\mathbf{y}}(t) = -\dot{W}(t)\mathbf{y}(t) - \gamma\Phi\big(W(t)\mathbf{y}(t) - \mathbf{u}(t)\big) + \dot{\mathbf{u}}(t), \qquad (13.5)$$

or equivalently,

$$\dot{\mathbf{y}}(t) = -M(t)\dot{W}(t)\mathbf{y}(t) - \gamma M(t)\Phi\big(W(t)\mathbf{y}(t) - \mathbf{u}(t)\big) + M(t)\dot{\mathbf{u}}(t), \qquad (13.6)$$

where, being the reciprocal of a capacitance parameter, design parameter $\gamma > 0 \in \mathbb{R}$ should be set as large as the hardware would permit or selected appropriately for simulative or experimental purposes. In addition, $M(t) = W^{-1}(t)$, and $\Phi(\cdot) : \mathbb{R}^{n+m} \to \mathbb{R}^{n+m}$ denotes an activation-function vector array. More specifically, the array $\Phi(\cdot)$ is made of $(n+m)$ monotonically increasing odd activation functions denoted by $\phi(\cdot)$. Besides, the neurons' connections and network architecture of CTZD model (13.5) are depicted in Figure 13.1. Correspondingly, Figure 13.2 shows the block diagram of such a neural-dynamic model.

Furthermore, it is worth mentioning here that, when using linear activation functions, CTZD model (13.5) reduces to the following linear one:

$$W(t)\dot{\mathbf{y}}(t) = -\big(\dot{W}(t) + \gamma W(t)\big)\mathbf{y}(t) + \big(\dot{\mathbf{u}}(t) + \gamma\mathbf{u}(t)\big). \qquad (13.7)$$

In general, for CTZD models (13.5) and (13.7), we have the following theorem [117, 132] which guarantees their global exponential convergence. Due to analysis similarity, the corresponding proof of the theorem is omitted and can be generalized from [102,104,107,122,129] by taking into account the Lyapunov function candidate and using Lyapunov theory or the theory of ordinary differential equations.

Theorem 21 *Considering strictly convex QP (13.1)-(13.2). If a monotonically increasing odd activation function array* $\Phi(\cdot)$ *is used, then state vector* $\mathbf{y}(t)$ *of CTZD (13.5), starting from any initial state* $\mathbf{y}(0) \in \mathbb{R}^{n+m}$, *converges to the unique theoretical solution* $\mathbf{y}^*(t)$ *of linear system (13.3). In addition, if the linear activation functions are used, then the state vector* $\mathbf{y}(t)$ *of CTZD (13.7) globally exponentially converges to the unique theoretical solution* $\mathbf{y}^*(t)$ *of linear system (13.3) with convergence rate* γ. *The first n elements of* $\mathbf{y}^*(t)$ *constitute the optimal solution* $\mathbf{x}^*(t)$ *to time-varying QP (13.1)–(13.2).*

13.3 DTZD Models

For the purpose of potential hardware implementation via digital circuits, it may be preferable to discretize CTZD model (13.5) by using Euler forward-difference rule [1,3,22,25,32,40,51,57,68]. The DTZD models are thus generalized and developed for solving online time-varying QP problem (13.1)–(13.2) in this section.

13.3.1 With $\dot{W}(t)$ and $\dot{\mathbf{u}}(t)$ Known

To discretize CTZD model (13.5) for solving online time-varying QP problem (13.1)–(13.2), we firstly refer to Euler forward-difference rule:

$$\dot{\mathbf{y}}(t = k\tau) \approx (\mathbf{y}((k+1)\tau) - \mathbf{y}(k\tau))/\tau,$$

where $\tau > 0$ denotes the sampling gap, and update index $k = 0, 1, 2, \cdots$. In general, we denote $\mathbf{y}_k = \mathbf{y}(t = k\tau)$ for presentation convenience. In addition, $W(t)$, $\dot{W}(t)$, $\mathbf{u}(t)$ and $\dot{\mathbf{u}}(t)$ (which are assumed to be known in this situation) are discretized by the standard sampling method. For convenience and also for consistency with $\mathbf{y}_k$, we use W_k, $\dot{W}_k$, $\mathbf{u}_k$ and $\dot{\mathbf{u}}_k$ standing for $W(t = k\tau)$, $\dot{W}(t = k\tau)$, $\mathbf{u}(t = k\tau)$, and $\dot{\mathbf{u}}(t = k\tau)$, respectively. Besides, $M_k = W_k^{-1}$. Thus, the DTZD model with $\dot{W}(t)$ and $\dot{\mathbf{u}}(t)$ known (i.e., the DTZDK model) can be derived from CTZD model (13.6) as

$$(\mathbf{y}_{k+1} - \mathbf{y}_k)/\tau = -M_k \dot{W}_k \mathbf{y}_k - \gamma M_k \Phi(W_k \mathbf{y}_k - \mathbf{u}_k) + M_k \dot{\mathbf{u}}_k,$$

which can be further formulated as

$$\mathbf{y}_{k+1} = (I - \tau M_k \dot{W}_k)\mathbf{y}_k - h M_k \Phi(W_k \mathbf{y}_k - \mathbf{u}_k) + \tau M_k \dot{\mathbf{u}}_k, \tag{13.8}$$

where $\mathbf{y}_k$ corresponds to the kth update or sample of $\mathbf{y}(t = k\tau)$, with $h = \tau\gamma > 0$ denoting the step size again and with $\Phi(\cdot)$ defined as before. In addition, τ should be set appropriately small for better convergence and precision purposes. When using linear activation functions, the above DTZDK model (13.8) reduces to

$$\mathbf{y}_{k+1} = ((1-h)I - \tau M_k \dot{W}_k)\mathbf{y}_k + M_k(h\mathbf{u}_k + \tau\dot{\mathbf{u}}_k). \tag{13.9}$$

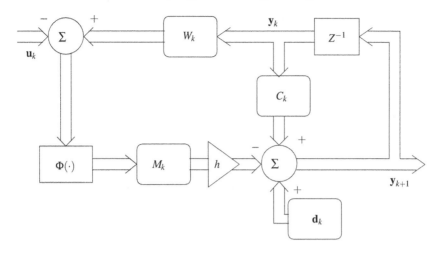

FIGURE 13.3 Block diagram realizing DTZD models.

13.3.2 With $\dot{W}(t)$ and $\dot{\mathbf{u}}(t)$ Unknown

Generally speaking, there exist some science and engineering fields, where it may be difficult to know the analytical form or numerical value of $\dot{W}(t)$ and $\dot{\mathbf{u}}(t)$. Thus, we investigate the DTZD model with $\dot{W}(t)$ and $\dot{\mathbf{u}}(t)$ unknown. In this situation, $\dot{W}(t)$ and $\dot{\mathbf{u}}(t)$ are generally estimated from signals $W(t)$ and $\mathbf{u}(t)$, e.g., by employing Euler backward-difference rule:

$$\dot{W}(t = k\tau) \approx (W(k\tau) - W((k-1)\tau))/\tau,$$
$$\dot{\mathbf{u}}(t = k\tau) \approx (\mathbf{u}(k\tau) - u((k-1)\tau))/\tau,$$

where τ is defined the same as before. Thus, from the proposed CTZD model (13.6), we can derive the DTZD model with $\dot{W}(t)$ and $\dot{\mathbf{u}}(t)$ unknown (i.e., the DTZDU model) for solving time-varying QP problem (13.1)-(13.2) as

$$(\mathbf{y}_{k+1} - \mathbf{y}_k)/\tau = -M_k(W_k - W_{k-1})\mathbf{y}_k/\tau - \gamma M_k \Phi(W_k\mathbf{y}_k - \mathbf{u}_k) + M_k(\mathbf{u}_k - \mathbf{u}_{k-1})/\tau,$$

which can be further formulated as

$$\mathbf{y}_{k+1} = M_k W_{k-1}\mathbf{y}_k - h M_k \Phi(W_k\mathbf{y}_k - \mathbf{u}_k) + M_k(\mathbf{u}_k - \mathbf{u}_{k-1}), \tag{13.10}$$

where h and $\Phi(\cdot)$ are defined as before. In addition, if we use the linear activation function array, the above DTZDU model (13.10) can be rewritten as

$$\mathbf{y}_{k+1} = (M_k W_{k-1} - hI)\mathbf{y}_k + (h+1)M_k\mathbf{u}_k - M_k\mathbf{u}_{k-1}. \tag{13.11}$$

Note that, from the above Euler backward-difference rule, we can neither obtain $\dot{W}(t = 0)$ nor $\dot{\mathbf{u}}(t = 0)$, since t is defined within $[0, +\infty)$ and W_{-1} and $\mathbf{u}_{-1}$ are undefined. So, in this situation, we choose $\dot{W}(t = 0) = 0$ and $\dot{\mathbf{u}}(t = 0) = 0$ or any other values to start the DTZDU model (13.10). Furthermore, Figure 13.3 depicts the block diagram of the DTZD models, where $C_k = I - \tau M_k\dot{W}_k$ and $\mathbf{d}_k = \tau M_k\dot{\mathbf{u}}_k$ for DTZDK model (13.8), or $C_k = M_k W_{k-1}$ and $\mathbf{d}_k = M_k(\mathbf{u}_k - \mathbf{u}_{k-1})$ for DTZDU model (13.10).

13.3.3 With $\dot{W}(t)$ and $\dot{\mathbf{u}}(t)$ Partially Known

Alternatively, there are another two situations, i.e., with $\dot{W}(t)$ unknown but $\dot{\mathbf{u}}(t)$ known, and with $\dot{W}(t)$ known but $\dot{\mathbf{u}}(t)$ unknown. In view of the similarity to the above DTZD models, we just show the DTZD models with $\dot{W}(t)$ and $\dot{\mathbf{u}}(t)$ partially known (i.e., the DTZDP models) as below:

$$\mathbf{y}_{k+1} = M_k W_{k-1} \mathbf{y}_k - h M_k \Phi(W_k \mathbf{y}_k - \mathbf{u}_k) + \tau M_k \dot{\mathbf{u}}_k, \tag{13.12}$$

$$\mathbf{y}_{k+1} = (I - \tau M_k \dot{W}_k) \mathbf{y}_k - h M_k \Phi(W_k \mathbf{y}_k - \mathbf{u}_k) + M_k(\mathbf{u}_k - \mathbf{u}_{k-1}). \tag{13.13}$$

13.4 Illustrative Examples

The previous two sections have proposed the CTZD and DTZD models for solving the time-varying convex QP problem subject to a time-varying linear-equality constraint. In this section, computer-simulation and numerical-experiment results are provided for substantiating the efficacy of the ZD models.

Example 13.1 Consider the time-varying convex QP problem subject to a time-varying linear-equality constraint with the following coefficients:

$$P(t) = \begin{bmatrix} 4 + \cos(2t) & \sin(0.5t) & \sin(0.5t)/2 & \sin(0.5t)/3 \\ \sin(0.5t) & 6 + \sin(3t) & \sin(0.5t) & \sin(0.5t)/2 \\ \sin(0.5t)/2 & \sin(0.5t) & 8 & \sin(0.5t) \\ \sin(0.5t)/3 & \sin(0.5t)/2 & \sin(0.5t) & 10 + \cos(t) \end{bmatrix},$$

$$\mathbf{q}(t) = - \begin{bmatrix} 1.5\sin(2t) \\ 1.5\sin(2t + 0.5\pi) \\ 1.5\sin(2t + \pi) \\ 1.5\sin(2t + 1.5\pi) \end{bmatrix}, \quad A(t) = \begin{bmatrix} \cos(t) \\ \cos(t - \pi/3) \\ \cos(t - 2\pi/3) \\ \cos(t - \pi) \end{bmatrix}^{\mathrm{T}}, \quad \mathbf{b}(t) = 1.5\sin(2t).$$

The simulative and numerical results are presented in Figures 13.4 through 13.6 and Table 13.1. Specifically, Figure 13.4(a) illustrates the simulative results of CTZD model (13.5) using linear activation functions. As shown in it, starting from an initial state randomly generated within $[-2, 2]$, neural state $\mathbf{x}(t)$ of CTZD model (13.5) converges to the theoretical solution of time-varying QP problem (13.1)–(13.2). From this figure, we confirm the theoretical result in Theorem 21 of Section 13.2. On the other hand, as presented previously, we have discretized the CTZD model (13.5) for the potential implementation on digital circuits or computers, and three DTZD models [i.e., DTZDK model (13.8), DTZDU model (13.10), and DTZDP model (13.12)] are generalized and developed for solving the time-varying QP problem (13.1)–(13.2). Figures 13.4(b) through 13.6 and Table 13.1 illustrate the numerical-experiment results that are of DTZDK model (13.8) with different values of step size h and sampling gap τ. As seen from Figure 13.4(b), state $\mathbf{x}_k$ of the proposed DTZDK model (13.8) converges to the theoretical solution. From Figures 13.5 and 13.6, we further see that good performance of DTZDK model (13.8) can be achieved by using appropriate step size h and sampling gap τ. Moreover, the results illustrated in Table 13.1 show the relationship between the maximal steady-state residual error $\max_{k \to \infty} \|\mathbf{e}_k\|_2 = \max_{k \to \infty} \|W_k \mathbf{y}_k - \mathbf{u}_k\|_2$ and the sampling gap τ, which is in an $O(\tau^2)$ manner. That is to say, the steady-state residual error of $\|\mathbf{e}_k\|_2$ reduces by 100 times when the sampling gap τ decreases by 10 times, which implies that τ can be selected appropriately small to satisfy effectively the usual precision we need in practice. Thus, we can have the important conclusion that the maximal steady-state residual errors of DTZDK model (13.8) are of order $O(\tau^2)$. These illustrate the efficacy of DTZDK model (13.8) on time-varying QP problem (13.1)–(13.2) solving.

The numerical results of DTZDU model (13.10) and DTZDP model (13.12) are also depicted in Table 13.1. From this table, we can observe that the maximal steady-state residual errors of these

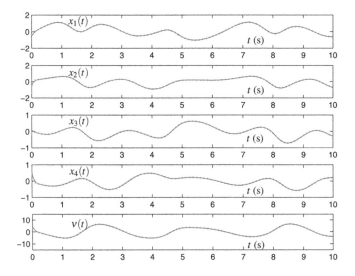

(a) By CTZD model (13.5) with $\gamma = 10$

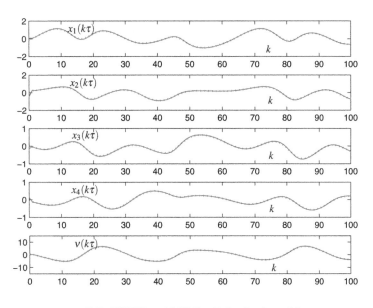

(b) By DTZDK model (13.8) with $h = 1$ and $\tau = 0.1$

FIGURE 13.4 Neural states of ZD models solving time-varying QP problem (13.1)–(13.2) for Example 13.1, where ZD solutions and theoretical solution are denoted by solid curves and dotted curves, respectively.

two models also show an $O(\tau^2)$ manner. Besides, DTZDU model (13.10) performs slightly less favorably than DTZDK model (13.8), in view of the former using the estimated time-derivative information of time-varying coefficients. In addition, using the accurate time-derivative information partially (e.g., with $\dot{\mathbf{u}}_k$ known and $\dot{W}_k$ estimated), DTZDP model (13.12) works better than DTZDU model (13.10) but less favorably than DTZDK model (13.8), which can also be seen from Table 13.1.

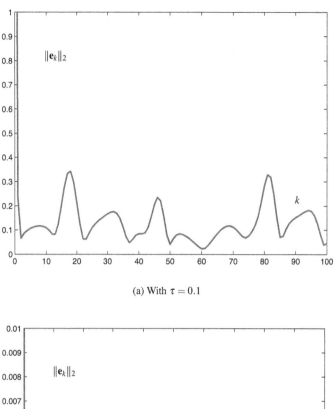

(a) With $\tau = 0.1$

(b) With $\tau = 0.01$

FIGURE 13.5 Residual errors of DTZDK model (13.8) solving time-varying QP problem (13.1)–(13.2), with $h = 1$ for Example 13.1.

Example 13.2 In the above example, the time-varying coefficients are of the sinusoidal form. As another illustrative example, we consider the following time-varying coefficients in the exponential form:

$$P(t) = \begin{bmatrix} 8 & 2\exp(-0.5t) \\ 2\exp(-0.5t) & 10 \end{bmatrix}, \ \mathbf{q}(t) = \begin{bmatrix} \exp(-0.5t) \\ \exp(-t) \end{bmatrix},$$

$$A(t) = \begin{bmatrix} 0.8\exp(-t) & 1.5 \end{bmatrix}, \ \mathbf{b}(t) = 0.5\exp(-2t).$$

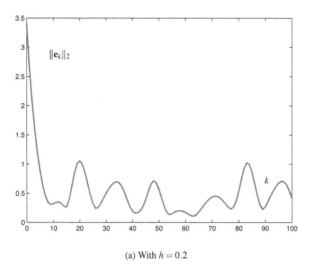

(a) With $h = 0.2$

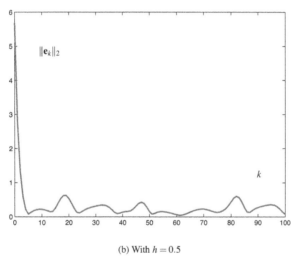

(b) With $h = 0.5$

FIGURE 13.6 Residual errors of DTZDK model (13.8) solving time-varying QP problem (13.1)–(13.2), with $\tau = 0.1$ for Example 13.1.

TABLE 13.1 Maximal Steady-State Residual Errors of DTZD Models When Solving Time-Varying QP Problem (13.1)–(13.2) for Example 13.1

Discrete-time ZD model	Step size h	Maximal steady-state residual error			
		$\tau = 0.1000$	$\tau = 0.0100$	$\tau = 0.0010$	$\tau = 0.0001$
DTZDK model (13.8)	$h = 0.2$	1.0539	1.6324×10^{-2}	1.6562×10^{-4}	1.6566×10^{-6}
	$h = 0.5$	0.6226	6.6159×10^{-3}	6.6259×10^{-5}	6.6264×10^{-7}
	$h = 0.8$	0.4263	4.1397×10^{-3}	4.1412×10^{-5}	4.1415×10^{-7}
	$h = 1.0$	0.3424	3.3119×10^{-3}	3.3130×10^{-5}	3.3132×10^{-7}
DTZDU model (13.10)	$h = 0.2$	1.2149	1.8936×10^{-2}	1.9221×10^{-4}	1.9226×10^{-6}
	$h = 0.5$	0.7281	7.6761×10^{-3}	7.6895×10^{-5}	7.6904×10^{-7}
	$h = 0.8$	0.4914	4.8020×10^{-3}	4.8060×10^{-5}	4.8065×10^{-7}
	$h = 1.0$	0.3987	3.8425×10^{-3}	3.8448×10^{-5}	3.8452×10^{-7}
DTZDP model (13.12)	$h = 0.2$	1.0788	1.6898×10^{-2}	1.7186×10^{-4}	1.7190×10^{-6}
	$h = 0.5$	0.6317	6.8618×10^{-3}	6.8755×10^{-5}	6.8761×10^{-7}
	$h = 0.8$	0.4350	4.2944×10^{-3}	4.2972×10^{-5}	4.2976×10^{-7}
	$h = 1.0$	0.3532	3.4358×10^{-3}	3.4378×10^{-5}	3.4381×10^{-7}

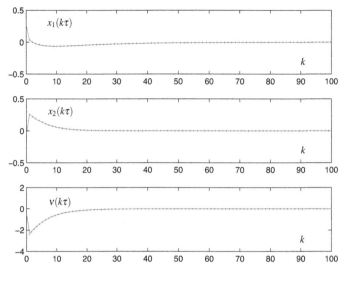

(a) By DTZDK model (13.8) with $h = 1$

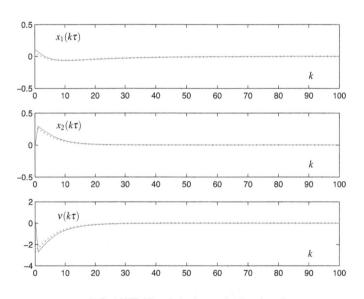

(b) By MATLAB optimization routine "quadprog"

FIGURE 13.7 States of DTZDK model (13.8) and MATLAB® optimization routine "quadprog" with $\tau = 0.1$ when solving time-varying QP problem (13.1)–(13.2) for Example 13.2.

TABLE 13.2 MSSREs of DTZD Models with $h = 1$ and MATLAB® Optimization Routine "Quadprog" Solving Time-Varying QP Problem (13.1)–(13.2) for Example 13.2

Discrete-time solver	Maximal steady-state residual error			
	$\tau = 0.1000$	$\tau = 0.0100$	$\tau = 0.0010$	$\tau = 0.0001$
DTZDK (13.8)	4.5046×10^{-3}	3.1653×10^{-5}	3.0557×10^{-7}	3.0450×10^{-9}
DTZDU (13.10)	7.1685×10^{-3}	5.1472×10^{-5}	4.9815×10^{-7}	4.9653×10^{-9}
DTZDP (13.12)	5.5743×10^{-3}	3.7570×10^{-5}	3.6106×10^{-7}	3.5968×10^{-9}
"quadprog"	4.5118×10^{-2}	4.0497×10^{-3}	4.0111×10^{-4}	4.0073×10^{-5}

It follows from Equation (13.3) that we have

$$W(t) = \begin{bmatrix} 8 & 2\exp(-0.5t) & 0.8\exp(-t) \\ 2\exp(-0.5t) & 10 & 1.5 \\ 0.8\exp(-t) & 1.5 & 0 \end{bmatrix},$$

$$\mathbf{u}(t) = \begin{bmatrix} -\exp(-0.5t) & -\exp(-t) & 0.5\exp(-2t) \end{bmatrix}^{\mathrm{T}}.$$

Being the numerical-experiment results, Figure 13.7(a) shows the state trajectories of DTZDK model (13.8) with $\tau = 0.1$ for time-varying QP problem (13.1)–(13.2) solving. From the subfigure, we see that state $\mathbf{x}_k$ of the proposed DTZDK model (13.8) converges to the theoretical solution of the time-varying QP problem. For comparative purposes, we also solve the above time-varying QP problem by using MATLAB optimization routine "quadprog," of which the initially less good but finally acceptable result is depicted in Figure 13.7(b). Besides, the numerical results of these two methods with different sampling gap τ are presented in Table 13.2. As seen from the table and other simulation results, the maximal steady-state residual errors of MATLAB optimization method show an $O(\tau)$ manner (when applied to time-varying QP), while those of DTZD models (i.e., DTZDK, DTZDU, and DTZDP models) are all of order $O(\tau^2)$. In addition, better convergence and precision can be achieved by DTZD models for the same time-varying QP problem solving, as compared with MATLAB optimization routine. This example also illustrates the efficacy and advantages of the proposed DTZD models.

13.5 Summary

In this chapter, following Zhang *et al.*'s design method, we have generalized and investigated a special class of continuous-time neural dynamics (i.e., CTZD) for real-time solution of the time-varying convex QP problem subject to a time-varying linear equality. Moreover, for the purposes of potential implementation on digital circuits or computers, the DTZD models (i.e., DTZDK, DTZDU, and DTZDP models) have been proposed and developed for such a time-varying quadratic-programming problem solving. Computer-simulation and numerical-experiment results of two examples have further illustrated the efficacy and advantages of the proposed ZD models on time-varying QP problem solving.

Chapter 14

Simulative and Experimental Application to Robot Arms

Abstract

As mentioned in Chapter 13, the quadratic programming (QP) subject to linear equality constraints has been theoretically analyzed and widely applied in various scientific and engineering areas. Therefore, in this chapter, we present an application of the proposed continuous-time zeroing dynamics (CTZD) model to the kinematic control of redundant robot arms via time-varying QP formulation and solution. Computer simulations performed on a four-link robot arm illustrate the superiority of the zeroing dynamics (ZD) model, as compared with the gradient dynamics (GD) one. Moreover, robotic experiments conducted on a six degrees-of-freedom (DOF), motor-driven, push-rod redundant robot arm substantiate the physical realizability and effectiveness of the ZD model.

14.1 Problem Formulation and Reformulation

As we know, the relation $\mathbf{f}(\cdot)$ between end-effector position vector $\mathbf{r}(t) \in \mathbb{R}^m$ and joint-angle variable vector $\theta(t) \in \mathbb{R}^n$ for a redundant robot arm is written readily as follows [125, 136]:

$$\mathbf{r}(t) = \mathbf{f}(\theta(t)). \tag{14.1}$$

In addition, differentiating Equation (14.1) with respect to time t, we have a pointwise linear relation between Cartesian velocity $\dot{\mathbf{r}}(t)$ and joint velocity $\dot{\theta}(t)$ as

$$J(\theta(t))\dot{\theta}(t) = \dot{\mathbf{r}}(t), \tag{14.2}$$

where $J(\theta(t)) \in \mathbb{R}^{m \times n}$ is Jacobian matrix defined as $J(\theta(t)) = \partial \mathbf{f}(\theta)/\partial \theta$. In terms of inverse kinematics [i.e., given $\mathbf{r}(t)$, to solve for $\theta(t)$], Equation (14.2) is under-determined (i.e., with $m < n$), admitting an infinite number of feasible solutions. Besides, note that, in Section 9.6, we present a pseudoinverse-type solution to the above Equation (14.2), and exploit the ZD method to solve for $J^+(t)$ in real time and in a simulated parallel manner so as to expedite the computational process. In this chapter, an optimization method is developed to solve Equation (14.2), which can make the inverse-kinematic solution repetitive [125, 136]. That is to say, this method can solve the joint-angle-drift problem of redundant robot arms.

14.1.1   Drift-Free Inverse Kinematics

As discussed in Section 9.6, the conventional solution to the inverse-kinematic problem (14.2) is the pseudoinverse-type solution, e.g., depicted in Equation (9.31). However, the pseudoinverse-type solution may not be repetitive [125, 136]. In other words, a closed path of the end-effector does not yield the closed trajectories in the joint space. Such a joint-angle-drift phenomenon will be undesirable for cyclic motion planning and control. To make the inverse-kinematic solution repetitive, the minimization of the joint displacement between current state $\theta(t)$ and initial state $\theta(0)$ can be exploited [125, 136]. In the formulation, the problem/scheme together with the performance index is

$$\text{minimize} \quad (\dot{\theta}(t) + \mathbf{d}(t))^{\mathrm{T}}(\dot{\theta}(t) + \mathbf{d}(t))/2 \tag{14.3}$$

$$\text{subject to} \quad J(\theta(t))\dot{\theta}(t) = \dot{\mathbf{r}}(t), \tag{14.4}$$

where $\mathbf{d}(t) = \varsigma(\theta(t) - \theta(0))$ with ς being a positive design parameter used to scale the magnitude of the robot arm's response to such a joint displacement.

14.1.2   QP Problem Reformulation

The above drift-free repetitive motion planning (RMP) scheme (14.3)–(14.4) of robot arm can be reformulated as a time-varying QP subject to a time-varying linear-equality constraint, i.e.,

$$\text{minimize} \quad \mathbf{x}^{\mathrm{T}}(t)P(t)\mathbf{x}(t)/2 + \mathbf{q}^{\mathrm{T}}(t)\mathbf{x}(t) \tag{14.5}$$

$$\text{subject to} \quad A(t)\mathbf{x}(t) = \mathbf{b}(t), \tag{14.6}$$

where $\mathbf{x}(t) = \dot{\theta}(t) \in \mathbb{R}^n$, $P(t) = I \in \mathbb{R}^{n \times n}$, $\mathbf{q}(t) = \mathbf{d}(t) = \varsigma(\theta(t) - \theta(0)) \in \mathbb{R}^n$, $A(t) = J(t) \in \mathbb{R}^{m \times n}$, and $\mathbf{b}(t) = \dot{\mathbf{r}}(t) \in \mathbb{R}^m$. Due to the positive-definiteness of matrix $P(t) = I$, QP (14.5)–(14.6) is strictly convex, which guarantees the solution uniqueness of the time-varying problem. Thus, based on the preliminary results on equality-constrained QP problem (discussed in Chapter 13 and references therein), we have its related Lagrangian $L(\mathbf{x}, v, t) = \mathbf{x}^{\mathrm{T}}(t)P(t)\mathbf{x}(t)/2 + \mathbf{q}^{\mathrm{T}}(t)\mathbf{x}(t) + v^{\mathrm{T}}(t)(A(t)\mathbf{x}(t) - \mathbf{b}(t))$, where $v(t) \in \mathbb{R}^m$ denotes Lagrange multiplier vector. As known, QP (14.5)–(14.6) can be solved by zeroing out the partial-derivative equations:

$$\begin{cases} \partial L(\mathbf{x}, v, t)/\partial \mathbf{x} = P(t)\mathbf{x}(t) + \mathbf{q}(t) + A^{\mathrm{T}}(t)v(t) = 0, \\ \partial L(\mathbf{x}, v, t)/\partial v = A(t)\mathbf{x}(t) - \mathbf{b}(t) = 0. \end{cases}$$

The above equations can further be written as

$$W(t)\mathbf{y}(t) = \mathbf{u}(t), \tag{14.7}$$

where

$$W(t) = \begin{bmatrix} P(t) & A^{\mathrm{T}}(t) \\ A(t) & 0 \end{bmatrix} \in \mathbb{R}^{(n+m)\times(n+m)},$$

$$\mathbf{y}(t) = \begin{bmatrix} \mathbf{x}(t) \\ v(t) \end{bmatrix} \in \mathbb{R}^{n+m}, \ \mathbf{u}(t) = \begin{bmatrix} -\mathbf{q}(t) \\ \mathbf{b}(t) \end{bmatrix} \in \mathbb{R}^{n+m}.$$

14.2 Solution Models

In this section, ZD and GD models are both developed for the real-time solution of the above time-varying QP (14.5)–(14.6) comparatively.

14.2.1 ZD Model

To solve the above QP (14.5)–(14.6) via time-varying linear system (14.7), we firstly define the following vector-valued error function:

$$\mathbf{e}(t) = W(t)\mathbf{y}(t) - \mathbf{u}(t) \in \mathbb{R}^{n+m}, \tag{14.8}$$

The ZD design formula is then adopted [102, 104, 107, 117, 118, 122, 129, 134]:

$$\dot{\mathbf{e}}(t) = -\gamma \Phi(\mathbf{e}(t)), \tag{14.9}$$

where design parameter $\gamma > 0$, being the reciprocal of a capacitance parameter, should be set as large as the hardware would permit or selected appropriately for simulative and experimental purposes. In addition, $\Phi(\cdot) : \mathbb{R}^{n+m} \to \mathbb{R}^{n+m}$ denotes an activation function array. The array $\Phi(\cdot)$ is made of $(n+m)$ monotonically increasing odd activation functions denoted by $\phi(\cdot)$. For example, the power-sigmoid activation function with $\xi = 4$ and $p = 5$ is used in this chapter.

$$\phi(e_i) = \begin{cases} e_i^p, & \text{if } |e_i| \geqslant 1; \\ \dfrac{1+\exp(-\xi)}{1-\exp(-\xi)} \dfrac{1-\exp(-\xi e_i)}{1+\exp(-\xi e_i)}, & \text{otherwise.} \end{cases}$$

By expanding the above ZD design formula (14.9) with the definition Equation (14.8), the following implicit dynamic equation of a ZD model can readily be constructed for solving the time-varying QP (14.5)–(14.6):

$$W(t)\dot{\mathbf{y}}(t) = -\dot{W}(t)\mathbf{y}(t) - \gamma\Phi\big(W(t)\mathbf{y}(t) - \mathbf{u}(t)\big) + \dot{\mathbf{u}}(t). \tag{14.10}$$

14.2.2 GD Model

By following the gradient-descent method [102, 117, 118, 134], the design of a GD model can be generalized and presented to solve the time-varying QP (14.5)–(14.6) via time-varying linear system (14.7). Specifically, we first define a scalar-valued norm-based energy function $\mathscr{E}(\mathbf{y}, t) = \|W(t)\mathbf{y}(t) - \mathbf{u}(t)\|_2^2/2$. Note that a minimum point of $\mathscr{E}$ is achieved with $\mathscr{E}(\mathbf{y}^*, t) = 0$, if and only if $\mathbf{y}^*(t)$ is the theoretical solution to (14.7), i.e., $\mathbf{y}^*(t) = W^{-1}(t)\mathbf{u}(t)$. The derivative of $\mathscr{E}(\mathbf{y}, t)$, with respect to $\mathbf{y}(t) \in \mathbb{R}^{n+m}$, is then derived simply as $\partial\mathscr{E}(\mathbf{y}, t)/\partial\mathbf{y} = W^{\mathrm{T}}(t)(W(t)\mathbf{y}(t) - \mathbf{u}(t)) \in \mathbb{R}^{n+m}$. Finally, according to the GD design formula $\dot{\mathbf{y}}(t) = -\gamma\partial\mathscr{E}(\mathbf{y}, t)/\partial\mathbf{y}$ with $\gamma > 0$ as before, the dynamic equation of the linear-activation form of a GD model for real-time solution of the time-varying QP and time-varying linear system is shown as follows:

$$\dot{\mathbf{y}}(t) = -\gamma W^{\mathrm{T}}(t)(W(t)\mathbf{y}(t) - \mathbf{u}(t)). \tag{14.11}$$

According to the nonlinear-activation-function technique [102,104,117,122,129,134], we can have the following generalized nonlinear-activation form of the GD model:

$$\dot{\mathbf{y}}(t) = -\gamma W^{\mathrm{T}}(t)\Phi(W(t)\mathbf{y}(t) - \mathbf{u}(t)), \tag{14.12}$$

where activation function array Φ is defined as before (e.g., discussed in more details in Chapters 7 and 8).

14.3   Computer Simulations

In this section, the presented RMP scheme (14.3)–(14.4) is applied to a four-link planar robot arm. The end-effector of the robot is expected to track a triangle path and a circle path in the simulations. The resultant time-varying QP (14.5)–(14.6) is solved by ZD model (14.10) and GD model (14.12). Note that the computing environment related to the simulations of this chapter is MATLAB$^{\circledR}$ R2008a performed on a personal digital computer with a Core(TM) Duo E4500 2.20GHz CPU, 2GB memory, and a Windows XP Professional operating system.

14.3.1   Triangle-Path Tracking

In this simulation example, the four-link robot arm's end-effector is expected to move along a desired isosceles-right-triangle path with the right-angle side length being 0.8 m. The motion duration is 30 s, and the initial joint state is $\theta(0) = [3\pi/4, -\pi/2, -\pi/4, \pi/6]^{\mathrm{T}}$ rad. Parameters ς and γ are set to be 4 and 10^5, respectively. The simulation results are depicted correspondingly in Figure 14.1, Table 14.1, Figures 14.2 and 14.3 as well as Table 14.2.

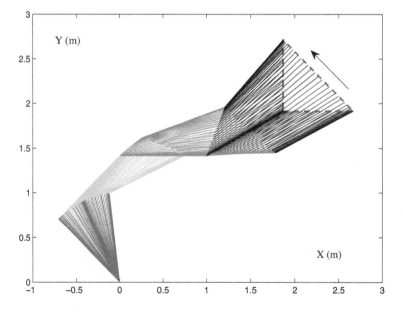

FIGURE 14.1 ZD-synthesized motion trajectories of four-link planar robot arm when its end-effector tracks an isosceles right-triangle path.

TABLE 14.1 Maximum Position and Velocity Errors of Four-Link Planar Robot Arm's End-Effector Synthesized by ZD Model (14.10) and GD Model (14.12)

	Triangle path		Circle path	
	Position error	**Velocity error**	**Position error**	**Velocity error**
ZD model	8.27×10^{-7} m	4.83×10^{-9} m/s	8.69×10^{-7} m	1.15×10^{-7} m/s
GD model	2.46×10^{-5} m	4.73×10^{-6} m/s	1.58×10^{-4} m	1.75×10^{-4} m/s

Specifically, synthesized by RMP scheme (14.3)–(14.4) using ZD model (14.10), Figure 14.1 illustrates the motion trajectories of the four-link planar robot arm operating in the 2-dimensional space (which is actually a plane). The arrow appearing in Figure 14.1 shows the motion direction. It is worth mentioning here that the motion trajectories synthesized by RMP scheme (14.3)–(14.4) using GD model (14.12) are very similar to those in Figure 14.1 and thus omitted. From Figure 14.1 and Table 14.1, we see that the simulated trajectories of the robot end-effector are very close to the desired isosceles-right-triangle path, with the maximum Cartesian position error being 8.27×10^{-7} m by using the ZD model and being 2.46×10^{-5} m by using the GD model (note that the error of using the ZD model is about 30 times smaller than that of the GD model).

As further shown in Figure 14.2, the solution of ZD model (14.10) is repetitive in the sense that the final state and initial state of the four-link planar robot arm coincide very well with each other. Note that, due to result similarity, the corresponding figure of GD model (14.12) is omitted. Therefore, these simulative observations show us evidently that this scheme can solve the joint-angle-drift problem very well. Furthermore, as seen from Table 14.1 and Figure 14.3, the maximum position and velocity errors synthesized by the ZD model (i.e., 8.27×10^{-7} m and 4.83×10^{-9} m/s, respectively) are much smaller than those by the GD model, when the end-effector tracks the triangle path. From Table 14.2, we observe additionally that the simulation time of the robot using ZD is also smaller than that using GD. Evidently, the ZD model is superior to the GD model in terms of inverse kinematic control, as the former achieves higher solution precision with less convergence/simulation time.

14.3.2 Circle-Path Tracking

As another illustrative example, the trajectory of the four-link robot arm's end-effector is expected to be a circle, with the radius being 0.8 m. The task duration is 10 s, and $\theta(0) = [0, \pi/12, \pi/24, -\pi/24]^{\mathrm{T}}$ rad. Design parameters ς and γ are set the same as those in the triangle example. The simulation results are correspondingly shown in Figures 14.4 and 14.5, in addition to Tables 14.1 and 14.2.

Specifically, the motion trajectories of the four-link planar robot arm synthesized by RMP scheme (14.3)–(14.4) are illustrated in Figure 14.4. As shown in Figure 14.4 and Table 14.1, the trajectory of the robot's end-effector is sufficiently close to the desired circle path, with the maximum Cartesian position error being 8.69×10^{-7} m by using the ZD model and being 1.58×10^{-4} m by using the GD model (note that the error of using ZD is about 182 times smaller than that of using GD). As shown in Figure 14.5 further, the solution of the RMP scheme is repetitive. Therefore, these simulative observations show us evidently that this scheme can solve the joint-angle-drift problem very well and again. Note that the corresponding figures of position errors by using ZD and GD models are omitted due to result similarity. From Table 14.1, we observe again that, for this circle path tracking task, the maximum position and velocity errors synthesized by the ZD model are also much smaller than those by the GD model. Besides, Table 14.2 shows confirmedly that the simulation time of the robot using ZD is smaller again, as compared with that using GD. These observations verify once more the ZD superiority for the robot inverse-kinematic problem solving.

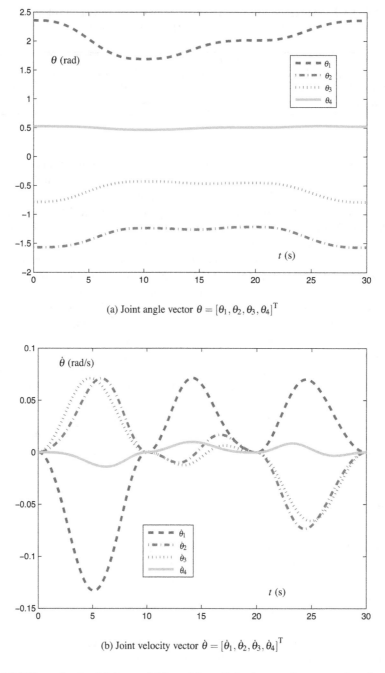

(a) Joint angle vector $\theta = [\theta_1, \theta_2, \theta_3, \theta_4]^T$

(b) Joint velocity vector $\dot{\theta} = [\dot{\theta}_1, \dot{\theta}_2, \dot{\theta}_3, \dot{\theta}_4]^T$

FIGURE 14.2 ZD-synthesized joint variables of four-link planar robot arm when its end-effector tracks a triangle path.

In summary, the above computer-simulation results based on the four-link planar robot arm have illustrated the efficacy of the presented RMP scheme as well as the ZD and GD models. More importantly, the simulation results show the superiority of the ZD method and model, as compared with the GD ones. Evidently, the ZD method and model methodologically exploit the time-derivative

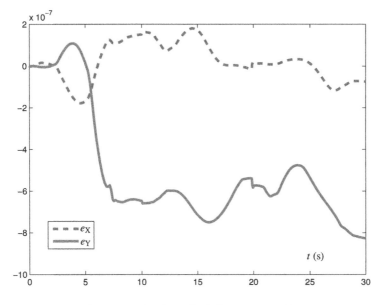

(a) Position error within $[-10, 2] \cdot 10^{-7}$ m by ZD model (14.10)

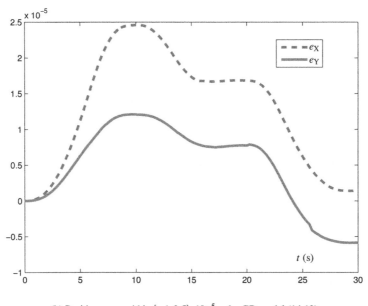

(b) Position error within $[-1, 2.5] \cdot 10^{-5}$ m by GD model (14.12)

FIGURE 14.3 ZD-synthesized and GD-synthesized position errors of four-link planar robot arm's end-effector when tracking a triangle path.

information of coefficients during the process of real-time time-varying problem solving, which is the reason why the ZD model globally exponentially converges to the exact solution of time-varying QP (14.5)–(14.6). In contrast, the GD model does not utilize such important information, and thus is less effective on the inverse kinematic control of the four-link robot arm.

TABLE 14.2 Simulation Time in Seconds (s) of Four-Link Planar Robot Arm Synthesized by ZD Model (14.10) and GD Model (14.12)

	Neural model	Testing index							
		1	2	3	4	5	6	7	8
Triangle path	ZD	1.31 s	1.32 s	1.32 s	1.32 s	1.33 s	1.33 s	1.33 s	1.34 s
	GD	6.29 s	5.55 s	5.48 s	5.92 s	5.80 s	5.70 s	5.74 s	5.76 s
Circle path	ZD	1.31 s	1.28 s	1.32 s	1.32 s	1.32 s	1.34 s	1.49 s	1.53 s
	GD	1.81 s	1.77 s	1.80 s	1.80 s	1.80 s	1.84 s	1.80 s	1.83 s

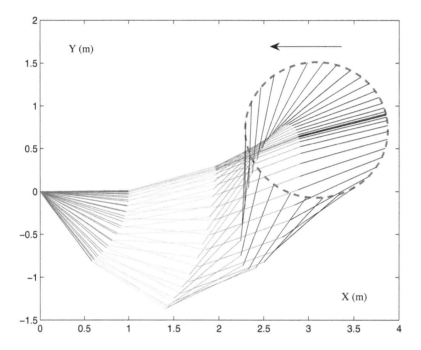

FIGURE 14.4 ZD-synthesized motion trajectories of four-link planar robot arm when its end-effector tracks a circle path.

14.4 Hardware Experiments

In the above computer simulations, we assume that everything is ideal and satisfies simulative requests. However, in practice (e.g., experiments), there are various difficulties. To keep the completeness of the research, a six-DOF redundant robot arm (or to say, six-link planar robot arm) hardware system is developed, investigated, and shown for further verification purposes. The whole robot arm system is mainly composed of a robot arm and a host computer. Figure 14.6 depicts this robot hardware system and its 3-dimensional model. Note that the host computer is a personal digital computer with a Pentium Dual-Core E5300 2.60GHz CPU, 4GB memory, and a Windows XP Professional operating system, which sends instructions to the robot arm's motion-control module. Here and now, in order to verify the presented RMP scheme (14.3)–(14.4) and the ZD model, we perform experiments by using the six-DOF redundant robot arm to track a straight-line segment path and an ellipse path in the 2-dimensional horizontal work-plane.

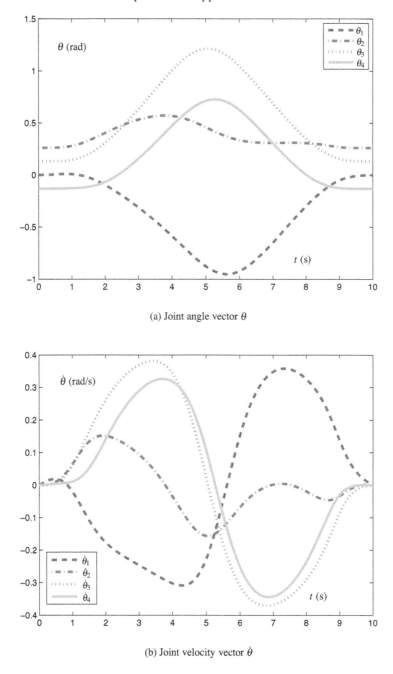

(a) Joint angle vector θ

(b) Joint velocity vector $\dot{\theta}$

FIGURE 14.5 ZD-synthesized joint variables of four-link planar robot arm when its end-effector tracks a circle path.

As the first experiment, the end-effector is expected to move forward and then backward along a straight-line segment path with length 0.05 m, where initial joint state $\theta(0) = [\pi/12, \pi/12, \pi/12, \pi/12, \pi/12, \pi/12]^T$ rad, and design parameters $\varsigma = 6$ and $\gamma = 10^4$. The task execution is shown in Figure 14.7; i.e., the end-effector of the robot arm moves smoothly and draws

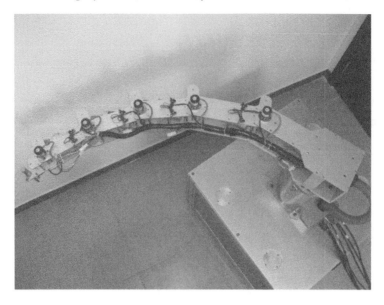

(a) Actual robot arm

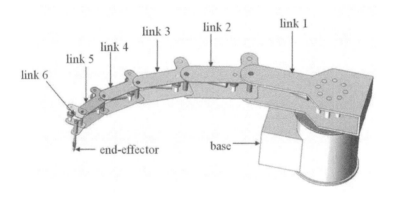

(b) 3-dimensional model

FIGURE 14.6 Actual six-DOF redundant robot arm in authors' laboratory and its 3-dimensional model. *Reproduced from Y. Zhang, Y. Wang, L. Jin, et al., Simulations and experiments of ZNN for online quadratic programming applied to manipulator inverse kinematics, Figure 6, Proceedings of the Third International Conference on Information Science and Technology, pp. 23–25, 2013. ©IEEE 2013. With kind permission of IEEE.*

a line successfully. Thus, this experiment illustrates well that the presented RMP scheme (14.3)–(14.4) and the ZD model are effective on the redundant robot arm's repetitive motion planning and control.

As another illustrative experiment, the trajectory of the end-effector is expected to be an ellipse, with the major radius being 0.02 m and the minor radius being 0.01 m. The initial joint state and

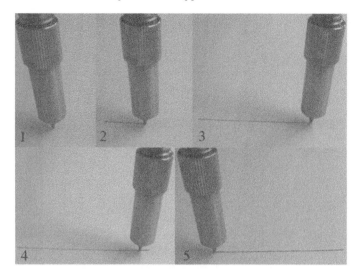

(a) Snapshots of straight-line segment task execution

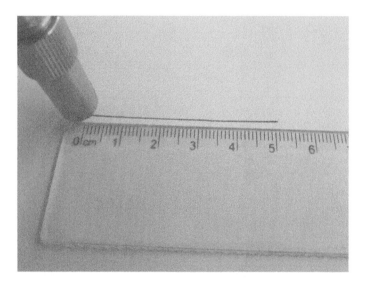

(b) Measurement of experimental result

FIGURE 14.7 Straight-line segment path tracking experiment of six-DOF redundant robot arm synthesized by RMP scheme (14.3)–(14.4) using ZD model (14.10). *Reproduced from Y. Zhang, Y. Wang, L. Jin, et al., Simulations and experiments of ZNN for online quadratic programming applied to manipulator inverse kinematics, Figure 7, Proceedings of the Third International Conference on Information Science and Technology, pp. 23–25, 2013. ©IEEE 2013. With kind permission of IEEE.*

design parameters are the same as those in the straight-line experiment. Figure 14.8 illustrates that the six-DOF robot synthesized by the presented RMP scheme (14.3)–(14.4) using the ZD model moves smoothly and draws an ellipse successfully during the task execution. It is illustrated that the solution can be readily implemented on hardware.

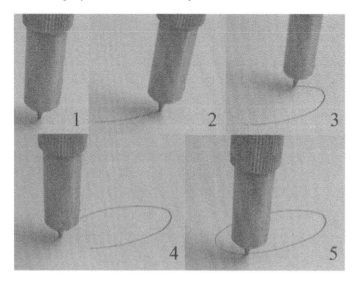

(a) Snapshots of ellipse task execution

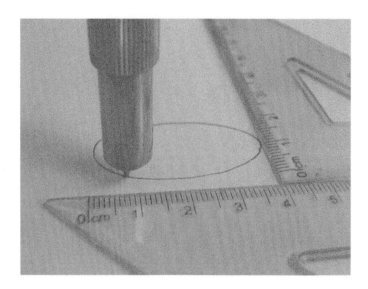

(b) Measurement of experimental result

FIGURE 14.8 Ellipse path tracking experiment of six-DOF redundant robot arm synthesized by RMP scheme (14.3)–(14.4) using ZD model (14.10). *Reproduced from Y. Zhang, Y. Wang, L. Jin, et al., Simulations and experiments of ZNN for online quadratic programming applied to manipulator inverse kinematics, Figure 8, Proceedings of the Third International Conference on Information Science and Technology, pp. 23–25, 2013. ©IEEE 2013. With kind permission of IEEE.*

In summary, the above experimental results based on the six-DOF planar robot arm have further illustrated the physical realizability and effectiveness of the RMP scheme (14.3)–(14.4) using the ZD model. Note that this ZD- and QP-based method can also be applied to robot arms operating in 3-dimensional space.

14.5 Summary

In this chapter, the ZD model has been developed, investigated, and applied to the real-time time-varying QP problem solving of robot arms. Being different from the conventional GD model, the ZD model has utilized the time-derivative information of the QP problem and thus has achieved higher solution precision with less convergence time. Computer simulations performed on a four-link planar robot arm have substantiated the superiority of the ZD model for the real-time solution of the time-varying QP problem, as compared with the GD one. Moreover, practical experiments have been conducted on an actual six-DOF planar redundant robot arm, which has substantiated the physical realizability and effectiveness of the ZD model.

Part VI

Time-Varying Inequality Solving

Chapter 15

Linear Inequality Solving

Abstract

In this chapter, we firstly propose a novel continuous-time zeroing dynamics (CTZD) model for real-time solution of the scalar-valued time-varying linear inequality. In addition, three different activation functions are exploited in the CTZD model. For potential hardware implementation, the corresponding discrete-time zeroing dynamics (DTZD) models are derived and proposed as well. Then, such zeroing dynamics (ZD) models are further extended for solving systems of time-varying linear inequalities in real time. For comparative purposes, the gradient dynamics (GD) model is also developed and exploited for solving systems of time-varying linear inequalities. Computer-simulation and numerical-experiment results further verify and illustrate the efficacy, novelty, and superiority of such ZD models for solving time-varying linear inequalities.

15.1 Introduction

Many problems encountered in science and technology involve solving a large package of linear inequalities, e.g., system identification [80], computer tomography [71], automatic control [81, 96], and signal restoration [6, 45, 61]; and it is often required to obtain the solution in real time. Currently, most of the reported algorithms are just for solving the constant linear inequalities [17, 42, 52, 93]. However, in this situation, faster convergence rate and more stringent parameters setting are often required with respect to the variational rate of time-varying coefficients [102, 104, 107, 117, 118, 122, 129, 134]. Thus, it may cause serious restrictions on physical implementation or sacrifice the solution precision.

In recent decades, because of the parallel distributed nature and the suitability for hardware implementation, neural dynamics have widely arisen in scientific computation and optimization, drawing extensive interests and investigation of researchers [17, 42, 52, 93, 102, 104, 107, 117, 118, 122, 129, 134]. In addition, due to the in-depth research on neural dynamics, numerous solvers have been proposed, developed, and investigated for such problems solving [17, 42, 52, 93]. For example, Cichocki and Bargiela [17] developed three continuous-time neural dynamics for solving linear inequality systems. These neural dynamics have penalty parameters that decrease to zero as time t tends to infinity in order to get better solution. Labonte [42] presented a class of discrete-time neural dynamics for solving the linear inequalities that implemented each of the different versions of the aforementioned relaxation-projection method. Xia *et al.* [93] presented two types of neural dynamics (designed based on gradient-based methods), i.e., continuous-time and discrete-time ones, for solving the linear inequality and equality systems, with rigorous proofs on the global convergence of such neural dynamics given. It is worth pointing out that most of the reported neural-dynamic models are related to gradient-based methods or designed theoretically for solving constant linear inequalities. As we know, the GD is based on the gradient-descent method to minimize a norm-based lower-bounded energy function, which has already been employed comprehensively to solve such constant problems [102, 104, 107, 117, 118, 122, 129, 134]. However, when such a GD model is applied to the case of having time-varying coefficients, a faster convergence rate is often required, as compared with the variational rate of the time-varying coefficients. On the other hand, time-varying linear inequalities have a wide application in the motion planning and control (MPC) of robot arms; e.g., time-varying linear inequalities were proposed, introduced, and investigated by Zhang *et al.* for robotic arms' obstacle-avoidance MPC, which are used to generate escape velocities of variable magnitude, driving the affected links away from obstacles. Thus, robot arms can avoid obstacles successfully [94, 116, 136].

Differing from the conventional GD model, a special class of neural dynamics (ZD) has been formally proposed by Zhang *et al.* for time-varying or constant problems solving, such as time-varying convex quadratic programming [117], Sylvester equation solving [107], matrix inversion [99, 102, 104, 118]. In view of the superiority of the above models, new ZD models depicted in implicit dynamic equations, are presented, investigated and analyzed for real-time solution of the scalar-valued and vector-valued time-varying linear inequalities in this chapter [94, 116]. For the situation of the scalar-valued time-varying linear inequality solving, three different activation functions are exploited in the ZD models, and the corresponding DTZD models are derived and proposed for numerical experimentation or potential hardware implementation. For the situation of the vector-valued time-varying linear inequality solving, a general ZD model is presented, investigated, and analyzed with theoretical analysis and results provided. The conventional GD model is also developed and exploited for comparative purposes.

15.2 Time-Varying Linear Inequality

In order to lay the basis for further discussion, the CTZD model is proposed first for real-time solution of the scalar-valued time-varying linear inequality in this section. Detailed design procedure and theoretical analysis are presented in this section as well.

15.2.1 CTZD Model

Without loss of generality, let us consider the following scalar-valued time-varying linear inequality problem [116]:

$$f(x(t),t) = a(t)x(t) - b(t) \leqslant 0 \in \mathbb{R}, \ t \in [0,+\infty). \tag{15.1}$$

For monitoring and controlling the process of solving the scalar-valued time-varying linear inequality (15.1), let us define the following error function $e(t)$ [for which there exists at least one trajectory of $x(t)$, such that $a(t)x(t) - b(t) \leqslant 0$ and simultaneously $e(t) = 0$]:

$$e(t) = (\max\{0, f(x(t),t)\})^2 / 2. \tag{15.2}$$

Then, by following Zhang *et al.*'s neural-dynamic design method (i.e., zeroing dynamics design method) [99,102,104,118], let us choose the following time-derivative $\dot{e}(t)$, such that $e(t)$ converges to zero:

$$\dot{e}(t) = \frac{de(t)}{dt} = -\gamma\phi(e(t)), \tag{15.3}$$

where design parameter $\gamma > 0$ should be set as large as the hardware would permit or selected appropriately for simulative or experimental purposes, and $\phi(\cdot) : \mathbb{R} \to \mathbb{R}$ denotes a general monotonically-increasing odd activation function that can be linear or nonlinear. As shown in Figure 15.1, three types of activation functions can be exploited, i.e., linear activation function (discussed in almost every chapter of this book), hyperbolic sine activation function (mainly discussed in Chapter 12.3.2), and power-sum activation function (mainly discussed in Chapter 7). Note that, by the authors' previous work and experiences as well as the previous chapters of this book, different choices of design parameter γ and activation function $\phi(\cdot)$ can affect the convergence performance of the proposed ZD models.

15.2.1.1 CTZD Model Using Linear Activation Function

In this subsection, we exploit the linear activation function $\phi(e(t)) = e(t)$ in the CTZD model design. Thus, we obtain

$$\dot{e}(t) = -\gamma e(t). \tag{15.4}$$

Considering Equation (15.2), we have

$$\dot{e}(t) = (\dot{a}(t)x(t) + a(t)\dot{x}(t) - \dot{b}(t))\max\{0, f(x(t),t)\}. \tag{15.5}$$

Thus, combining Equations (15.2), (15.4), and (15.5), we have

$$a(t)\dot{x}(t) = -\dot{a}(t)x(t) + \dot{b}(t) - \frac{\gamma}{2}\max\{0, f(x(t),t)\}. \tag{15.6}$$

With $a(t) \neq 0$ assumed [which means that $a(t)$ keeps positive or negative all the time], we further obtain

$$\dot{x}(t) = \frac{1}{a(t)}\left(-\dot{a}(t)x(t) + \dot{b}(t) - \frac{\gamma}{2}\max\{0, f(x(t),t)\}\right). \tag{15.7}$$

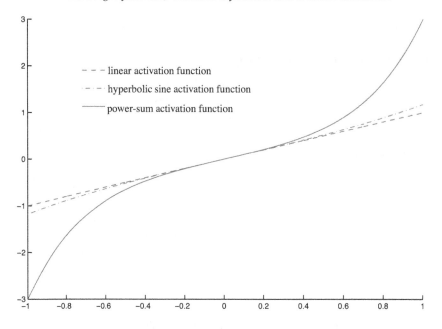

FIGURE 15.1 Three different activation functions exploited in ZD models.

Therefore, we obtain the CTZD model using a linear activation function for solving the scalar-valued time-varying linear inequality; i.e., CTZD model (15.7). Here, $x(t)$ denotes the neural state of the aforementioned CTZD model (15.7). Besides, because the proposed CTZD model (15.7) makes use of the time-derivative information of the coefficients, the smooth convergence performance can be achieved for solving exactly the scalar-valued time-varying linear inequality problem (15.1). That is, we have the theorem.

Theorem 22 *Consider smoothly time-varying scalars $a(t) \neq 0 \in R$ and $b(t) \in \mathbb{R}$ in inequality (15.1). For CTZD model (15.7) using a linear activation function, starting from any initial value $x(0) \in \mathbb{R}$, neural state $x(t)$ of CTZD model (15.7) exponentially converges to the theoretical time-varying solution set of linear inequality (15.1).*

Proof The well-known Lyapunov theory [78, 100, 123, 131, 132, 136] is used to simply prove the validity of the above theorem. We first define a Lyapunov function candidate as

$$v(x(t),t) = (\max\{0, f(x(t),t)\})^2 /2 \geqslant 0,$$

where $v(x(t),t) = 0$ for any $x(t)$ inside the theoretical time-varying solution set of (15.1), and $v(x(t),t) > 0$ for any $x(t)$ outside the theoretical time-varying solution set of inequality (15.1). Then we derive its time derivative along the trajectory of CTZD model (15.7) as

$$
\begin{aligned}
\dot{v}(x(t),t) &= \frac{\mathrm{d}v(x(t),t)}{\mathrm{d}t} \\
&= \left(\dot{a}(t)x(t) + a(t)\dot{x}(t) - \dot{b}(t)\right)\max\{0, f(x(t),t)\} \\
&= -\gamma(\max\{0, f(x(t),t)\})^2 /2 = -\gamma v(x(t),t) \leqslant 0, \qquad (15.8)
\end{aligned}
$$

which guarantees the negative-definiteness of $\dot{v}(x(t),t)$. In addition, from (15.8), we have $v(x(t),t) = v(x(0),0)\exp(-\gamma t)$. As $v(x(t),t) = (\max\{0, a(t)x(t) - b(t)\})^2 /2 = e(t)$, we have

$$e(t) = e(0)\exp(-\gamma t). \qquad (15.9)$$

That is to say, starting from any initial value $x(0) \in \mathbb{R}$, neural state $x(t)$ of the proposed CTZD model (15.7) exponentially converges to the theoretical time-varying solution set of (15.1) with $e(t)$ exponentially converging to 0 (as time t tends to ∞). The proof is thus complete.

15.2.1.2 CTZD Model Using Nonlinear Activation Functions

In this subsection, we exploit two types of nonlinear activation functions for solving the scalar-valued time-varying linear inequality:

1) the hyperbolic sine activation function with constant integer parameter $\xi \geqslant 1$: $\phi(e) = (\exp(\xi e) - \exp(-\xi e))/2$ (see also Chapter 12.3.2); and,

2) the power-sum activation function with constant integer parameter $N \geqslant 2$: $\phi(e) = \sum_{\kappa=1}^{N} e^{2\kappa-1}$ (see also Chapter 7).

Thus, from Equations (15.2) and (15.3), we can obtain the following nonlinearly activated CTZD model:

$$(a(t)\max\{0, f(x(t),t)\})\dot{x}(t) = -\gamma\phi\left((\max\{0, f(x(t),t)\})^2/2\right)$$
$$- (\dot{a}(t)x(t) + \dot{b}(t))\max\{0, f(x(t),t)\},$$

and its improved division-free explicit model [for $a(t) > 0$ as an example]:

$$\dot{x}(t) = -\gamma\phi\left((\max\{0, f(x(t),t)\})^2/2\right) - (\dot{a}(t)x(t) + \dot{b}(t))\max\{0, f(x(t),t)\}. \tag{15.10}$$

It is worth mentioning here that, once neural state $x(t)$ goes inside the theoretical solution set of inequality (15.1), according to the expressions of the nonlinear activation functions used in this chapter, the right-hand side of Equation (15.10) equals zero, and $\dot{x}(t) = 0$, which elegantly avoids the division-by-zero situation.

As a result, we obtain CTZD model (15.10) using nonlinear activation functions for solving the scalar-valued time-varying linear inequality. Similar to CTZD model (15.7), due to the utilization of the time-derivative information of the coefficients, good convergence performance of CTZD model (15.10) can be achieved as well.

15.2.2 DTZD Model

At the beginning of this subsection, we give the following consideration: because of derivation similarity, we only use the linear activation function to derive the DTZD model for solving the scalar-valued time-varying linear inequality in this subsection. The derivation of the DTZD models using nonlinear activation functions can also be done, e.g., from CTZD model (15.10), but left to interested readers (together with its numerical experiments) to complete as a topic of exercise.

For solving inequality (15.1), we use the well-known Euler forward-difference rule to estimate $\dot{x}(t)$ as follows [1, 3, 22, 25, 32, 40, 51, 57, 68]:

$$\dot{x}(t = k\tau) \approx (x((k+1)\tau) - x(k\tau))/\tau,$$

where $\tau > 0$ denotes the sampling gap of $x(t)$, and $k = 0, 1, 2, \cdots$. Throughout this chapter, we denote $x_k = x(t = k\tau)$ for short. Accordingly, $a_k = a(t = k\tau)$, $b_k = b(t = k\tau)$, $\dot{a}_k = \dot{a}(t = k\tau)$, and $\dot{b}_k = \dot{b}(t = k\tau)$. Then, the DTZD model for solving scalar-valued time-varying linear inequality (15.1) is derived from CTZD model (15.7) as follows:

$$\frac{x_{k+1} - x_k}{\tau} = \frac{1}{a_k}\left(-\dot{a}_k x_k + \dot{b}_k - \frac{\gamma}{2}\max\{0, f(x_k, k\tau)\}\right). \tag{15.11}$$

Defining $h = \tau\gamma > 0$ as the step size, from the above Equation (15.11), we can obtain the DTZD model with $\dot{a}_k$ and $\dot{b}_k$ known (i.e., the DTZDK model):

$$x_{k+1} = \left(1 - \tau\frac{\dot{a}_k}{a_k}\right)x_k - \frac{h}{2a_k}\max\{0, f(x_k, k\tau)\} + \frac{\tau}{a_k}\dot{b}_k, \qquad (15.12)$$

where x_k represents the kth sampling (or to say, updating) value of the solution $x(t)$, i.e., $x(t = k\tau)$. In addition, the sampling gap τ should be set appropriately small for simulative or experimental purposes, and the step size h should take on a suitable value as well. Furthermore, it is worth mentioning that different choices of sampling gap τ and step size h can lead to different convergence performance of DTZDK model (15.12).

However, it may be difficult to get the value of $\dot{a}_k$ or $\dot{b}_k$ directly in certain real-world applications. Therefore, we use Euler backward-difference rule to estimate $\dot{a}_k$ and $\dot{b}_k$:

$$\dot{a}_k \approx (a_k - a_{k-1})/\tau, \ \dot{b}_k \approx (b_k - b_{k-1})/\tau.$$

In this paragraph, let us consider the situation with $\dot{a}_k$ and $\dot{b}_k$ both unknown. According to the aforementioned DTZDK model (15.12), the following DTZD model with $\dot{a}_k$ and $\dot{b}_k$ unknown (i.e., the DTZDU model) can be derived:

$$x_{k+1} = \frac{a_{k-1}}{a_k}x_k - \frac{h}{2a_k}\max\{0, f(x_k, k\tau)\} + \frac{\Delta_{b_k}}{a_k}, \qquad (15.13)$$

with $\Delta_{b_k} = b_k - b_{k-1}$. Thus, DTZDU model (15.13) with $\dot{a}_k$ and $\dot{b}_k$ both unknown for solving the scalar-valued time-varying linear inequality (15.1) is obtained.

In addition, there are another two situations with derivatives partially known; i.e., 1) $\dot{a}_k$ is unknown but $\dot{b}_k$ known, and 2) $\dot{a}_k$ is known but $\dot{b}_k$ unknown. Because of derivation similarity, we just show the DTZD models (i.e., DTZDP models) as below:

$$x_{k+1} = \frac{a_{k-1}}{a_k}x_k - \frac{h}{2a_k}\max\{0, f(x_k, k\tau)\} + \frac{\tau}{a_k}\dot{b}_k,$$

$$x_{k+1} = \left(1 - \tau\frac{\dot{a}_k}{a_k}\right)x_k - \frac{h}{2a_k}\max\{0, f(x_k, k\tau)\} + \frac{\Delta_{b_k}}{a_k}.$$

About the detailed discussion, performance, and efficacy of the DTZDP-type models, please see also Chapter 13 on time-varying quadratic programming. On the other hand, the numerical experiments on the above DTZDP models can be done readily, but left intentionally to interested readers to complete as a topic of exercise.

15.3 Constant Linear Inequality

In the above section, we have mentioned that most of the reported algorithms for solving a linear inequality are just for the constant linear inequality. The constant linear inequality solving can be viewed as a special case of the real-time solution of the scalar-valued time-varying linear inequality. In this section, we exploit the ZD method to solve the constant linear inequality problem in a smooth-solution and solution-extendable manner, which is in the form of

$$f(x) = ax - b \leqslant 0 \in \mathbb{R}, \qquad (15.14)$$

where $a \neq 0 \in \mathbb{R}$ and $b \in \mathbb{R}$ are constant scalars, while $x \in \mathbb{R}$ is the unknown to be obtained. The simplified CTZD (S-CTZD) model and the simplified DTZD (S-DTZD) model are thus investigated for the solution of constant linear inequality problem (15.14) in the following subsections, respectively.

15.3.1 S-CTZD Model

We exploit the linear activation function and two types of nonlinear activation functions in the S-CTZD model to solve (15.14) in this subsection.

Situation 1 Using linear activation function

Since a and b are both constant scalars, CTZD model (15.7) reduces to the following S-CTZD model for solving constant linear inequality problem (15.14):

$$\dot{x}(t) = -\frac{\gamma}{2a}\max\{0, f(x)\}. \tag{15.15}$$

Thus, we obtain S-CTZD model (15.15) using a linear activation function for constant linear inequality solving.

Situation 2 Using nonlinear activation functions

In this situation, $\phi(\cdot)$ can be the hyperbolic sine activation function or power-sum activation function. Deriving from CTZD model (15.10), we have

$$\dot{x}(t) = -\gamma\phi\left((\max\{0, f(x)\})^2/2\right). \tag{15.16}$$

Therefore, S-CTZD model (15.16) using a nonlinear activation function for solving the constant linear inequality is obtained.

Consider constant scalars $a \neq 0 \in \mathbb{R}$ and $b \in \mathbb{R}$ in (15.14). It follows from the presented theorem and related results that, for S-CTZD model (15.15) and S-CTZD model (15.16), starting from randomly generated initial value $x(0) \in \mathbb{R}$, neural states $x(t)$ of S-CTZD models converge to the boundary or inside of the theoretical solution set of (15.14).

15.3.2 S-DTZD Model

In this subsection, we use the well-known Euler forward-difference rule to estimate $\dot{x}(t)$. Therefore, the simplified DTZD model for constant linear inequality solving (i.e., the S-DTZD model) is derived. Specifically, as a and b are both constant, the S-DTZD model is derived from DTZD model (15.12):

$$x_{k+1} = x_k - \frac{h}{2a}\max\{0, f(x_k)\}, \tag{15.17}$$

where $h = \tau\gamma > 0$ denotes the step size and should take on a suitable value [similar to that in DTZDK model (15.12)]. In addition, different choices of the step size h can affect the convergence performance of S-DTZD model (15.17).

15.4 Illustrative Examples

In the previous two sections, the CTZD model, the DTZD models and their simplified models have been presented for solving the scalar-valued time-varying (or constant) linear inequality. In this section, some illustrative examples are provided for substantiating the efficacy of the aforementioned ZD models. Firstly, we define the initial solution set $\mathbb{S}(0) = \{x^*(0)|a(0)x^*(0) - b(0) \leqslant 0\}$ for the

CTZD models or $\mathbb{S}(0) = \{x_0^* | a_0 x_0^* - b_0 \leqslant 0\}$ for the DTZD models. In order to observe the convergence process, the initial value is set to be outside the theoretical initial solution set $\mathbb{S}(0)$ in all of the following six examples, and the proposed ZD models for solving the scalar-valued time-varying linear inequality can guarantee that neural states $x(t)$ of ZD models converge to the boundary of the theoretical solution set of the scalar-valued linear inequality. In addition, it is easy to see that, if starting inside the theoretical initial solution set $\mathbb{S}(0)$, neural states $x(t)$ of ZD models will keep inside the theoretical solution set all the time.

Note that the following two scalar-valued inequalities are used to verify the efficacy and superiority of the proposed ZD models:

$$f_1(x(t),t) = (\sin(2t) + 4)x(t) - (\sin(5t) + 7) \leqslant 0; \tag{15.18}$$

$$f_2(x) = 4x - 7 \leqslant 0. \tag{15.19}$$

Correspondingly, we have theoretical initial solution sets $\mathbb{S}(0) = \{x^*(0) | x^*(0) \leqslant 1.75\}$ and $\mathbb{S}(0) = \{x_0^* | x_0^* \leqslant 1.75\}$, respectively, for (15.18) and (15.19). For observation and comparison conveniences, we set initial values $x(0)$ and x_0 to be 2, which is outside the theoretical initial solution set $\mathbb{S}(0)$ in the following examples.

Example 15.1 In this example, CTZD model (15.7) is exploited for the real-time solution of scalar-valued time-varying linear inequality (15.18). In Figure 15.2(a), the solid curve denotes the time-varying trajectory of neural state $x(t)$ of CTZD model (15.7), while the dash-dotted curve denotes the upper bound of the theoretical solution set of (15.18). In addition, the trajectory of $x(t)$ is fairly smoothly time-varying. Figure 15.2(b) illustrates the convergence behavior of error function $|e(t)|$. The simulation results illustrate the exponential convergence of CTZD model (15.7). Furthermore, Figure 15.3 displays the convergence behaviors of error function $|e(t)|$ during the process of solving scalar-valued time-varying problem (15.18) with different values of γ. As seen from Figures 15.2 and 15.3, when the value of γ increases from 1 to 100, the proposed CTZD model (15.7) achieves a much faster convergence rate. Therefore, we can draw a conclusion that the proposed CTZD model (15.7) can achieve magnificent performance for solving scalar-valued time-varying linear inequality (15.1).

Example 15.2 In order to further verify the efficacy and superiority of CTZD model (15.10), nonlinear activation functions are exploited in this example. First, we set $\xi = 2$ for the hyperbolic sine activation function and $p = 5$ for the power-sum activation function. The simulation results using the three activation functions with $\gamma = 1$ are shown in Figure 15.4. We observe from the figure that, using hyperbolic sine activation function or power-sum activation function, CTZD model (15.10) can achieve a much faster convergence performance than CTZD model (15.7) using a linear activation function.

It follows from the above two examples that the CTZD models can solve effectively the scalar-valued time-varying linear inequality problem no matter whether linear or nonlinear activation functions are used.

Example 15.3 As another illustrative example, we exploit the aforementioned DTZDK model (15.12) to solve (15.18). In this example, without loss of generality, we set $\tau = 0.1$ and $h = 1$. Figure 15.5(a) displays the time-varying trajectory of discrete-time neural state x_k of DTZDK model (15.12), denoted by the solid curve, as well as the upper bound of the theoretical solution set of (15.18), denoted by the dash-dotted curve. It is seen from Figure 15.5(b) that discrete-time neural state x_k of DTZDK model (15.12) converges well to the theoretical time-varying solution set of (15.18) within 10 iterations.

Example 15.4 As a supplementary example, we investigate the different convergence performance, of the aforementioned DTZDK model (15.12) with different values of τ and h in this example. In

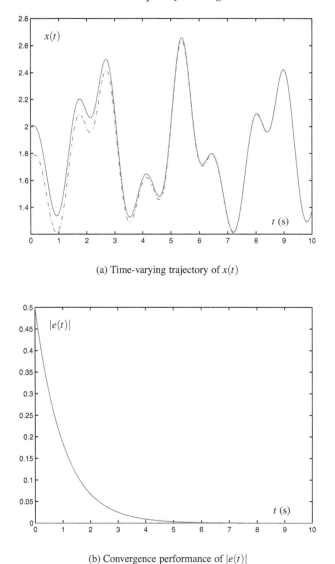

(a) Time-varying trajectory of $x(t)$

(b) Convergence performance of $|e(t)|$

FIGURE 15.2 Solving (15.18) via CTZD model (15.7) with $\gamma = 1$.

Figure 15.6, the convergence characteristics of DTZDK model (15.12) with different values of h are shown. As seen from the figure, when the value of h increases from 0.1 to 0.5, the convergence becomes much faster, i.e., from about 0.6 s (60 iterations) to about 0.12 s (12 iterations). Evidently, step size h plays an important role in the proposed DTZDK model (15.12). Moreover, Figure 15.7 illustrates the convergence characteristics of DTZDK model (15.12) with different values of τ. As shown in this figure, the convergence time of DTZDK model (15.12) reduces from about 0.05 s (5 iterations) to about 0.005 s (5 iterations) when the value of τ decreases from 0.01 s (i.e., 10 ms) to 0.001 s (i.e., 1 ms). Therefore, we can conclude that sampling gap τ and step size h play important roles in DTZDK model (15.12).

Example 15.5 In this example, $\dot{a}_k$ and $\dot{b}_k$ are both assumed unknown. Thus, DTZDU model (15.13) is exploited to solve (15.18). Being the same as those in previous examples, in Figure 15.8(a), the

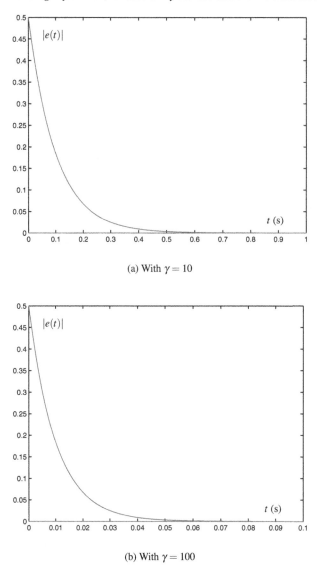

(a) With $\gamma = 10$

(b) With $\gamma = 100$

FIGURE 15.3 Solving (15.18) via CTZD model (15.7) with different γ.

solid curve denotes the time-varying trajectory of discrete-time neural state x_k of DTZDU model (15.13) while the dash-dotted curve denotes the upper bound of the theoretical solution set of (15.18). With $\tau = 0.1$ and $h = 1$, the fast convergence of DTZDU model (15.13) is achieved, as seen from Figure 15.8(b). That is to say, by using Euler backward-difference rule to estimate $\dot{a}_k$ and $\dot{b}_k$, DTZDU model (15.13) can solve the scalar-valued time-varying linear inequality problem accurately and efficiently.

In summary of Examples 15.3 through 15.5, the efficacy and superiority of the proposed DTZD models are illustrated and verified well. Besides, a suitably small sampling gap τ and a suitably large step size h should be chosen, respectively, for the higher solution precision and the faster convergence we may need in practice.

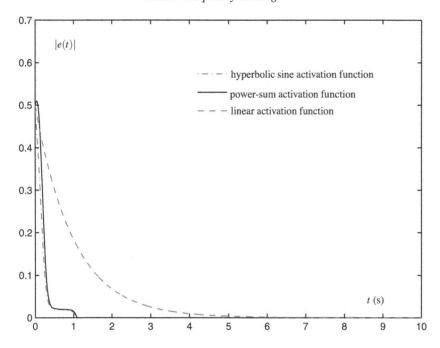

FIGURE 15.4 Solving (15.18) via CTZD models (15.7) and (15.10).

Example 15.6 As a special case of the scalar-valued time-varying linear inequality solving, the simplified ZD models are exploited to solve the constant linear inequality (15.19) in this example. In Figure 15.9(a), which shows the convergence of $|e(t)|$, S-CTZD models (15.15) and (15.16) are exploited to solve the constant linear inequality problem (15.19). Specifically, Figure 15.9(a) implies that, starting from an initial state outside the theoretical initial solution set $\mathbb{S}(0)$, neural states $x(t)$ of the proposed S-CTZD models [i.e., (15.15) and (15.16)] exponentially converges to the theoretical solution set of (15.19). From Figure 15.9(b), which is synthesized by S-DTZD model (15.17) with different values of h, we can observe that, by increasing the value of h, a much faster convergence rate of the proposed S-DTZD model (15.17) can be achieved. In addition, the proposed S-DTZD model (15.17) solves well the constant linear inequality problem (15.14) in an efficient and accurate manner.

In summary, from the above six examples, we can draw the conclusion that the proposed CTZD and DTZD models are all effective and efficient on solving scalar-valued time-varying linear inequality (15.1) and its corresponding constant problem (15.14).

15.5 System of Time-Varying Linear Inequalities

In this section, let us consider the following problem of time-varying linear inequality system (or termed time-varying linear inequalities):

$$A(t)\mathbf{x}(t) \leqslant \mathbf{b}(t) \in \mathbb{R}^n, \tag{15.20}$$

where $A(t) \in \mathbb{R}^{n \times n}$ and $\mathbf{b}(t) \in \mathbb{R}^n$ are smoothly time-varying; and let us define the time-varying theoretical solution set $\mathbb{S}(t) = \{\mathbf{x}(t) | \mathbf{x}(t) \in \mathbb{R}^n \text{ solves } (15.20)\}$. For presentation convenience, we

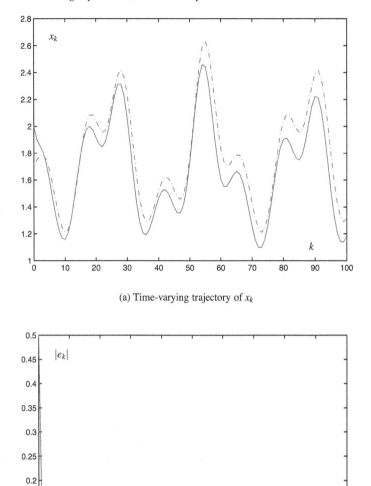

(a) Time-varying trajectory of x_k

(b) Convergence performance of $|e_k|$

FIGURE 15.5 Solving (15.18) via DTZDK model (15.12) with $\tau = 0.1$ and $h = 1$.

define the residual error as

$$\mathbf{y}(t) = A(t)\mathbf{x}(t) - \mathbf{b}(t) \in \mathbb{R}^n,$$

and $\mathbf{y}(t) = [y_1(t), y_2(t), \cdots, y_n(t)]^{\mathrm{T}}$. The ZD (including CTZD and DTZD) models are then presented and analyzed for solving time-varying linear inequality system (15.20) in the following subsections.

Our objective is to find the unknown $\mathbf{x}(t) \in \mathbb{R}^n$ in real time t, such that the above time-varying linear inequality system (15.20) always holds true. For time-varying linear inequality system solv-

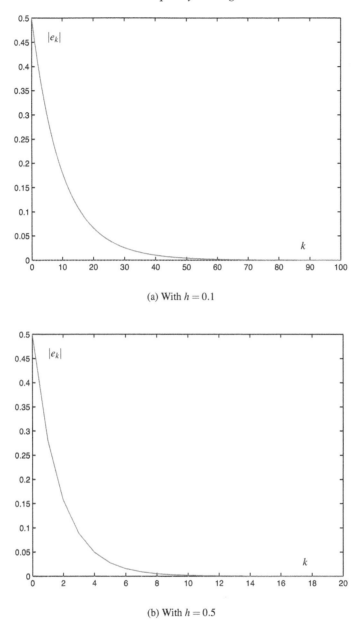

(a) With $h = 0.1$

(b) With $h = 0.5$

FIGURE 15.6 Solving (15.18) via DTZDK model (15.12) with $\tau = 0.01$ and different h.

ing, unlike constant linear inequality system, it needs to guarantee the existence of time-varying solution set $\mathbb{S}(t)$ to the time-varying linear inequality system. Otherwise, time-varying solution $\mathbf{x}(t)$ will not be continuous or even does not exist. In order to guarantee the existence of time-varying solution $\mathbf{x}(t)$, and also for simplicity and clarity, we limit the discussion for the situation that coefficient matrix $A(t)$ is a nonsingular square matrix for any time instant $t \in [0, +\infty)$, although the extension for solving more general time-varying linear inequalities is possible.

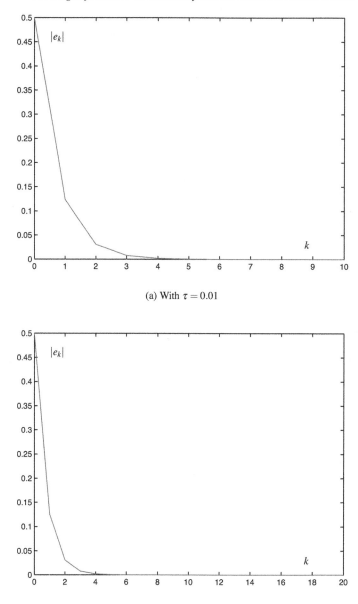

(a) With $\tau = 0.01$

(b) With $\tau = 0.001$

FIGURE 15.7 Solving (15.18) via DTZDK model (15.12) with $h = 1$ and different τ.

15.5.1   CTZD Model

To monitor and control the process of solving a time-varying linear inequality system, we first define the following vector-valued error function (instead of the scalar-valued energy function usually used in gradient-based dynamic approaches):

$$\mathbf{e}(t) = [e_1(t), e_2(t), \cdots, e_n(t)]^{\mathrm{T}},$$

and $e_i(t) = (\max\{0, y_i(t)\})^2/2$, with $i = 1, 2, \cdots, n$.

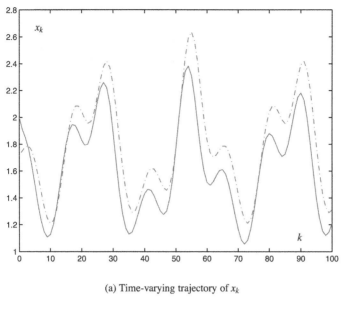

(a) Time-varying trajectory of x_k

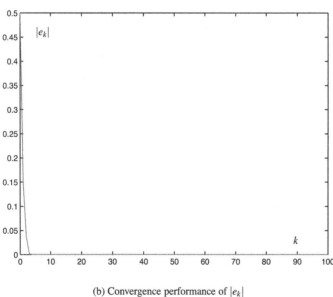

(b) Convergence performance of $|e_k|$

FIGURE 15.8 Solving (15.18) via DTZDU model (15.13) with $\tau = 0.1$ and $h = 1$.

Secondly, the differential equation for the error function $\mathbf{e}(t)$ has to be constructed, such that $\mathbf{e}(t)$ converges to zero. Specifically, we choose the time derivative $\dot{\mathbf{e}}(t)$ via the following exponent-type ZD design formula (which has been used in most chapters of the book) so that $\mathbf{e}(t)$ exponentially converges to zero:

$$\frac{d\mathbf{e}(t)}{dt} = -\gamma \mathbf{e}(t), \tag{15.21}$$

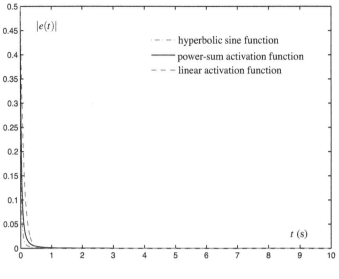

(a) Via S-CTZD models (15.15) and (15.16)

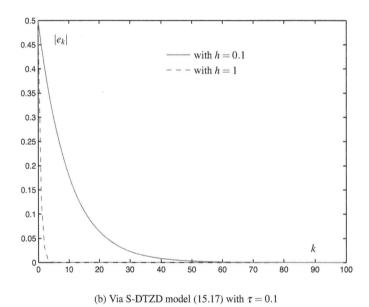

(b) Via S-DTZD model (15.17) with $\tau = 0.1$

FIGURE 15.9 Solving (15.19) via simplified ZD models.

which is equivalently written as

$$
\begin{bmatrix} \dot{e}_1(t) \\ \dot{e}_2(t) \\ \vdots \\ \dot{e}_n(t) \end{bmatrix} = -\frac{\gamma}{2} \begin{bmatrix} (\max\{0, y_1(t)\})^2 \\ (\max\{0, y_2(t)\})^2 \\ \vdots \\ (\max\{0, y_n(t)\})^2 \end{bmatrix},
$$

with constant γ denoting a positive design parameter used to scale the convergence rate of the CTZD model. Expanding the above design formula, we obtain that, $\forall i \in \{1, 2, \cdots, n\}$,

$$\max\{0, y_i(t)\}\dot{y}_i(t) = -\frac{\gamma}{2}(\max\{0, y_i(t)\})^2, \tag{15.22}$$

where, if $\max\{0, y_i(t)\} = 0$, then $y_i(t) \leqslant 0$, which implies that the ith subsystem of time-varying linear inequality system (15.20) holds true and that $\dot{y}_i(t)$ should be zero. So, the above equation (15.22) is equivalent to

$$\dot{y}_i(t) = -\frac{\gamma}{2}\max\{0, y_i(t)\}, \quad \forall i \in \{1, 2, \cdots, n\}.$$

So, we have

$$\begin{bmatrix} \dot{y}_1(t) \\ \dot{y}_2(t) \\ \vdots \\ \dot{y}_n(t) \end{bmatrix} = -\frac{\gamma}{2}\begin{bmatrix} \max\{0, y_1(t)\} \\ \max\{0, y_2(t)\} \\ \vdots \\ \max\{0, y_n(t)\} \end{bmatrix} = -\frac{\gamma}{2}\max\{0, \mathbf{y}(t)\}.$$

In view of $y_i(t) = [A(t)\mathbf{x}(t) - \mathbf{b}(t)]_i$, we know

$$\dot{y}_i(t) = [\dot{A}(t)\mathbf{x}(t) + A(t)\dot{\mathbf{x}}(t) - \dot{\mathbf{b}}(t)]_i, \quad \forall i = 1, 2, \cdots, n,$$

which leads to the following dynamic equation of a CTZD model for solving time-varying linear inequality system (15.20):

$$A(t)\dot{\mathbf{x}}(t) = -\dot{A}(t)\mathbf{x}(t) + \dot{\mathbf{b}}(t) - \frac{\gamma}{2}\max\{0, A(t)\mathbf{x}(t) - \mathbf{b}(t)\}, \tag{15.23}$$

where $\mathbf{x}(t) \in \mathbb{R}^n$, starting from an initial state $\mathbf{x}(0) \in \mathbb{R}^n$, denotes the state vector. Generally speaking, similar to traditional neural-dynamic approaches, design parameter γ, being the reciprocal of a capacitance parameter in the hardware implementation, should be set as large as the hardware would permit, or selected appropriately for simulative or experimental purposes. In addition, to construct the analog hardware of CTZD model (15.23), we can use and connect analog adders, multipliers, integrators, and limiters, which are usually based on operational amplifiers [8, 54, 81, 94, 132]. The circuit schematic of CTZD model (15.23) is thus depicted in Figure 15.10, and it can be seen that CTZD model (15.23) per iteration contains $4n$ addition operations, $3n^2 + n$ multiplication operations, n comparison operations [as for function $\max(\cdot)$], and n integrator operations. Therefore, the computational complexity of CTZD model (15.23) is of $O(n^2)$ operations.

While the above proposes a CTZD model for solving time-varying linear inequality system (15.20), theoretical analysis is presented for it via the following theorem.

Theorem 23 *For CTZD model (15.23), with a smoothly time-varying nonsingular coefficient matrix $A(t)$ and a smoothly time-varying coefficient vector $\mathbf{b}(t)$ in (15.20),*

1) *if initial state $\mathbf{x}(0) \in \mathbb{R}^n$ is outside the initial solution set $\mathbb{S}(0)$ of (15.20), the two-norm based error function $\|e(t)\|_2$ is globally exponentially convergent to zero with convergence rate $\gamma > 0$; and,*

2) *if initial state $\mathbf{x}(0) \in \mathbb{R}^n$ is inside the initial solution set $\mathbb{S}(0)$ of (15.20), the two-norm based error function $\|e(t)\|_2$ is always equal to zero.*

That is, CTZD model (15.23) generates an exact time-varying solution of (15.20).

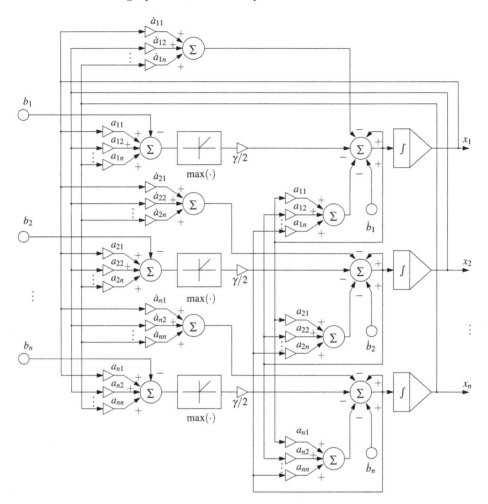

FIGURE 15.10 Circuit schematic [94] realizing CTZD model (15.23) via derivative feedback, where CTZD model (15.23) is rewritten as $\dot{\mathbf{x}}(t) = (-A(t) + I)\dot{\mathbf{x}}(t) - \dot{A}(t)\mathbf{x}(t) + \dot{\mathbf{b}}(t) - \gamma \max\{0, A(t)\mathbf{x}(t) - \mathbf{b}(t)\}/2$ for modeling purposes.

Proof Consider CTZD model (15.23), which is derived from ZD design formula (15.21). From design formula (15.21), we obtain

$$\mathbf{e}(t) = \mathbf{e}(0)\exp(-\gamma t),$$

which means that, as $t \to \infty$, $\mathbf{e}(t) \to 0$ globally and exponentially. Specifically, there are two situations as follows.

1) If any randomly generated initial state $\mathbf{x}(0)$ is outside the initial solution set $\mathbb{S}(0)$ of (15.20), due to $y_i(0) = [A(0)\mathbf{x}(0) - \mathbf{b}(0)]_i > 0$, and $e_i(0) = \max\{0, y_i(0)\}^2/2 = y_i^2(0)/2$, then $e_i(0) > 0$, $\exists i \in \{1, 2, \cdots, n\}$, and $\|\mathbf{e}(0)\|_2 > 0$. Therefore, the norm-based error function $\|\mathbf{e}(t)\|_2 = \|\mathbf{e}(0)\|_2\exp(-\gamma t)$ is globally exponentially convergent to zero, with convergence rate $\gamma > 0$. In addition, it is evident that $\|\mathbf{e}(t)\|_2$ is a convex function with respect to x for any fixed time instant t, and $\|\mathbf{e}(t)\|_2 = 0$ is equivalent to having solved time-varying linear inequality system (15.20). This implies that state vector $\mathbf{x}(t) \in \mathbb{R}^n$ is convergent to time-varying solution set $\mathbb{S}(t)$ in a global and exponential manner.

2) If any randomly generated initial state $\mathbf{x}(0)$ is inside the initial solution set $\mathbb{S}(0)$ of (15.20),

evidently, we obtain $e_i(0) = 0$, $\forall i \in \{1, 2, \cdots, n\}$, and $\mathbf{e}(0) = 0$. Therefore, no matter how time t evolves, the error $\|\mathbf{e}(t)\|_2$ is always equal to zero, which guarantees residual error $\mathbf{y}(t) \leqslant 0$. This implies that state vector $\mathbf{x}(t) \in \mathbb{R}^n$ will always stay inside the time-varying solution set $\mathbb{S}(t)$ of (15.20) in this situation.

The proof is thus complete.

15.5.2 GD Model and Comparison

For comparative purposes, the GD model is also developed and exploited for the real-time solution of the above-presented time-varying linear inequality system. However, similar to most gradient-based numerical algorithms and neural-dynamic schemes aforementioned, the GD model and its related methods are designed intrinsically for problems with constant coefficient matrices or vectors. The GD design procedure is now presented below for solving (15.20).

That is, following the conventional GD design method, we first define a scalar-valued norm-based energy function:

$$\mathscr{E}(t) = \frac{1}{2}\sum_{i=1}^{n}\max\{0, [A(t)\mathbf{x}(t) - \mathbf{b}(t)]_i\}^2 = \frac{1}{2}\|\max\{0, A(t)\mathbf{x}(t) - \mathbf{b}(t)\}\|_2^2.$$

Secondly, we can get its gradient information:

$$\frac{\partial\mathscr{E}(t)}{\partial\mathbf{x}} = \frac{1}{2}\frac{\partial\|\max\{0, A(t)\mathbf{x}(t) - \mathbf{b}(t)\}\|_2^2}{\partial\mathbf{x}} = A^{\mathrm{T}}(t)\max\{0, A(t)\mathbf{x}(t) - \mathbf{b}(t)\}.$$

Finally, a typical continuous-time learning rule based on the negative-gradient information leads to the following dynamic equation of a GD model:

$$\dot{\mathbf{x}}(t) = -\gamma\frac{\partial\mathscr{E}(t)}{\partial\mathbf{x}} = -\gamma A^{\mathrm{T}}(t)\max\{0, A(t)\mathbf{x}(t) - \mathbf{b}(t)\}, \tag{15.24}$$

where $\mathbf{x}(t)$, starting from an initial state $\mathbf{x}(0) \in \mathbb{R}^n$, is the state vector. Design parameter $\gamma > 0$ is defined the same as that in CTZD model (15.23).

It is worth comparing CTZD model (15.23) and GD model (15.24), both of which are exploited for time-varying linear inequality system (15.20) solving. Their differences lie in the following facts (with more detailed comparison of this type presented in Section 9.4).

1) The design of CTZD model (15.23) is based on the elimination of every element of vector-valued error function $\mathbf{e}(t) = [e_1(t), e_2(t), \cdots, e_n(t)]^{\mathrm{T}}$. In contrast, the design of GD model (15.24) is based on the elimination of scalar-valued norm-based energy function $\mathscr{E}(t) = \|\max\{0, A(t)\mathbf{x}(t) - \mathbf{b}(t)\}\|_2^2/2$ as a whole.

2) The design of CTZD model (15.23) is based on a new method for the time-varying problem solving. Thus, the CTZD state globally exponentially converges to time-varying solution set $\mathbb{S}(t)$ of (15.20) in an error-free manner. In contrast, the design of GD model (15.24) is based on the negative-gradient method intrinsically for handling the linear inequality problem with constant coefficients only.

3) CTZD model (15.23) systematically exploits the time-derivative information of $A(t)$ and $\mathbf{b}(t)$ [i.e., $\dot{A}(t)$ and $\dot{\mathbf{b}}(t)$] during its real-time solving process. This is one reason why CTZD model (15.23) exactly converges to time-varying solution set $\mathbb{S}(t)$ of (15.20). In contrast, GD model (15.24) has not exploited such important time-derivative information, and thus may not be effective enough in solving the problem of time-varying linear inequalities. These points can also be seen from the ensuing simulation section, Section 15.6.

4) CTZD model (15.23) and its design method, which make good use of the time-derivative information, belong to a predictive approach and thus they are more effective. In contrast, belonging to the conventional tracking approach, GD model (15.24) and its method act by adapting to the change of matrix $A(t)$ and vector $\mathbf{b}(t)$ in a posterior passive manner, and thus theoretically may not stay inside the time-varying solution set $\mathbb{S}(t)$ of (15.20) exactly.

15.5.3 DTZD Models

In this subsection, for the purpose of digital-hardware or digital-computer implementation, the discrete-time model of the presented CTZD model (15.23) are generalized and developed for the real-time solution of time-varying linear inequality system (15.20). In order to discretize CTZD model (15.23), we refer to the following Euler forward-difference rule again: $\dot{\mathbf{x}}(t = k\tau) \approx (\mathbf{x}((k+1)\tau) - \mathbf{x}(k\tau))/\tau$, where $\tau > 0$ denotes the sampling gap (i.e., $\tau = t_{k+1} - t_k$) and k denotes the update index (with $k = 0, 1, 2, \cdots$). In general, we denote $\mathbf{x}_k = \mathbf{x}(t = k\tau)$ for presentation convenience. In addition, $A(t)$ and $\dot{A}(t)$ (which is assumed to be known now) are discretized by the same standard sampling method. For convenience and also for consistency with $\mathbf{x}_k$, we use $A_k = A(t = k\tau)$, $\dot{A}_k = \dot{A}(t = k\tau)$, $\mathbf{b}_k = \mathbf{b}(t = k\tau)$, $\dot{\mathbf{b}}_k = \dot{\mathbf{b}}(t = k\tau)$, and $V_k = A^{-1}(t = k\tau)$. Thus, CTZD model (15.23) is discretized as

$$A_k(\mathbf{x}_{k+1} - \mathbf{x}_k)/\tau = -\dot{A}_k\mathbf{x}_k + \dot{\mathbf{b}}_k - \frac{\gamma}{2}\max\{0, A_k\mathbf{x}_k - \mathbf{b}_k\},$$

which is further written as

$$A_k(\mathbf{x}_{k+1} - \mathbf{x}_k) = -\tau\dot{A}_k\mathbf{x}_k + \tau\dot{\mathbf{b}}_k - \frac{h}{2}\max\{0, A_k\mathbf{x}_k - \mathbf{b}_k\}, \tag{15.25}$$

with $h = \tau\gamma > 0$ denoting the step size.

As matrix $A(t)$ is nonsingular at any time instant t, A_k is invertible at each update (or to say, sample) with $k = 0, 1, 2, \cdots$. Therefore, the above DTZD model (15.25) can be rewritten in the following form:

$$\mathbf{x}_{k+1} = \mathbf{x}_k - \tau V_k\dot{A}_k\mathbf{x}_k + \tau V_k\dot{\mathbf{b}}_k - \frac{h}{2}V_k\max\{0, A_k\mathbf{x}_k - \mathbf{b}_k\}. \tag{15.26}$$

For presentation convenience, in this chapter, the above discrete-time model (15.26) is called the DTZD model with $\dot{A}(t)$ and $\dot{\mathbf{b}}(t)$ known, i.e., the DTZDK model.

As we also know, it may be difficult to obtain the numerical value or analytical form of $\dot{A}(t)$ and $\dot{\mathbf{b}}(t)$ directly in certain real-world applications. Thus, it is worth investigating the DTZD model with $\dot{A}(t)$ and $\dot{\mathbf{b}}(t)$ unknown. In this situation, $\dot{A}(t)$ and $\dot{\mathbf{b}}(t)$ can be estimated from $A(t)$ and $\mathbf{b}(t)$, respectively, by employing again Euler backward-difference rule: $\dot{A}_k \approx (A_k - A_{k-1})/\tau$. Then, from DTZDK model (15.26), we derive the following DTZD model with $\dot{A}(t)$ and $\dot{\mathbf{b}}(t)$ unknown (i.e., the DTZDU model) for solving time-varying linear inequality system (15.20):

$$\mathbf{x}_{k+1} = \mathbf{x}_k - V_k(A_k - A_{k-1})\mathbf{x}_k + V_k(\mathbf{b}_k - \mathbf{b}_{k-1}) - \frac{h}{2}V_k\max\{0, A_k\mathbf{x}_k - \mathbf{b}_k\}, \tag{15.27}$$

where update index $k = 0, 1, 2, \cdots$. Similar to the discussions in Subsections 2.2.2, 12.3.2, and 13.3.2, note that, from the above Euler backward-difference rule, we cannot obtain $\dot{A}(0)$, since t starts from 0 s and A_{-1} is undefined. In this situation, $A_{-1} = A_0$ [i.e., $\dot{A}(0) = 0$] can be set to start DTZDU model (15.27).

For further investigation, the simplified DTZD model without using the time derivative $\dot{A}(t)$ and $\dot{\mathbf{b}}(t)$ (i.e., the S-DTZD model) can be presented below for solving time-varying linear inequality system (15.20):

$$\mathbf{x}_{k+1} = \mathbf{x}_k - \frac{h}{2}V_k\max\{0, A_k\mathbf{x}_k - \mathbf{b}_k\}. \tag{15.28}$$

15.6 Illustrative Examples

In Section 15.5, the CTZD, DTZD, S-DTZD, and GD models have been presented for solving the time-varying linear inequality system. In this section, some illustrative computer-simulation and numerical-experiment examples are provided for substantiating the efficacy and superiority of the aforementioned ZD models.

15.6.1 About CTZD and GD Models

Now let us consider time-varying linear inequality system (15.20) with coefficients being as follows:

$$A(t) = \left[\begin{array}{cc} -\sin(10t) & \cos(10t) \\ \cos(10t) & \sin(10t) \end{array} \right], \quad \mathbf{b}(t) = \left[\begin{array}{c} \sin(10t) \\ \cos(10t) \end{array} \right].$$

For illustration and comparison, both CTZD model (15.23) and GD model (15.24) are exploited for solving the time-varying linear inequality system. Considering that different initial states of the CTZD model may result in different convergence performance, we investigate the following two situations.

Situation 1 Initial state outside initial solution Set $\mathbb{S}(0)$

If initial state $\mathbf{x}(0)$ is outside the initial solution set $\mathbb{S}(0)$ of (15.20), by applying CTZD model (15.23) to solving the time-varying linear inequality system (15.20) with $\gamma = 1$, the dynamic performance of the state vector $\mathbf{x}(t)$ can be observed from Figure 15.11(a). Using GD model (15.24) under the same conditions, as seen from Figure 15.11(b), the change rate of the GD state vector $\mathbf{x}(t)$ is small, which shows that GD model (15.24) is designed intrinsically for solving constant problems. To help readers understand better, Figure 15.12 shows the transient behavior of the residual error $\mathbf{y}(t) = A(t)\mathbf{x}(t) - \mathbf{b}(t)$ synthesized by CTZD model (15.23) and GD model (15.24) for the real-time solution of time-varying linear inequality system (15.20). It is seen from Figure 15.12(a) that residual error $\mathbf{y}(t) = A(t)\mathbf{x}(t) - \mathbf{b}(t)$ synthesized by CTZD model (15.23) exponentially decreases to zero within 6 s. In contrast, Figure 15.12(b) shows that residual error $\mathbf{y}(t) = A(t)\mathbf{x}(t) - \mathbf{b}(t)$ synthesized by GD model (15.24) is oscillating around zero all the time, which means that the GD solution cannot satisfy the requirement $A(t)\mathbf{x}(t) - \mathbf{b}(t) \leqslant 0$. In summary, CTZD model (15.23) is more effective on the real-time solution of time-varying linear inequalities [as compared with GD model (15.24)].

To further monitor and show the solution process of CTZD model (15.23) and GD model (15.24), transient behaviors of the error and energy functions are preferable measure choices. It is seen from Figure 15.13(a) that, by applying CTZD model (15.23) to solving time-varying linear inequality system (15.20), its error function $\|\mathbf{e}(t)\|_2$ exponentially converges to zero within around 6 s. In contrast, it follows from Figure 15.13(b) that, using GD model (15.24) for the real-time solution of (15.20), the energy function $\mathscr{E}(t)$ is always rather large, which implies that the GD solution cannot converge well to the time-varying solution set of (15.20). Once more, in the authors' opinion, the reason why ZD model (15.23) has a better performance on solving the time-varying linear inequality system lie in the time-derivative information of time-varying coefficient in (15.20) being fully utilized by CTZD model (15.23). Note that this point is further explained in the ensuing remark (i.e., Remark 5) at the end of this subsection. Thus, we can confirm that CTZD model (15.23) is more effective on the time-varying linear inequality system solving.

In addition, it is worth pointing out that, as shown in Figure 15.14(a), the convergence time of CTZD model (15.23) can be expedited from 6 s to 0.6 s, when design parameter γ increases from 1 to 10. Furthermore, as design parameter γ is set to be 100, the convergence time is almost close to zero, which is shown in Figure 15.14(b). For better understanding the exponential convergence

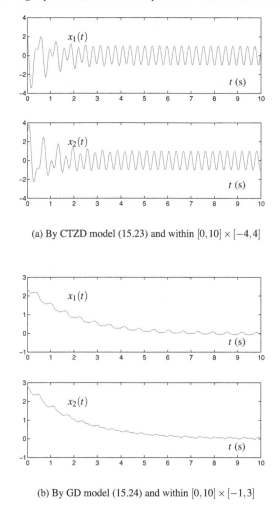

(a) By CTZD model (15.23) and within $[0, 10] \times [-4, 4]$

(b) By GD model (15.24) and within $[0, 10] \times [-1, 3]$

FIGURE 15.11 Neural states $\mathbf{x}(t)$ starting from $\mathbf{x}(0) \notin \mathbb{S}(0)$.

property of CTZD model (15.23), some more numerical experiments on the average convergence time of CTZD model (15.23) with respect to different γ and prescribed precision are conducted, where totally 120 initial states $\mathbf{x}(0)$ outside the initial solution set $\mathbb{S}(0)$ are generated randomly. The results are shown in Table 15.1. As seen from the table, CTZD model (15.23) has better precision as γ increases. So, design parameter γ plays an important role in CTZD model (15.23) and should be selected appropriately large to satisfy the convergence rate needed in this situation.

Situation 2 Initial state inside initial solution set $\mathbb{S}(0)$

As analyzed in Section 15.5, if initial state $\mathbf{x}(0)$ is inside the initial solution set $\mathbb{S}(0)$, for CTZD model (15.23), we know that its error function remains zero without an appreciable convergence process. But, for GD model (15.24), we only know that the initial state makes the time-varying linear inequality system (15.20) satisfied; so, the energy function equals zero in the beginning (but one may not know how its trajectory changes afterwards). To illustrate and verify the above situations, in Figure 15.15, we show the error and energy functions synthesized by CTZD model (15.23) and GD

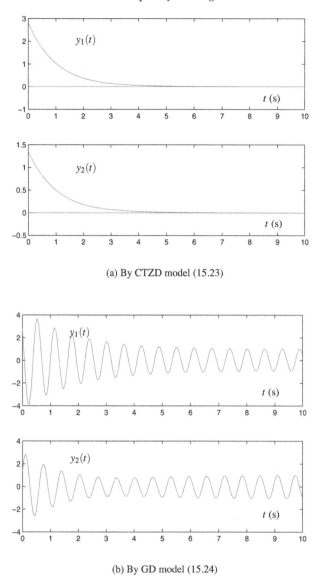

(a) By CTZD model (15.23)

(b) By GD model (15.24)

FIGURE 15.12 Residual errors $\mathbf{y}(t) = A(t)\mathbf{x}(t) - \mathbf{b}(t)$ starting with $\mathbf{x}(0) \notin \mathbb{S}(0)$.

model (15.24) for solving time-varying linear inequality system (15.20), respectively. As seen from Figure 15.15(a), by applying CTZD model (15.23) to time-varying linear inequality system (15.20), the error function $\|\mathbf{e}(t)\|_2$ is always equal to zero. In contrast, it follows from Figure 15.15(b) that, using GD model (15.24) for the real-time solution of (15.20) with design parameter $\gamma = 1$, energy function $\mathscr{E}(t)$ equals zero at first, and then it starts to oscillate, which means that the solution of GD model (15.24) cannot stay inside the time-varying solution set of (15.20), even if initial state $\mathbf{x}(0)$ is inside it. These computer-simulation results have further substantiated the presented theoretical analysis.

Furthermore, with the above two situations included, the planar trajectories of state vector $\mathbf{x}(t)$ of CTZD model (15.23) at different time instants, starting from 150 randomly generated initial states

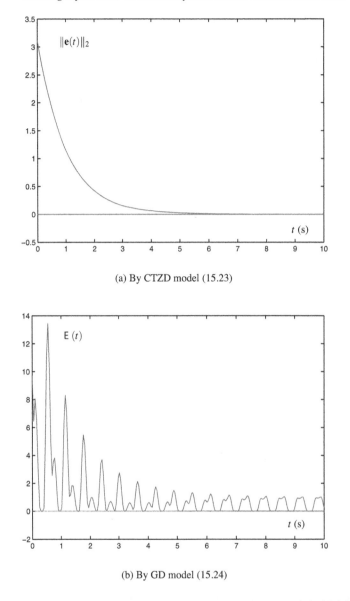

(a) By CTZD model (15.23)

(b) By GD model (15.24)

FIGURE 15.13 Error and energy functions starting with $\mathbf{x}(0) \notin \mathbb{S}(0)$.

in $[-10, 10]^2$, are shown in Figure 15.16. Although solution set $\mathbb{S}(t)$ is different (i.e., time-varying, dynamic) at different time instants, evidently the trajectories all converge to or stay inside $\mathbb{S}(t)$.

In summary, from the above computer-simulation and numerical-experiment results, we can conclude that the solution of GD model (15.24) cannot converge to the time-varying solution set of (15.20) from any initial state $\mathbf{x}(0)$. In contrast, the solution of CTZD model (15.23) has a much better convergence performance. In details, if the initial state is outside the time-varying solution set $\mathbb{S}(t)$, the CTZD solution globally exponentially converges to $\mathbb{S}(t)$; otherwise, the CTZD solution always stays inside $\mathbb{S}(t)$, which is our desired result. Thus, we can conclude that CTZD model (15.23) is much more effective and efficient for the real-time solution of the time-varying linear inequality system, as compared with GD model (15.24).

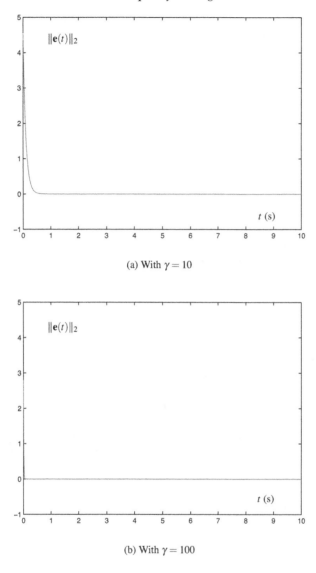

(a) With $\gamma = 10$

(b) With $\gamma = 100$

FIGURE 15.14 Error functions $\|\mathbf{e}(t)\|_2$ synthesized by CTZD model (15.23) with different γ.

Remark 5 Before ending this subsection, it is worth studying further (and also explain) the effect of using time-derivative information in CTZD model (15.23). When constant matrix $A \in \mathbb{R}^{n \times n}$ and constant vector $\mathbf{b} \in \mathbb{R}^n$ are considered, due to time-derivatives $\dot{A}(t) = 0$ and $\dot{\mathbf{b}}(t) = 0$ in this case, CTZD model (15.23) reduces to $A\dot{\mathbf{x}}(t) = -\gamma \max\{0, A\mathbf{x}(t) - \mathbf{b}\}/2$. The latter model can solve the system of constant linear inequalities and can be seen as a special case of CTZD model (15.23). When this model is applied to the system of time-varying linear inequalities, it can be written as below (and termed the simplified CTZD model, i.e., the S-CTZD model):

$$A(t)\dot{\mathbf{x}}(t) = -\frac{\gamma}{2}\max\{0, A(t)\mathbf{x}(t) - \mathbf{b}(t)\}. \tag{15.29}$$

Evidently, the difference between CTZD model (15.23) and S-CTZD model (15.29) lies in the time-derivative information of time-varying coefficients in (15.20) being utilized or not. As illustrated above, CTZD model (15.23) is effective on solving the time-varying linear inequality system. In

TABLE 15.1 CTZD Model (15.23) Convergence
Comparison for Solving Time-Varying Linear Inequalities with
Randomly Generated 10 Initial States

	Average convergence time in seconds		
Precision	$\gamma = 10^4$	$\gamma = 10^6$	$\gamma = 10^8$
$\|\mathbf{e}(t)\|_2 < 10^{-2}$	5.92×10^{-4}	5.96×10^{-6}	6.11×10^{-8}
$\|\mathbf{e}(t)\|_2 < 10^{-4}$	1.14×10^{-3}	1.06×10^{-5}	1.06×10^{-7}
$\|\mathbf{e}(t)\|_2 < 10^{-6}$	1.87×10^{-3}	1.67×10^{-5}	1.72×10^{-7}
$\|\mathbf{e}(t)\|_2 < 10^{-8}$	2.32×10^{-3}	1.98×10^{-5}	2.03×10^{-7}

order to test the reason why CTZD model (15.23) has a better performance on solving the time-varying linear inequality system, S-CTZD model (15.29) is exploited for solving the time-varying linear inequality system under the same conditions. If initial state $\mathbf{x}(0)$ is outside the initial solution set $\mathbb{S}(0)$ of (15.20), as shown in Figure 15.17(a), error function $\|\mathbf{e}(t)\|_2$ is quite large instead of converging to zero, which means that S-CTZD model (15.29) (without using the time-derivative information) is not effective on solving the time-varying linear inequality system. On the other hand, if initial state $\mathbf{x}(0)$ is inside the initial solution set $\mathbb{S}(0)$, as seen from Figure 15.17(b), error function $\|\mathbf{e}(t)\|_2$ starts from zero and then begins to oscillate, which implies that the solution of S-CTZD model (15.29) cannot stay inside the time-varying solution set of (15.20), even if initial state $\mathbf{x}(0)$ is inside it. The simulation results are similar to those of GD model (15.24). Therefore, we can conclude that the time-derivative information of time-varying coefficients in (15.20) plays an important role in CTZD model (15.23). That is why CTZD model (15.23) has a better performance on solving the time-varying linear inequality system. Theoretically speaking, the time-derivative information incorporates the variation trend of time-varying coefficients and plays a predictive role in the process of solving the time-varying linear inequality system, and thus CTZD model (15.23) is more effective.

15.6.2 About DTZD Models

In this subsection, numerical studies are performed to verify the efficacy of the DTZD models [i.e., DTZDK model (15.26), DTZDU model (15.27), and S-DTZD model (15.28)] for online solution of time-varying linear inequality system (15.20). The corresponding residual error is defined for the DTZD models as

$$\|\mathbf{e}_k\|_2 = \|\max\{0, A_k\mathbf{x}_k - \mathbf{b}_k\}\|_2.$$

Besides, the minimum number of updates (MNU) required for converging to steady state, and the maximal steady-state residual error (MSSRE) (discussed previously in Chapters 2 through 4, 12, and 13), are captured to measure the efficacy of the DTZD models for online solution of time-varying linear inequality system (15.20), which are based on the following practical criterion:

$$\| \|\mathbf{e}_{k+1}\|_2 - \|\mathbf{e}_k\|_2 | / \max\{\|\mathbf{e}_{k+1}\|_2, \text{eps}\} < 10^{-5}, \tag{15.30}$$

where the "eps" denotes a very small positive constant, e.g., $2^{-52} \approx 2.2204 \times 10^{-16}$ (which is used in MATLAB$^{\circledR}$ and termed "eps" to denote the spacing of floating point numbers, and which, more specifically, is the distance from 1.0 to the next larger double-precision number). As a result, by introducing such a small constant, we can elegantly avoid the division-by-zero situation. When criterion (15.30) is satisfied for the first time during the ZD solving process, we have MNU $= k + 1$, and MSSRE $= \max\{\|\mathbf{e}_j\|_2\}$ with $j = k+1, k+2, k+3, \cdots, \text{int}(T/\tau)$, with $T > 0$ denoting the actual task duration.

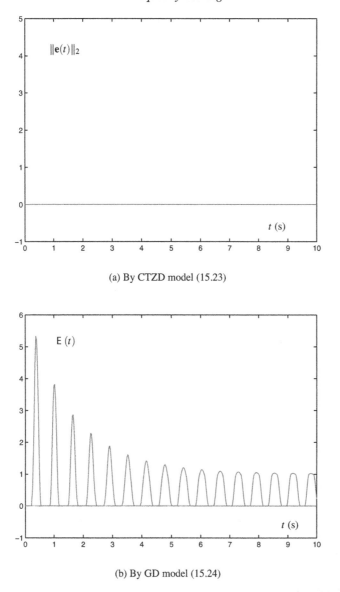

(a) By CTZD model (15.23)

(b) By GD model (15.24)

FIGURE 15.15 Error and energy functions starting with $\mathbf{x}(0) \in \mathbb{S}(0)$.

15.6.2.1 About DTZDK Model

In this subsection, DTZDK model (15.26) is investigated by using different initial state $\mathbf{x}_0 \notin \mathbb{S}(0)$, different step size h, and different sampling gap τ. The coefficients of time-varying linear inequality system (15.20), $A(t)$ and $\mathbf{b}(t)$, are set to be $[\sin(t), \cos(t); -\cos(t), \sin(t)] \in \mathbb{R}^{2 \times 2}$ and $[\sin(t); \cos(t)]$, respectively (note that the MATLAB presentation is used). Additionally, it is worth noting that the situation of using initial state $\mathbf{x}_0 \in \mathbb{S}(0)$ is omitted in this subsection, because it seems trivial (i.e., the residual error of the DTZD model will theoretically always equal zero).

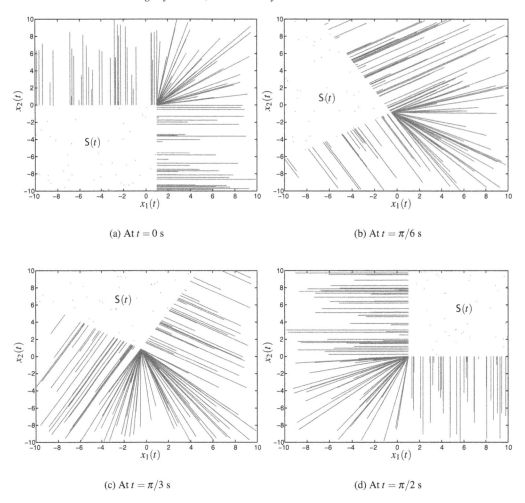

FIGURE 15.16 Planar trajectories of $\mathbf{x}(t) \in [-10, 10]^2$ solved by CTZD model (15.23) starting from 150 randomly generated initial states for handling time-varying linear inequality system (15.20) at different time instants.

With different initial state $\mathbf{x}_0$ In this case, step size h and sampling gap τ are fixed to be 1 and 0.01, respectively. Besides, the initial state of DTZDK model (15.26) is set as $\mathbf{x}_0 = [10; 10]$ or $\mathbf{x}_0 = [5; 5]$. Figure 15.18 shows the corresponding numerical results, which illustrate the efficacy of DTZDK model (15.26) for the online solution of time-varying linear inequality system (15.20).

With different step size h   In this case, the sampling gap is fixed to be $\tau = 0.01$ and the initial state of DTZDK model(15.26) is set as $\mathbf{x}_0 = [10; 10]$. The corresponding numerical results are shown in Table 15.2. As seen from the table, the MSSREs of DTZDK model (15.26) with different values of h are all small enough, which illustrate the effectiveness of DTZDK model (15.26). In addition, as shown in Table 15.2, the MNUs are relatively small (i.e., less than 113), which show the high convergence speed of (15.26). Thus, the good performance of DTZDK model (15.26) (i.e., with relatively small MNUs and MSSREs) is achieved by choosing an appropriate value of step size $h > 0$; especially $h \geqslant 2.1$ is preferred (which is quite different from the criterion of $0 < h < 2$ used in most chapters of the book).

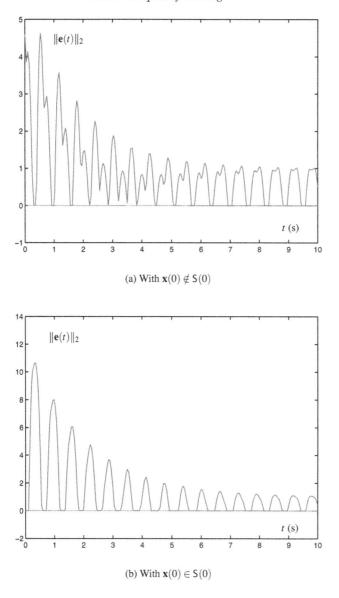

(a) With $\mathbf{x}(0) \notin S(0)$

(b) With $\mathbf{x}(0) \in S(0)$

FIGURE 15.17 Error functions synthesized by S-CTZD model (15.29).

With different sampling gap τ In this case, step size h is fixed (e.g., 1 or 2) and the initial state of DTZDK model (15.26) is set as $\mathbf{x}_0 = [10; 10]$. The corresponding numerical results are shown in Table 15.3. From the table, we see that the relationship between the MSSRE and sampling gap τ follows an $O(\tau^2)$ manner. That is to say, the MSSRE decreases by 100 times when sampling gap τ decreases by 10 times, which implies that τ can be chosen practically small, such that the solution precision required in practice is satisfied readily. Thus, we can draw an important conclusion that the MSSRE of DTZDK model (15.26) is generally of order $O(\tau^2)$. Besides, the above numerical

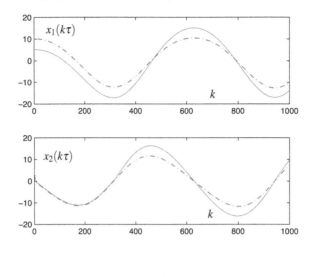

(a) Neural states

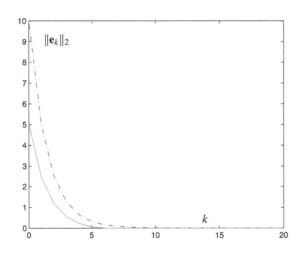

(b) Residual errors

FIGURE 15.18 Numerical results of DTZDK model (15.26) with $h = 1$ and $\tau = 0.01$, starting from $\mathbf{x}_0 = [10; 10]$ (corresponding to dashed-dotted curves) or $\mathbf{x}_0 = [5; 5]$ (corresponding to solid curves) for online solution of time-varying linear inequality system (15.20).

results illustrate well the efficacy of DTZDK model (15.26) for online solution of time-varying linear inequality system (15.20).

15.6.2.2   About DTZDU Model

In this subsection, DTZDU model (15.27) with different initial state $\mathbf{x}_0$, different step size h, and different sampling gap τ is applied to online solution of time-varying linear inequality system (15.20).

TABLE 15.2 Numerical Results Synthesized by DTZDK Model (15.26) with Different h

h	τ	MNU	MSSRE
0.2	0.01	112	2.177×10^{-2}
0.4	0.01	83	1.111×10^{-2}
0.6	0.01	81	7.438×10^{-3}
0.8	0.01	81	5.585×10^{-3}
1.0	0.01	26	4.471×10^{-3}
1.5	0.01	30	2.982×10^{-3}
2	0.01	29	2.236×10^{-3}
2.1	0.01	3	0
2.5	0.01	3	0
3	0.01	3	0
3.5	0.01	3	0
4	0.01	3	0
5	0.01	3	0
10	0.01	3	0
100	0.01	3	0
1000	0.01	3	0

TABLE 15.3 Numerical Results Synthesized by DTZDK Model (15.26) with Different τ

h	τ	MNU	MSSRE
1	0.1	38	5.909×10^{-1}
1	0.01	26	4.471×10^{-3}
1	0.001	79	4.007×10^{-5}
1	0.0001	197	3.908×10^{-7}
2	0.1	37	3.107×10^{-1}
2	0.01	29	2.236×10^{-3}
2	0.001	111	2.172×10^{-5}
2	0.0001	126	1.995×10^{-7}

The coefficients of time-varying linear inequality system (15.20) are as below:

$$A(t) = \begin{bmatrix} (\sin(t)\ln(t+1))/(t+1) & \cos(t)\exp(-t/10) \\ \cos(t)\exp(-t/20) & \sin(t)\exp(-t/15) \end{bmatrix}, \ \mathbf{b}(t) = \begin{bmatrix} t\sin(10t) \\ \cos(10t) \end{bmatrix}.$$

With different initial state $\mathbf{x}_0$ In this case, step size h and sampling gap τ are fixed to be the same as before, i.e., 1 and 0.01, respectively. In addition, initial state $\mathbf{x}_0$ of DTZDU model (15.27) is still set as $\mathbf{x}_0 = [10; 10]$ or $\mathbf{x}_0 = [5; 5]$. The corresponding numerical results are shown in Figure 15.19. As seen from the figure, the MNUs of DTZDU model (15.27) are very small, which illustrate the efficacy of DTZDU model (15.27).

With different step size h In this case, the sampling gap is fixed to be $\tau = 0.01$ and the initial state of DTZDU model(15.27) is set as $\mathbf{x}_0 = [10; 10]$. The corresponding numerical results are shown in Table 15.4, which illustrate the efficacy of DTZDU model (15.27) for online solution of time-varying linear inequality system (15.20). Similar to the situation of the DTZDK model, the

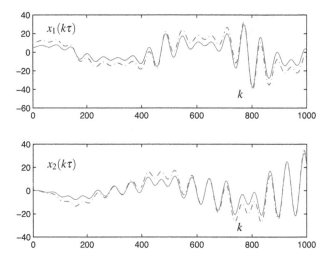

(a) Neural states

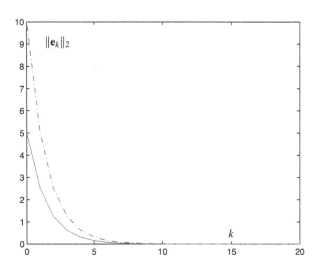

(b) Residual errors

FIGURE 15.19 Numerical results of DTZDU model (15.27) with $h = 1$ and $\tau = 0.01$, starting from $\mathbf{x}_0 = [10; 10]$ (corresponding to dashed-dotted curves) or $\mathbf{x}_0 = [5; 5]$ (corresponding to solid curves) for online solution of time-varying linear inequality system (15.20).

excellent performance of DTZDU model (15.27) (i.e., with relatively small MNUs and MSSREs) is achieved by choosing an appropriate value of step size h (e.g., $h > 2$), as shown in Table 15.4. Note that, as compared with Table 15.2, the best precision of the DTZD solution (i.e., with MNU= 3, MSSRE= 0) is also achieved in this case (i.e., with $h > 2$ and $\tau = 0.01$).

TABLE 15.4 Numerical Results
Synthesized by DTZDU Model (15.27)
with Different h

h	τ	MNU	MSSRE
0.5	0.01	44	6.401×10^{-2}
1	0.01	42	4.129×10^{-2}
1.5	0.01	42	3.276×10^{-2}
2	0.01	3	3.076×10^{-2}
2.5	0.01	3	0
3	0.01	3	0
3.5	0.01	3	0
5	0.01	3	0
10	0.01	3	0
100	0.01	3	0
1000	0.01	3	0

TABLE 15.5 Numerical Results
Synthesized by DTZDU Model
(15.27) with Different τ

h	τ	MNU	MSSRE
1	0.01	42	4.129×10^{-2}
1	0.001	25	6.420×10^{-4}
1	0.0001	33	6.495×10^{-6}
2	0.01	3	3.076×10^{-2}
2	0.001	3	3.512×10^{-4}
2	0.0001	3	3.584×10^{-6}

With different sampling gap τ In this case, step size h is fixed (e.g., 1 or 2) and the initial state of DTZDU model (15.27) is set as $\mathbf{x}_0 = [10; 10]$. The corresponding numerical results are shown in Table 15.5, which illustrate as well the efficacy of DTZDU model (15.27) for online solution of time-varying linear inequality system (15.20). The $O(\tau^2)$ pattern of the MSSRE is also found.

15.6.2.3 About S-DTZD Model

In this subsection, S-DTZD model (15.28) is investigated by using different initial state $\mathbf{x}_0$, different step size h, and different sampling gap τ. The coefficients of time-varying linear inequality system (15.20) are

$$A(t) = \begin{bmatrix} t & \exp(-t/10) \\ \exp(-t/20) & \exp(-t/15) \end{bmatrix}, \; \mathbf{b}(t) = \begin{bmatrix} \exp(-t/10) \\ t \end{bmatrix}.$$

With different initial state $\mathbf{x}_0$ In this case, step size h and sampling gap τ are fixed to be 1 and 0.01, respectively. In addition, the initial state of S-DTZD model (15.28) is set as $\mathbf{x}_0 = [10; 10]$ or $\mathbf{x}_0 = [5; 5]$. Figure 15.20 shows the corresponding numerical results, illustrating the efficacy of S-DTZD model (15.28) on solving time-varying linear inequality system (15.20) with the above coefficients $A(t)$ and $\mathbf{b}(t)$.

With different step size h In this case, the sampling gap is fixed to be $\tau = 0.01$, and the initial state of S-DTZD model (15.28) is set as $\mathbf{x}_0 = [10; 10]$. The corresponding numerical results are shown in Table 15.6. Compared with DTZDK model (15.26) and DTZDU model (15.27), S-DTZD model

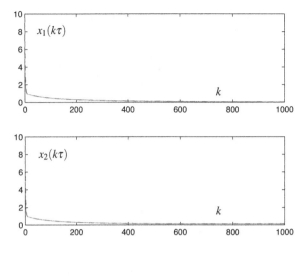

(a) Neural states

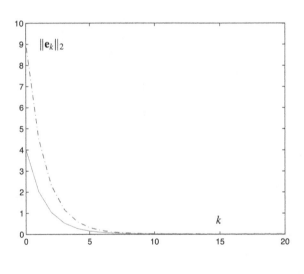

(b) Residual errors

FIGURE 15.20 Numerical results of S-DTZD model (15.28) with $h = 1$ and $\tau = 0.01$, starting from $\mathbf{x}_0 = [10; 10]$ (corresponding to dashed-dotted curves) or $\mathbf{x}_0 = [5; 5]$ (corresponding to solid curves) for online solution of time-varying linear inequality system (15.20).

does not use the time-derivative information of coefficients $A(t)$ and $\mathbf{b}(t)$. Therefore, S-DTZD model (15.28) can be applied to the situation with slow variational rates of time-varying coefficients.

With different sampling gap τ   In this case, step size h is fixed (e.g., 1 or 2 again) and the initial state of S-DTZD model (15.28) is set as $\mathbf{x}_0 = [10; 10]$. The corresponding numerical results are shown in Table 15.7. From the table, we can conclude that the MSSRE of S-DTZD model (15.28) is of order $O(\tau)$, relatively worse than order $O(\tau^2)$ [of which DTZDK model (15.26) and DTZDU

TABLE 15.6 Numerical Results Synthesized by S-DTZD Model (15.28) with Different h

h	τ	MNU	MSSRE
0.5	0.01	39	3.195×10^{-2}
1	0.01	21	1.753×10^{-2}
1.5	0.01	13	1.230×10^{-2}
2	0.01	4	9.910×10^{-3}
2.5	0.01	3	0
3	0.01	3	0
3.5	0.01	3	0
5	0.01	3	0
10	0.01	3	0
100	0.01	3	0
1000	0.01	3	0

TABLE 15.7 Numerical Results Synthesized by S-DTZD Model (15.28) with Different τ

h	τ	MNU	MSSRE
1	0.1	not captured	not captured
1	0.01	21	1.753×10^{-2}
1	0.001	21	1.997×10^{-3}
1	0.0001	20	2.256×10^{-4}
2	0.1	not captured	not captured
2	0.01	4	9.910×10^{-3}
2	0.001	3	9.991×10^{-4}
2	0.0001	3	9.998×10^{-5}

model (15.27) are]. That is to say, the performance of S-DTZD model (15.28) can also be improved to some extent by decreasing the sampling gap τ to achieve the precision requirement in some engineering applications. This, in some sense, shows the efficacy of S-DTZD model (15.28) and the superiority of DTZDK model (15.26) and DTZDU model (15.27).

In summary, the above numerical results (i.e., Figures 15.18 through 15.20 and Tables 15.2–15.7) illustrate well the efficacy of the proposed DTZD models for online solution of time-varying linear inequality system (15.20). Besides, please note that, for all DTZD models proposed in this chapter, the exactly error-free solution of time-varying linear inequality system (15.20) can be achieved with $h > 2$, as shown in Tables 15.2, 15.4, and 15.6. This is a really interesting discovery.

15.7 Summary

In this chapter, we have proposed and investigated the novel CTZD and DTZD models for solving scalar-valued and vector-valued time-varying linear inequalities in real time. For solving the scalar-valued time-varying linear inequality, we have exploited the linear activation function, hyperbolic sine activation function, and power-sum activation function in the CTZD and DTZD models. For solving online vector-valued time-varying linear inequalities, the GD model has also been exploited for comparative purposes. Through illustrative examples with verification and comparison, we have substantiated the efficacy and superiority of the CTZD and DTZD models for solving scalar-valued and vector-valued time-varying linear inequalities.

Chapter 16

System of Time-Varying Nonlinear Inequalities Solving

Abstract

To further solve the system of time-varying nonlinear inequalities, this chapter proposes two new ZD models. The first model is based on the conventional Zhang *et al.*'s neural-dynamic design method (i.e., the conventional zeroing dynamics method), and is termed a conventional ZD (CZD) model. The other one is based on a novel variant of the conventional Zhang *et al.*'s neural-dynamic design method, and is termed a modified ZD (MZD) model. The theoretical analysis of both CZD and MZD models is presented to show their excellent convergence performance. Compared with the CZD model for solving the system of time-varying nonlinear inequalities, it is discovered that the MZD model incorporates the CZD model as its special case [i.e., the MZD model using linear activation functions (MZDL) reduces to the CZD model exactly]. Besides, the MZD model using power-sum activation functions (MZDP) possesses superior convergence performance to the CZD model. Moreover, computer-simulation results illustrate and substantiate the theoretical analysis and efficacy of both CZD and MZD models for solving the system of time-varying nonlinear inequalities.

16.1 Introduction

Systems of nonlinear inequalities in the form of $\mathbf{f}(x) \leqslant 0$ are widely encountered in science and engineering fields [46, 77, 92, 100]. For example, Singh [77] presented a stability inequality for nonlinear discrete-time systems with slope-restricted nonlinearity. In [92], nonlinear convex programming problems subject to nonlinear inequality constraints were addressed. Li and Liao [46] derived the means quare stability conditions for stochastic delayed recurrent neural networks, and such conditions are all expressed in terms of linear matrix inequalities. Zhang [100] converted a pure robotic issue into a mathematical problem of solving a set of nonlinear equations and inequalities during the task execution. It is worth pointing out that systems of nonlinear inequalities may make great contributions to the development of mathematics and related fields to some extent, because researchers in various fields may use some famous nonlinear inequalities (e.g., Cauchy-Schwarz inequality, Chebyshev inequality, and Markov inequality) as a powerful tool to prove some theorems and solve some practical problems [95].

Therefore, on one hand, considerable efforts have been devoted to the analysis of systems of nonlinear inequalities. For example, Lee *et al.* [43] considered some existence results for vector nonlinear inequalities without any monotonicity assumption. Robinson [69] developed a necessary and sufficient condition for the stability of the solution set of a system of linear inequalities (or to say, linear inequality system) over a closed convex set in a Banach space, when the functions defining the inequalities are subject to small perturbation. On the other hand, many methods, including Newton's approach and its improved algorithms, have been developed for the numerical solution of systems of nonlinear inequalities [11, 14, 19, 33, 58, 70]. However, since the computing time for solving nonlinear inequalities greatly depends on the dimension and the structure of the problem, these numerical algorithms may only work well for small-scale constant problems solving and may encounter serious speed bottleneck due to the serial nature of the digital computer employed.

In recent decades, due to the in-depth research on neural dynamics, some solvers have been proposed, developed, and investigated for solving the problems of nonlinear inequalities [92, 100]. For example, Zhang [100] presented a primal neural network (or to say, primal neural-dynamics) for solving a system of nonlinear equations and inequalities but with a large penalty parameter in order to get a good solution. Xia and Wang [92] presented a recurrent neural network for solving nonlinear convex programming problems subject to nonlinear inequality constraints on condition that the objective function is strictly convex and the constraint function is convex. However, most of the aforementioned neural networks (or neural dynamics) are related to gradient-based methods or designed theoretically for solving constant problems. As we know, the gradient dynamics (GD) is designed based on the gradient-decent method and is widely employed for solving constant problems. However, when such a GD method is applied to real-time solution of time-varying problems, a faster convergence rate is often required when compared with the variational rate of time-varying coefficients in real time. This may impose stringent restrictions on physical implementation or sacrifice the solution precision. Note that time-varying systems start to play a more and more important role in practical applications, which urgently requires effective tools (such as ZNN and ZD, thus emerging as required since 2001 and 2008, respectively) for solving such time-varying problems in real time.

In view of the superiority of the ZD method for various time-varying equality problems solving, two new ZD models are proposed and investigated for handling the system of time-varying nonlinear inequalities (or, to say, time-varying nonlinear inequality system) in real time [95]. The first model is based on the conventional Zhang *et al.*'s neural-dynamic design method (i.e., the conventional zeroing dynamics method), and is termed a conventional CZD model. The other one is based on a novel variant of the conventional Zhang *et al.*'s neural-dynamic design method, and is termed an MZD model. Specifically, in this chapter, the design method of the CZD model is based on a lower-bounded error function, while the MZD model is based on a lower-unbounded error function. Therefore, the error function of the CZD model remains zero (lower-bounded) for any initial state inside the initial solution set of the time-varying nonlinear inequality system; and the error function of the MZD model always equals the initial error (lower-unbounded) for any initial state under the same conditions [95]. It is worth pointing out that both CZD and MZD models have excellent convergence performance for any initial state outside the initial solution set. Simply put, both CZD and MZD models generate exact time-varying solutions. In addition, two types of activation functions (i.e., linear and power-sum activation functions) are investigated in the MZD model.

Compared with the CZD model for solving the time-varying nonlinear inequality system, it is discovered that the MZD model incorporates the CZD model as a special case [i.e., the MZD model using MZDL reduces to the CZD model exactly]. Besides, the MZD model using MZDP possesses superior convergence performance to the CZD model. Computer-simulation results further illustrate and substantiate the theoretical analysis and efficacy of both CZD and MZD models for solving the time-varying nonlinear inequality system.

16.2 Problem Formulation

In this chapter, we are concerned with the following solvable system of time-varying nonlinear inequalities [95]:

$$\mathbf{f}(\mathbf{x}(t),t) \leqslant 0 \in \mathbb{R}^n, \ t \in [0,+\infty), \tag{16.1}$$

where $\mathbf{f}(\mathbf{x}(t),t) = [f_1(\mathbf{x}(t),t), f_2(\mathbf{x}(t),t), \cdots, f_n(\mathbf{x}(t),t)]^T$ with $f_i(\mathbf{x}(t),t) : \mathbb{R}^n \times \mathbb{R}^+ \to \mathbb{R}$ ($i = 1,2,\cdots,n$) denoting a smooth nonlinear function mapping, and $\mathbf{x}(t) = [x_1(t), x_2(t), \cdots, x_n(t)]^T$ with $x_i(t)$ denoting its ith element. Let the time-varying solution set $\mathbb{S}(t) = \{\mathbf{x}(t) | \mathbf{x}(t) \in \mathbb{R}^n$ is a solution of (16.1)$\}$, with $\mathbb{S}(0)$ denoting its initial solution set. The main purpose of this chapter is to propose two new ZD models for solving the time-varying nonlinear inequality system (16.1) [i.e., finding the unknown vector $\mathbf{x}(t) \in \mathbb{S}(t)$ in real time t such that the above time-varying nonlinear inequality system holds true]. In the ensuing sections, both CZD and MZD models, together with the corresponding theoretical analysis, are presented and investigated for solving (16.1).

16.3 CZD Model and Convergence Analysis

In this section, by following the conventional Zhang *et al.*'s neural-dynamic design method, a new CZD model is proposed and investigated for the real-time solution of time-varying nonlinear inequality system (16.1).

16.3.1 Model Design and Formulation

By the conventional Zhang *et al.*'s neural-dynamic design method, the design procedure of CZD is presented as below.

First, a vector-valued lower-bounded error function can be defined as follows (instead of a scalar-valued norm-based energy function associated with GD):

$$\mathbf{e}(t) = [e_1(t), e_2(t), \cdots, e_n(t)]^T,$$

with $e_i(t) = (\max\{0, f_i(\mathbf{x}(t),t)\})^2/2$, $i = 1,2,\cdots,n$. Then, the following ZD design formula is applied [102, 104, 107, 117, 118, 122, 129, 134]:

$$\frac{\mathrm{d}\mathbf{e}(t)}{\mathrm{d}t} = -\gamma_1 \mathbf{e}(t), \tag{16.2}$$

which is equivalently written as

$$
\begin{bmatrix} \dot{e}_1(t) \\ \dot{e}_2(t) \\ \vdots \\ \dot{e}_n(t) \end{bmatrix} = -\frac{\gamma_1}{2} \begin{bmatrix} (\max\{0, f_1(\mathbf{x}(t),t)\})^2 \\ (\max\{0, f_2(\mathbf{x}(t),t)\})^2 \\ \vdots \\ (\max\{0, f_n(\mathbf{x}(t),t)\})^2 \end{bmatrix},
$$

with constant γ_1 denoting a positive design parameter used to scale the convergence rate of the CZD model. Expanding the above equation, we obtain that, $\forall i \in \{1, 2, \cdots, n\}$,

$$
\max\{0, f_i(\mathbf{x}(t),t)\} \dot{f}_i(\mathbf{x}(t),t) = -\frac{\gamma_1}{2} (\max\{0, f_i(\mathbf{x}(t),t)\})^2,
$$

where, if $\max\{0, f_i(\mathbf{x}(t),t)\} = 0$, then $f_i(\mathbf{x}(t),t) \leqslant 0$, which implies that the ith subsystem of time-varying nonlinear inequality system (16.1) holds true. So, the above equation is equivalent to

$$
\dot{f}_i(\mathbf{x}(t),t) = -\frac{\gamma_1}{2} \max\{0, f_i(\mathbf{x}(t),t)\}, \quad \forall i \in \{1, 2, \cdots, n\}.
$$

Thus, we have

$$
\begin{bmatrix} \dot{f}_1(\mathbf{x}(t),t) \\ \dot{f}_2(\mathbf{x}(t),t) \\ \vdots \\ \dot{f}_n(\mathbf{x}(t),t) \end{bmatrix} = -\frac{\gamma_1}{2} \begin{bmatrix} \max\{0, f_1(\mathbf{x}(t),t)\} \\ \max\{0, f_2(\mathbf{x}(t),t)\} \\ \vdots \\ \max\{0, f_n(\mathbf{x}(t),t)\} \end{bmatrix}.
$$

In view of

$$
\frac{\mathrm{d} f_i}{\mathrm{d} t} = \left(\frac{\partial f_i}{\partial x_1} \frac{\mathrm{d} x_1}{\mathrm{d} t} + \frac{\partial f_i}{\partial x_2} \frac{\mathrm{d} x_2}{\mathrm{d} t} + \cdots + \frac{\partial f_i}{\partial x_n} \frac{\mathrm{d} x_n}{\mathrm{d} t} \right) + \frac{\partial f_i}{\partial t},
$$

we can obtain the following dynamic equation of the CZD model for solving time-varying nonlinear inequality system (16.1):

$$
J(\mathbf{x}(t),t) \dot{\mathbf{x}}(t) = -\frac{1}{2} \gamma_1 \max\{0, \mathbf{f}(\mathbf{x}(t),t)\} - \frac{\partial \mathbf{f}}{\partial t}, \tag{16.3}
$$

where

$$
J(\mathbf{x}(t),t) = \begin{bmatrix} \frac{\partial f_1}{\partial x_1} & \frac{\partial f_1}{\partial x_2} & \cdots & \frac{\partial f_1}{\partial x_n} \\ \frac{\partial f_2}{\partial x_1} & \frac{\partial f_2}{\partial x_2} & \cdots & \frac{\partial f_2}{\partial x_n} \\ \vdots & \vdots & \ddots & \vdots \\ \frac{\partial f_n}{\partial x_1} & \frac{\partial f_n}{\partial x_2} & \cdots & \frac{\partial f_n}{\partial x_n} \end{bmatrix}, \quad \frac{\partial \mathbf{f}}{\partial t} = \begin{bmatrix} \frac{\partial f_1}{\partial t} \\ \frac{\partial f_2}{\partial t} \\ \vdots \\ \frac{\partial f_n}{\partial t} \end{bmatrix},
$$

and $\mathbf{x}(t)$, starting with an initial state $\mathbf{x}(0) \in \mathbb{R}^n$, denotes the state vector corresponding to a time-varying solution of (16.1). For simplicity and clarity, we confine the discussion to the situation that Jacobian matrix $J(\mathbf{x}(t),t)$ is a nonsingular matrix for any time instant $t \in [0, \infty)$, although the extension for solving more general situations is possible.

16.3.2   Convergence Analysis

The detailed theoretical analysis of CZD model (16.3) is presented in this subsection via the following theorem.

Theorem 24 *For CZD model (16.3),*

1) *if initial state $\mathbf{x}(0) \in \mathbb{R}^n$ is outside the initial solution set $\mathbb{S}(0)$ of (16.1), norm-based error function $\|\mathbf{e}(t)\|_2$ is globally and exponentially convergent to zero with convergence rate $\gamma_1 > 0$; and,*

2) *if initial state $\mathbf{x}(0) \in \mathbb{R}^n$ is inside the initial solution set $\mathbb{S}(0)$ of (16.1), norm-based error function $\|\mathbf{e}(t)\|_2$ is always equal to zero.*

That is, CZD model (16.3) generates an exact time-varying solution of (16.1), with an exponential convergence performance.

Proof Consider CZD model (16.3), which is derived from ZD design formula (16.2). From (16.2), we can obtain

$$\mathbf{e}(t) = \mathbf{e}(0)\exp(-\gamma_1 t),$$

which means that, as $t \to +\infty$, $\mathbf{e}(t) \to 0$ globally and exponentially. Specifically, there are two situations discussed as follows.

1) If any randomly generated initial state $\mathbf{x}(0) \in \mathbb{R}^n$ is outside the initial solution set $\mathbb{S}(0)$ of (16.1), then there exists at least $i \in \{1, 2, \cdots, n\}$ such that $f_i(\mathbf{x}(0), 0) > 0$ and

$$e_i(0) = \max\{0, f_i(\mathbf{x}(0), 0)\}^2/2 = f_i^2(\mathbf{x}(0), 0)/2 > 0.$$

In addition, according to the definition of $\mathbf{e}(t)$, we know $\|\mathbf{e}(0)\|_2 > 0$. Therefore, norm-based error function $\|\mathbf{e}(t)\|_2 = \|\mathbf{e}(0)\|_2 \exp(-\gamma_1 t)$ is globally and exponentially convergent to zero with convergence rate $\gamma_1 > 0$. Evidently, when $\|\mathbf{e}(t)\|_2 = 0$, time-varying nonlinear inequality system (16.1) is solved. This implies that state vector $\mathbf{x}(t) \in \mathbb{R}^n$ of CZD model (16.3) is correspondingly convergent to time-varying solution set $\mathbb{S}(t)$ of (16.1).

2) If any randomly generated initial state $\mathbf{x}(0) \in \mathbb{R}^n$ is inside the initial solution set $\mathbb{S}(0)$ of (16.1), evidently, we obtain $e_i(0) = 0$, $\forall i \in \{1, 2, \cdots, n\}$. Thus, $\mathbf{e}(0) = 0$. That is, no matter how time t evolves, $\|\mathbf{e}(t)\|_2$ is always equal to zero, which guarantees residual error $\mathbf{f}(\mathbf{x}(t), t)$ less than or equal to zero, i.e., $\mathbf{f}(\mathbf{x}(t), t) \leqslant 0$. This implies that state vector $\mathbf{x}(t) \in \mathbb{R}^n$ of CZD model (16.3) in this situation will always stay inside the time-varying solution set $\mathbb{S}(t)$ of (16.1).

The proof is thus complete.

16.4 MZD Model and Convergence Analysis

CZD model (16.3) is presented in the above section for handling time-varying nonlinear inequality system (16.1), which shows excellent convergence performance. In this section, by generalizing and extending the conventional Zhang *et al.*'s neural-dynamic design method, a novel design method and a MZD model are proposed and investigated, as well as the corresponding convergence analysis.

16.4.1 Model Design and Formulation

To monitor and control the solution process of time-varying nonlinear inequality system (16.1), we can first define a vector-valued lower-unbounded error function alternatively (i.e., not a lower-bounded one used in the above section):

$$\mathbf{e}(t) = [e_1(t), e_2(t), \cdots, e_n(t)]^{\mathrm{T}},$$

with $e_i(t) = f_i(\mathbf{x}(t), t)$ directly, $i = 1, 2, \cdots, n$.

Secondly, a differential equation for new error function $\mathbf{e}(t)$ is elaborately constructed via the following novel design formula such that time-varying nonlinear inequality system (16.1) holds true as $\mathbf{e}(t)$ exponentially converges to zero [which is evidently different from the conventional ZD design formula (16.2)]:

$$\frac{d\mathbf{e}(t)}{dt} = -\gamma_2 \text{JMP}\big(\mathbf{e}(0)\big) \diamond \Phi\big(\mathbf{e}(t)\big), \qquad (16.4)$$

where constant $\gamma_2 > 0$ is a design parameter used to scale the convergence rate of the MZD model, and $\text{JMP}(\cdot) : \mathbb{R}^n \to \mathbb{R}^n$ denotes the array of jump functions [95] with each defined as

$$\text{jmp}(b) = \begin{cases} 1, & \text{if } b > 0, \\ 0, & \text{if } b \leqslant 0. \end{cases}$$

The vector-multiplication operator $\diamond$ is proposed and defined as [95, 106]:

$$\mathbf{u} \diamond \mathbf{v} = \begin{bmatrix} u_1 v_1 \\ u_2 v_2 \\ \vdots \\ u_m v_m \end{bmatrix} \in \mathbb{R}^m.$$

Furthermore, $\Phi(\cdot) : \mathbb{R}^n \to \mathbb{R}^n$ denotes an array of activation functions. In general, any monotonically increasing odd activation function $\phi(\cdot)$, being an element of vector array $\Phi(\cdot)$, can be used for the construction of novel design formula (16.4) and its resultant MZD model. To do so, the following two types of activation functions are exploited and investigated in this chapter (as depicted comparatively in Figure 15.1):

1) Linear activation function $\phi(f_j) = f_j$

2) Power-sum activation function $\phi(f_j) = \sum_{\kappa=1}^{N} f_j^{2\kappa-1}$ with integer $N \geqslant 2$

Note that, as discussed in Remark 5.1 (of Chapter 5), nonlinearities always exist in practice, which is one of the main motivations for us to extend the conventional Zhang *et al.*'s neural-dynamic design method and investigate different activation function arrays. Even if the linear activation function array is used, the nonlinear phenomena may appear in its hardware implementation; e.g., in the form of saturation or inconsistency of the linear slopes in analog implementation, or in the form of truncation and round-off errors in digital implementation. Thus, the investigation of different activation function arrays (such as the power-sum activation function array) may give us some insights into the positive and negative effects of nonlinearities existing in the implementation of the linear activation function array. Besides, different choices of activation function array $\Phi(\cdot)$ lead to different convergence performance of novel design formula (16.4) and its resultant MZD model. Now we present the following two theorems for novel design formula (16.4) using linear and power-sum activation functions.

Theorem 25 *For novel design formula (16.4) using linear activation functions, i.e., $\dot{e}(t) = -\gamma_2 JMP\big(e(0)\big) \diamond e(t)$*

1) *if initial error function $e(0) > 0$ [i.e., each element of $e(0)$ is greater than zero], error function $e(t)$ is exponentially convergent to zero with convergence rate $\gamma_2 > 0$;*

2) *if initial error function $e(0) \leqslant 0$ [i.e., each element of $e(0)$ is less than or equal to zero], error function $e(t)$ always equals $e(0)$; and,*

3) *if some elements of initial error function $e(0)$ are greater than zero, the corresponding error functions $e_i(t)$ are exponentially convergent to zero, while the other error functions $e_j(t)$, with $j = 1, 2, \cdots, n$ and $j \neq i$, are always equal to their initial errors.*

Proof Consider design formula (16.4) using linear activation functions, from which a set of n decoupled differential equations can be written as follows: $\forall i \in \{1, 2, \cdots, n\}$:

$$\dot{e}_i(t) = -\gamma_2 \mathrm{jmp}\left(e_i(0)\right) e_i(t).$$

In view of the definition of jump function $\mathrm{jmp}(\cdot)$, we have

$$\dot{e}_i(t) = \begin{cases} -\gamma_2 e_i(t), & \text{if } e_i(0) > 0, \\ 0, & \text{if } e_i(0) \leqslant 0. \end{cases}$$

So, when $e_i(0) > 0$, $\dot{e}_i(t) = -\gamma_2 e_i(t)$, of which the solution is $e_i(t) = \exp(-\gamma_2 t) e_i(0)$; and, when $e_i(0) = 0$, $\dot{e}_i(t) = 0$, of which the solution is $e_i(t) = e_i(0)$. For vector-valued $\mathbf{e}(t)$, we have the following three situations.

1) If initial error function $\mathbf{e}(0) > 0$, then error function $\mathbf{e}(t) = \exp(-\gamma_2 t) \mathbf{e}(0)$, which exponentially converges to zero with convergence rate γ_2 [i.e., as $t \to +\infty$, $\mathbf{e}(t) \to 0$ exponentially].

2) If initial error function $\mathbf{e}(0) \leqslant 0$, then error function $\mathbf{e}(t)$ is always equal to $\mathbf{e}(0)$ [i.e., $\mathbf{e}(t) = \mathbf{e}(0)$ no matter how time t evolves].

3) If some elements of initial error function $\mathbf{e}(0)$ are greater than zero, the corresponding error functions $e_i(t) = \exp(-\gamma_2 t) e_i(0)$, which are exponentially convergent to zero with convergence rate γ_2, while the other error functions $e_j(t) = e_j(0)$, with $j = 1, 2, \cdots, n$ and $j \neq i$, i.e., they are always equal to their initial errors.

The proof is thus complete.

Theorem 26 *Consider novel design formula (16.4) using power-sum activation functions, i.e.,*

$$\dot{\mathbf{e}}(t) = -\gamma_2 JMP(\mathbf{e}(0)) \diamond \sum_{\kappa=1}^{N} \left(\mathbf{e}(t)\right)^{2\kappa-1},$$

where vector-square operation $(\mathbf{e})^2$ is defined as $\mathbf{e} \diamond \mathbf{e}$ and we can so define the vector-power operation [e.g., $(\mathbf{e})^{2\kappa-1}$].

1) *If initial error function $\mathbf{e}(0) > 0$, the superior convergence of error function $\mathbf{e}(t)$ can be achieved, as compared with the situation of Theorem 25.*

2) *If initial error function $\mathbf{e}(0) \leqslant 0$, error function $\mathbf{e}(t)$ always equals $\mathbf{e}(0)$.*

3) *If some elements of initial error function $\mathbf{e}(0)$ are greater than zero, the superior convergence of the corresponding elements of error function $\mathbf{e}(t)$ can be achieved, as compared with the situation of Theorem 25. Besides, the other elements of $\mathbf{e}(t)$ are always equal to their initial errors.*

Proof It can be generalized from the proof of Theorem 25 by considering $\sum_{\kappa=1}^{N} e_i^{2\kappa-1}(t) \geqslant e_i(t)$ for any $e_i(t) > 0$, $\forall i \in \{1, 2, \cdots, n\}$, and is thus omitted.

Thirdly, we now continue to discuss novel design formula (16.4), which can be further written as

$$\begin{bmatrix} \dot{e}_1(t) \\ \dot{e}_2(t) \\ \vdots \\ \dot{e}_n(t) \end{bmatrix} = -\gamma_2 \begin{bmatrix} \mathrm{jmp}\left(e_1(0)\right)\phi\left(e_1(t)\right) \\ \mathrm{jmp}\left(e_2(0)\right)\phi\left(e_2(t)\right) \\ \vdots \\ \mathrm{jmp}\left(e_n(0)\right)\phi\left(e_n(t)\right) \end{bmatrix}.$$

Expanding the above design formula, in view of $e_i(t) = f_i(\mathbf{x}(t), t)$, we obtain that, $\forall i \in \{1, 2, \cdots, n\}$,

$$\left(\frac{\partial f_i}{\partial x_1} \frac{\mathrm{d}x_1}{\mathrm{d}t} + \frac{\partial f_i}{\partial x_2} \frac{\mathrm{d}x_2}{\mathrm{d}t} + \cdots + \frac{\partial f_i}{\partial x_n} \frac{\mathrm{d}x_n}{\mathrm{d}t} \right) + \frac{\partial f_i}{\partial t} = -\gamma_2 \mathrm{jmp}\left(f_i(\mathbf{x}(0), 0)\right) \phi\left(f_i(\mathbf{x}(t), t)\right),$$

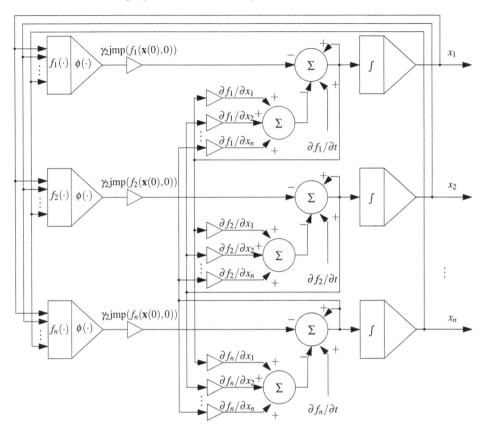

FIGURE 16.1 Circuit schematic realizing MZD model (16.5), where MZD model (16.5) is rewritten as $\dot{\mathbf{x}}(t) = (-J(\mathbf{x}(t),t) + I)\dot{\mathbf{x}}(t) - \gamma_2 \mathrm{JMP}(\mathbf{f}(\mathbf{x}(0),0)) \diamond \Phi(\mathbf{f}(\mathbf{x}(t),t)) - \partial \mathbf{f}/\partial t$ for modeling purposes.

which further leads to the following MZD model described in the form of a dynamic equation [which is also a differential equation with mass matrix $J(\mathbf{x}(t),t)$]:

$$J(\mathbf{x}(t),t)\dot{\mathbf{x}}(t) = -\gamma_2 \mathrm{JMP}\Big(\mathbf{f}(\mathbf{x}(0),0)\Big) \diamond \Phi\Big(\mathbf{f}(\mathbf{x}(t),t)\Big) - \frac{\partial \mathbf{f}}{\partial t}, \qquad (16.5)$$

where $J(\mathbf{x}(t),t)$ and $\partial \mathbf{f}/\partial t$ are defined the same as those in CZD model (16.3), and $\mathbf{x}(t) \in \mathbb{R}^n$, starting with an initial state $\mathbf{x}(0) \in \mathbb{R}^n$, denotes the state vector. As stated before, neural-dynamical models can be implemented physically by the designated hardware. In order to construct the analog circuit of MZD model (16.5), we can use and connect analog adders/subtractors, multipliers, integrators, limiters, activation functions and differentiators, which are usually based on operational amplifiers. The circuit schematic of MZD model (16.5) is thus depicted in Figure 16.1, which is an important and necessary step for the final hardware implementation of the neural-dynamical model. It is worth pointing out that, when MZD model (16.5) uses linear activation functions, it reduces to the following MZDL model:

$$J(\mathbf{x}(t),t)\dot{\mathbf{x}}(t) = -\gamma_2 \mathrm{JMP}\Big(\mathbf{f}(\mathbf{x}(0),0)\Big) \diamond \mathbf{f}(\mathbf{x}(t),t) - \frac{\partial \mathbf{f}}{\partial t}; \qquad (16.6)$$

and that, when MZD model (16.5) uses power-sum activation functions, it reduces to the following

MZDP model:

$$J(\mathbf{x}(t),t)\dot{\mathbf{x}}(t) = -\gamma_2 \text{JMP}\Big(\mathbf{f}(\mathbf{x}(0),0)\Big) \diamond \sum_{\kappa=1}^{N} \Big(\mathbf{f}(\mathbf{x}(t),t)\Big)^{2\kappa-1} - \frac{\partial \mathbf{f}}{\partial t}. \tag{16.7}$$

16.4.2 Convergence Analysis

The above subsection proposes MZD model (16.5) for solving time-varying nonlinear inequality system (16.1). In this subsection, the convergence results of MZD model (16.5) are presented, which are based on the use of the linear or power-sum activation function array.

Theorem 27 *For MZD model (16.5) using linear activation functions [i.e., MZDL model (16.6)],*

1) *if initial state $\mathbf{x}(0) \in \mathbb{R}^n$ is outside the initial solution set $\mathbb{S}(0)$ of (16.1), then there exists at least $i \in \{1,2,\cdots,n\}$, such that $e_i(t)$ is globally and exponentially convergent to zero with convergence rate $\gamma_1 > 0$, while other error functions $e_j(t)$, with $j = 1,2,\cdots,n$ and $j \neq i$, are always equal to $e_j(0)$; and*

2) *if initial state $\mathbf{x}(0) \in \mathbb{R}^n$ is inside the initial solution set $\mathbb{S}(0)$ of (16.1), error function $\mathbf{e}(t)$ is always equal to $\mathbf{e}(0)$.*

That is, MZDL model (16.6) generates an exact time-varying solution of (16.1), with an exponential convergence performance.

Proof It can be generalized from the proof of Theorem 25 by considering the equivalence between $\mathbf{x}(0) \notin \mathbb{S}(0)$ and the existence of at least one $i \in \{1,2,\cdots,n\}$, such that $e_i(0) > 0$, and is thus omitted due to analysis similarity.

Theorem 28 *For MZD model (16.5) using power-sum activation functions [i.e., MZDP model (16.7)],*

1) *if initial state $\mathbf{x}(0) \in \mathbb{R}^n$ is outside the initial solution set $\mathbb{S}(0)$ of (16.1), then there exists at least $i \in \{1,2,\cdots,n\}$ such that $e_i(t)$ has the superior convergence, as compared with the situation of Theorem 27, while the other error functions $e_j(t)$, with $j = 1,2,\cdots,n$ and $j \neq i$, are always equal to $e_j(0)$; and*

2) *if initial state $\mathbf{x}(0) \in \mathbb{R}^n$ is inside the initial solution set $\mathbb{S}(0)$ of (16.1), error function $\mathbf{e}(t)$ is always equal to $\mathbf{e}(0)$.*

That is, MZDP model (16.7) generates an exact time-varying solution of (16.1), with its convergence generally much faster than that of MZDL model (16.6).

Proof It can be generalized from the proofs of Theorems 25 through 27, and is thus omitted due to analysis similarity.

16.4.3 Comparison

In this subsection, we further discuss the relationship and differences between MZD model (16.5) and CZD model (16.3), both of which are exploited for solving time-varying nonlinear inequality system (16.1).

On one hand, there exist evident differences between MZD model (16.5) and CZD model (16.3). For example, the error functions, design formulas and the resultant models differ from each other [95]. In addition, MZD model (16.5) allows us to have many choices of activation functions. Thus,

MZD model (16.5) can be regarded as a more general form, while CZD model (16.3) has no choices of activation functions and can be seen as a specific form.

On the other hand, what link exists between them? To analyze it, we first focus on CZD model (16.3). We have two situations discussed as follows.

1) If $f_i(\mathbf{x}(0),0) > 0$, with $i \in \{1,2,\cdots,n\}$, then, according to the proof of Theorem 24, $e_i(t)$ of CZD model (16.3) is globally and exponentially convergent to zero with rate $\gamma_2 > 0$. That is, $f_i(\mathbf{x}(t),t)$ converges to zero as t increases. In this situation, $\max\{0,f_i(\mathbf{x}(t),t)\} = f_i(\mathbf{x}(t),t)$. One would thus have $\mathrm{jmp}\big(f_i(\mathbf{x}(0),0)\big) \diamond f_i(\mathbf{x}(t),t) = \max\{0,f_i(\mathbf{x}(t),t)\} = f_i(\mathbf{x}(t),t)$ when $f_i(\mathbf{x}(0),0) > 0$.

2) If $f_i(\mathbf{x}(0),0) \leqslant 0$, with $i \in \{1,2,\cdots,n\}$, then, according to the proof of Theorem 24, $e_i(t)$ of CZD model (16.3) is always equal to zero, which guarantees $f_i(\mathbf{x}(t),t) \leqslant 0$. So one would have $\mathrm{jmp}\big(f_i(\mathbf{x}(0),0)\big) \diamond f_i(\mathbf{x}(t),t) = \max\{0,f_i(\mathbf{x}(t),t)\} = 0$ when $f_i(\mathbf{x}(0),0) \leqslant 0$.

Therefore, we can conclude that $\mathrm{jmp}(f_i(\mathbf{x}(0),0)) \diamond f_i(\mathbf{x}(t),\ t) = \max\{0,f_i(\mathbf{x}(t),t)\}$, no matter what the initial value of $f_i(\mathbf{x}(0),0)$ is. That is,

$$\mathrm{JMP}(\mathbf{f}(\mathbf{x}(0),0)) \diamond \mathbf{f}(\mathbf{x}(t),t) = \max\{0,\mathbf{f}(\mathbf{x}(t),t)\}. \tag{16.8}$$

Now look at MZD model (16.5). When we use linear activation functions, it reduces to MZDL model (16.6). By comparison and in view of (16.8), it is further discovered that MZDL model (16.6) is exactly CZD model (16.3) for solving the time-varying nonlinear inequality system when $\gamma_2 = \gamma_1/2$. Note that design parameters γ_2 and γ_1 can be selected by users. So in the following simulation we set $\gamma_2 = \gamma_1/2$, such that MZDL model (16.6) is exactly CZD model (16.3). In summary, we can draw a conclusion that MZD model (16.5) incorporates CZD model (16.3) as its special case. In other words, in addition to the discovery on a general form of Newton iteration being given by ZD (as shown in most chapters of this book), we discover further that a general form of CZD model (16.3) can be given by MZD model (16.5) for solving the time-varying nonlinear inequality system.

16.5   Illustrative Example

In the previous three sections, both MZD model (16.5) and CZD model (16.3) have been presented for solving the time-varying nonlinear inequality system. The comparison between two such models has also been conducted there in addition to the theoretical analysis. Note that MZDL model (16.6) can be regarded as CZD model (16.3). Therefore, we only investigate MZDP model (16.7) and CZD model (16.3) in this section.

Now let us consider the following system of time-varying nonlinear inequalities:

$$\mathbf{f}(\mathbf{x}(t),t) = \begin{cases} \ln x_1(t) - 1/(t+1) \leqslant 0, \\ x_1(t)x_2(t) - \exp(1/(t+1))\sin(10t) \leqslant 0, \end{cases} \tag{16.9}$$

where evidently $x_1(t) > 0$. For illustration and comparison, both MZDP model (16.7) and CZD model (16.3) are exploited for solving time-varying nonlinear inequality system (16.9). Considering that different initial states of CZD and MZDP models may lead to different convergence performance, we investigate and illustrate the following three situations.

Situation 1   Consider randomly generated initial states $\mathbf{x}(0)$ outside the initial solution set $\mathbb{S}(0)$ of (16.9) such that $\mathbf{f}(\mathbf{x}(0),0) > 0$ completely. CZD and MZDP models are applied with $\gamma_2 = \gamma_1/2 = 1$

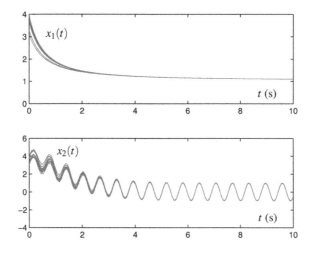

(a) By CZD model (16.3)

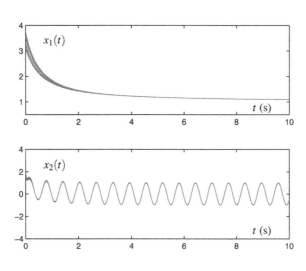

(b) By MZDP model (16.7)

FIGURE 16.2 Neural states $\mathbf{x}(t)$ starting from 10 randomly generated initial states $\mathbf{x}(0)$ with $f_1(\mathbf{x}(0),0) > 0$ and $f_2(\mathbf{x}(0),0) > 0$.

and with 10 randomly generated initial states $\mathbf{x}(0) \notin \mathbb{S}(t)$ to solving time-varying nonlinear inequality system (16.9). The corresponding time-varying solutions [i.e., the trajectories of state vector $\mathbf{x}(t)$] are obtained and illustrated in Figure 16.2. Comparing the results, we see that state vectors $\mathbf{x}(t)$ of CZD model (16.3) converge to their steady-state curves within 4 s, while those of MZDP model (16.7) converge faster [especially $x_2(t)$]. To help readers understand better, Figure 16.3 shows the transient behavior of residual errors $\mathbf{f}(\mathbf{x}(t),t)$ corresponding to the above time-varying solutions. It is seen from Figure 16.3(a) that residual errors $\mathbf{f}(x(t),t)$ of CZD model (16.3) decrease to zero within

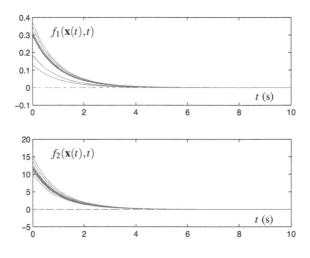

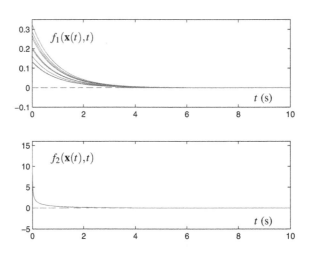

(a) By CZD model (16.3)

(b) By MZDP model (16.7)

FIGURE 16.3 Residual errors $\mathbf{f}(\mathbf{x}(t),t)$ starting with 10 randomly generated initial states $\mathbf{x}(0)$ for which $f_1(\mathbf{x}(0),0) > 0$ and $f_2(\mathbf{x}(0),0) > 0$.

4 s. In contrast, as seen from Figure 16.3(b), residual errors $\mathbf{f}(\mathbf{x}(t),t)$ of MZDP model (16.7) have superior convergence. These illustrate that both CZD and MZD models are effective on solving the system of time-varying nonlinear inequalities. In addition, MZDP model (16.7) is more effective, as compared with CZD model (16.3).

To further monitor and show the solution processes of the CZD and MZDP models, norm-based error function $\|\mathbf{e}(t)\|_2$ may be a preferable measure choice. From Figure 16.4, we observe again that both CZD and MZD are effective and efficient on solving the time-varying nonlinear inequality system; and that MZDP model (16.7) is more effective on solving the time-varying nonlinear in-

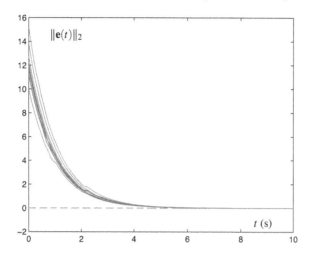

(a) By CZD model (16.3)

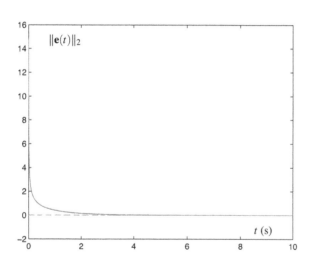

(b) By MZDP model (16.7)

FIGURE 16.4 Norm-based error functions $\|\mathbf{e}(t)\|_2$ starting with 10 randomly generated initial states $\mathbf{x}(0)$ for which $f_1(\mathbf{x}(0),0) > 0$ and $f_2(\mathbf{x}(0),0) > 0$.

equality system, as compared with CZD model (16.3). In addition, it is worth pointing out that, as shown in Figure 16.5, the convergence time of norm-based error function $\|\mathbf{e}(t)\|_2$ of MZDP model (16.7) (just take it as an example) can be expedited effectively; e.g., to 0.35 s or 0.0035 s (i.e., 3.5 ms), when design parameter γ_2 increases to 10 or 1000. So, design parameter γ_2 plays an important role in MZD model (16.5) and should be selected appropriately large to satisfy the convergence rate needed in this situation.

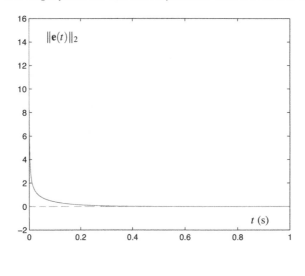

(a) With $\gamma_2 = 10$

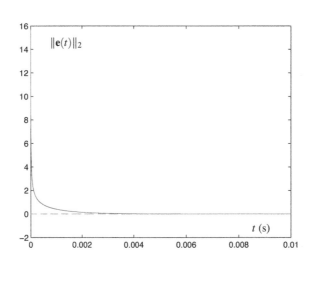

(b) With $\gamma_2 = 1000$

FIGURE 16.5 Norm-based error functions $\|\mathbf{e}(t)\|_2$ of MZDP model (16.7) with different γ_2 and starting with 10 randomly generated initial states $\mathbf{x}(0)$ for which $f_1(\mathbf{x}(0),0) > 0$ and $f_2(\mathbf{x}(0),0) > 0$.

Situation 2   In practice, it may be difficult to generate randomly initial state $\mathbf{x}(0)$, such that $\mathbf{f}(\mathbf{x}(0),0) > 0$ completely. Thus, it is worth investigating the performance of CZD and MZDP models for solving time-varying nonlinear inequality system (16.9), when randomly generated initial states $\mathbf{x}(0)$ are outside the initial solution set $\mathbb{S}(0)$ of (16.9) in a more general sense that some elements of $\mathbf{f}(\mathbf{x}(0),0)$ are greater than zero while the others are less than or equal to zero. The corresponding simulation results are illustrated in Figure 16.6, where $f_1(\mathbf{x}(0),0) < 0$ and $f_2(\mathbf{x}(0),0) > 0$. As shown in Figure 16.6(a) synthesized by CZD model (16.3), error functions $e_1(t)$ are always equal

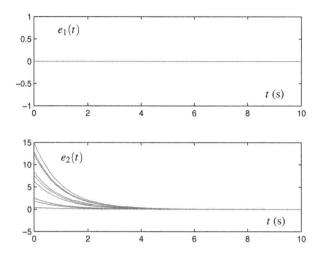

(a) By CZD model (16.3)

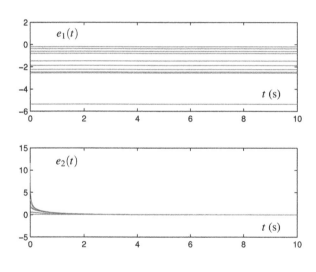

(b) By MZDP model (16.7)

FIGURE 16.6 Error functions $\mathbf{e}(t)$ starting with 10 randomly generated initial states $\mathbf{x}(0)$ for which $f_1(\mathbf{x}(0),0) < 0$ and $f_2(\mathbf{x}(0),0) > 0$.

to zero and error functions $e_2(t)$ converge to zero within 4 s. From Figure 16.6(b) synthesized by MZDP model (16.7), error functions $e_1(t)$ are always equal to their negative initial values $e_1(0) < 0$, and error functions $e_2(t)$ have much faster convergence than those of CZD model (16.3).

Situation 3 If initial states $\mathbf{x}(0)$ are inside the initial solution set $\mathbb{S}(0)$ [i.e., $\mathbf{f}(\mathbf{x}(0),0) \leqslant 0$], then equation (16.8) equals zero. Thus, MZD model (16.5) and CZD model (16.3) both reduce to $J(\mathbf{x}(t),t)\dot{\mathbf{x}}(t) = -\partial \mathbf{f}/\partial t$. However, because the definitions of their error functions differ from

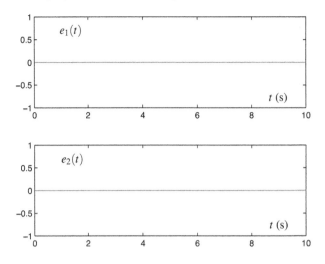

(a) By CZD model (16.3)

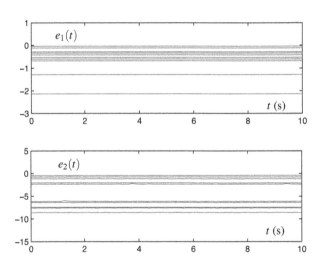

(b) By MZD model (16.5)

FIGURE 16.7 Error functions $\mathbf{e}(t)$ starting with 10 randomly generated initial states $\mathbf{x}(0)$ for which $f_1(\mathbf{x}(0),0) < 0$ and $f_2(\mathbf{x}(0),0) < 0$ [or to say, $\mathbf{x}(0) \in \mathbb{S}(0)$].

each other, the error functions of CZD model (16.3) remain zero without appreciable convergence process, and the error functions of MZD model (16.5) are equal to their initial values [i.e., $\mathbf{e}(0)$]. This point is substantiated well in Figure 16.7.

Finally, starting from 300 randomly generated initial states $\mathbf{x}(0)$ within $[0,4] \times [-10,10]$, the planar trajectories of state vector $\mathbf{x}(t)$ of MZDP model (16.7) at different time instants are shown

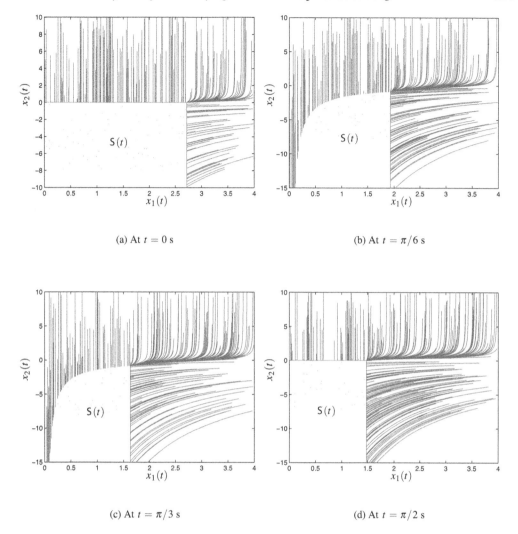

(a) At $t = 0$ s

(b) At $t = \pi/6$ s

(c) At $t = \pi/3$ s

(d) At $t = \pi/2$ s

FIGURE 16.8 Planar trajectories of $\mathbf{x}(t)$ solved by MZDP model (16.7) starting from 300 randomly generated initial states in $[0,4] \times [-10,10]$ at different time instants.

in Figure 16.8. Although time-varying solution set $\mathbb{S}(t)$ is different (i.e., dynamic) at different time instants, evidently the trajectories all converge to or stay inside $\mathbb{S}(t)$.

In summary, from the above simulation results, we can conclude that both MZDP model (16.7) and CZD model (16.3) are effective on solving the time-varying nonlinear inequality system, no matter whether initial state $\mathbf{x}(0)$ is inside or outside the initial solution set $\mathbb{S}(0)$; and that the solution of MZDP model (16.7) has a much better convergence performance when initial state $\mathbf{x}(0)$ is outside the initial solution set $\mathbb{S}(0)$, as compared with that of CZD model (16.3).

16.6   Summary

In this chapter, both CZD model (16.3) and MZD model (16.5) have been proposed for solving the time-varying nonlinear inequality system depicted in (16.1). The exponential convergence performance of the CZD and MZD models have been analyzed and presented. For MZD model (16.5), two types of activation functions have also been studied. Compared with CZD model (16.3) for solving the time-varying nonlinear inequality system, it has been discovered that MZD model (16.5) can be regarded as a generalized form of CZD model (16.3) for solving the time-varying nonlinear inequality system. Finally, an illustrative computer-simulation example and the corresponding numerical results have further verified and illustrated the efficacy of CZD model (16.3), MZD model (16.5), and their design methods for solving the time-varying nonlinear inequality system. Besides, the superiority of MZD model (16.5) has been substantiated, as compared with the CZD one.

Part VII

Application to Fractals

Chapter 17

Fractals Yielded via Static Nonlinear Equation

Abstract

In this chapter, a novel kind of fractal is yielded by using the complex-valued DTZD (CVDTZD) model to solve nonlinear equations in the complex domain. Such a CVDTZD model is designed based on the elimination of an indefinite complex-valued error function, instead of a square-based nonnegative energy function associated with the complex-valued discrete-time gradient-based dynamics (CVDTGD). Compared with the well-known (generalized) Newton fractals (i.e., the famous fractals generated by the well-known Newton iteration), we find that the novel fractals synthesized by the proposed CVDTZD model incorporate such Newton fractals as special cases. In addition, the fractals generated by the novel CVDTZD model are completely different from Newton fractals. The CVDTZD model using different types of activation functions can be seen as a new iterative algorithm to generate new fractals.

17.1   Introduction

The problem of generating fractals is an important and interesting issue in the field of computer geometry and graphics, and the theories and techniques on fractals have developed since 1970s. The fractals may show and reveal the chaotic phenomena appeared in the microscopic self-computing or self-organization natural or artificial architectures [13, 36, 56]. Generation of many fractals are closely related to conventional Newton iteration for solving certain types of nonlinear equations in the complex domain, such as complex root finding, complex polynomials solving, and complex-exponential function solving [87]. Solving different nonlinear equations may output different plots when generating fractals [9, 13, 65, 87].

It is worth mentioning that, in engineering applications, e.g., robotics, many problems can be abstracted as the time-varying mathematical problems (rather than constant mathematical problems), such as time-varying quadratic programming (QP) problems. As required, the ZD method was proposed intrinsically to solve the time-varying problems and can also be exploited to solve such time-varying engineering problems in robotics. For example, as shown in Chapters 13 and 14, the redundancy-resolution problems of robot arms can be abstracted as the time-varying QP problems, and turned into one of the applications of the ZD method, which may handle simultaneously the target-tracking and feedback problems of robot arms.

Furthermore, note that the ZD method, based on the indefinite error function, is quite different from the conventional algorithms and can avoid the lagging errors caused by the conventional algorithms for solving time-varying problems. The reason why ZD can achieve the excellent performance for solving time-varying problems is that the time-derivative information of time-varying coefficients involved in time-varying problems is utilized in the design procedure of the ZD models.

By discretizing the complex-valued CTZD (CVCTZD) model for nonlinear equation solving in the complex domain, we can obtain the corresponding CVDTZD model for the online problem solving [90,109]. Such a proposed CVDTZD model can be applied to generating new fractals, which are completely different from the fractals generated by the conventional or generalized Newton iteration [13,87].

17.2   Complex-Valued ZD Models

In this chapter, we consider the following nonlinear equation in the complex domain for generating new fractals [90]:

$$f(z) = 0 \in \mathbb{C}, \tag{17.1}$$

where $f(\cdot) : \mathbb{C} \to \mathbb{C}$ denotes a nonlinear complex mapping, and $z \in \mathbb{C}$ denotes the complex-valued scalar to be obtained. The well-known Newton iteration in the complex domain has been applied to computing the root of such a complex-valued nonlinear equation. The corresponding fractals can thus be generated by Newton iteration. In this chapter, we exploit a CVDTZD model to find out the root of the complex-valued nonlinear equation coinciding with obtaining some new fractals, which are quite different from Newton fractals. The following types of nonlinear equations are considered to be solved for yielding new fractals:

1) $f(z) = z^p - a$, where $p \geqslant 3$ is a positive integer parameter and a is a complex-valued constant scalar (please also see and compare it with Part I of the book);

2) $f(z) = \sin(z)$ or $f(z) = \cos(z)$;

3) $f(z) = \varphi(z)\exp(\psi(z))$ [87] or $f(z) = \exp(\psi(z))$ as the former's special case, where $\varphi(z)$ and $\psi(z)$ are complex functions with respect to complex scalar z.

17.2.1   CVCTZD Model

By following Zhang *et al.*'s design method, we can construct a CVCTZD model to solve nonlinear Equation (17.1). To monitor the solution process of nonlinear Equation (17.1), the following indefinite complex-valued error function is first defined (i.e., it can be positive, zero, negative, real, complex, bounded, or unbounded, even including lower-unbounded):

$$e(t) = f(z) \in \mathbb{C},$$

where, evidently, $z \in \mathbb{C}$ converges to theoretical solution z^* of nonlinear Equation (17.1) as error function $e(t)$ converges to zero. In order to make the error function diminish to zero, the following ZD design formula is employed:

$$\dot{e}(t) = -\gamma\phi(e(t)),$$

or exactly

$$\frac{\mathrm{d}f(z)}{\mathrm{d}t} = -\gamma\phi(f(z)), \tag{17.2}$$

where design parameter $\gamma > 0 \in \mathbb{R}$, which corresponds to the reciprocal of a capacitance parameter, should be set as large as the hardware would permit, or set appropriately large for simulative or experimental purposes [8]. Meanwhile, as shown and discussed in almost every chapter of the book, $\phi(\cdot) : \mathbb{C} \to \mathbb{C}$ is typically a monotonically increasing odd activation function. In addition, six types of activation functions (i.e., linear activation function, power activation function, power-sum activation function, bipolar sigmoid activation function, power-sigmoid activation function, and hyperbolic sine activation function) have been introduced and investigated in the ZD research (especially in this book).

Expanding ZD design formula (17.2), we can have the following complex-valued differential equation (termed as the so-called CVCTZD model):

$$\frac{\partial f(z)}{\partial z}\frac{\mathrm{d}z}{\mathrm{d}t} = -\gamma\phi(f(z)),$$

or, with $f'(z) = \partial f(z)/\partial z \neq 0$ assumed,

$$\dot{z}(t) = -\gamma\frac{\phi(f(z))}{f'(z)}, \tag{17.3}$$

where neural state $z(t) \in \mathbb{C}$, starting from a properly chosen initial condition $z(0) \in \mathbb{C}$, can converge to theoretical solution $z^*(t)$ of nonlinear Equation (17.1).

17.2.2 CVDTZD Model

In this subsection, we present the discrete-time form of CVCTZD model (17.3) to solve online nonlinear Equation (17.1), which can be viewed as the essential step for the new fractal generation. By applying the Euler forward-difference rule to CVCTZD model (17.3), we can obtain CVDTZD model as follows:

$$\frac{z_{k+1} - z_k}{\tau} = -\gamma\frac{\phi(f(z_k))}{f'(z_k)}, \text{ with } f'(z_k) \neq 0,$$

which can be further written as

$$z_{k+1} = z_k - h\frac{\phi(f(z_k))}{f'(z_k)}, \text{ with } f'(z_k) \neq 0, \tag{17.4}$$

where parameters $\tau > 0 \in \mathbb{R}$, $h = \tau\gamma > 0 \in \mathbb{R}$ and $k = 0, 1, 2, \cdots$ denote the sampling gap of $z(t)$, the step size, and the iteration index, respectively. Generally speaking (as shown in the related chapters of the book), different choices of step size h and activation function $\phi(\cdot)$ lead to different convergence performance of CVDTZD model (17.4) for solving nonlinear Equation (17.1).

Remark 6 During the generation of new fractals, if occasionally $f'(z_k)$ equals zero, a small constant is added to $f'(z_k)$ in the related computer code; e.g., $\alpha = 2^{-52} \approx 2.2204 \times 10^{-16}$ is added, which is termed "eps" in MATLAB® to denote the spacing of floating point numbers, and, more specifically, the distance from 1.0 to the next larger double precision number). As a result, the value of $f'(z_k)$ approximates to zero but does not equal zero (which guarantees the successful generation of new fractals). More importantly, by introducing such a small constant, we can elegantly avoid the division-by-zero situation, and the result of the visualization of new fractals will not be weakened.

Now, review CVDTZD model (17.4). If we utilize the linear activation function $\phi(e) = e$, CVDTZD model (17.4) reduces to

$$z_{k+1} = z_k - h\frac{f(z_k)}{f'(z_k)}. \tag{17.5}$$

If $h = 1$, CVDTZD model (17.4) further reduces to

$$z_{k+1} = z_k - \frac{f(z_k)}{f'(z_k)}, \tag{17.6}$$

which is exactly Newton iteration for solving the nonlinear equation in the complex domain. This implies that the conventional Newton iteration can be viewed as a special case of the general CVDTZD model (17.4). In addition, Newton iteration (17.6) can be applied to generating different fractals by solving different types of nonlinear equations, and the fractals yielded are thus called Newton fractals [87]. In the ensuing section, we show that the new fractals generated by CVDTZD model (17.4) for solving nonlinear equations are straightforward but completely different from Newton fractals [i.e., the fractals generated by the conventional (or generalized) Newton iteration].

Remark 7 In the authors' opinion, simplicity is beauty. As Occam's razor suggests, when you have two competing theories that make exactly the same predictions, the simpler one is the better one. From the resultant CVCTZD model (17.3) and CVDTZD model (17.4), we can see that, being simple, direct, and straightforward is actually the advantage of this work on the condition of achieving excellent results, and it is thus easy for readers to understand and use the presented ZD method to solve various time-varying problems (including the complex-valued nonlinear equation problem). On the other hand, exploiting different types of monotonically increasing odd activation functions is one of the characteristics of the method. Note that, by using the different types of activation functions, the ZD method can achieve interesting or excellent performances for solving various time-varying problems and simultaneously presenting the new fractals.

Remark 8 Note that the complex-valued ZD method can also be extended and then used for the learning, generalization, prediction, and classification of the back-propagation (BP) neural network, such as the complex-exponential Fourier neural network [135]. While the trigonometric functions of Fourier series are exploited as the activation functions, we can derive the weights-direct-determination method to decide the optimal neural-network weights based on the pseudoinverse formulation [135] (where pseudoinverse is also termed Moore–Penrose inverse and discussed in Chapter 9 of the book). This method remedies the weaknesses of the conventional BP neural network, such as small convergence rate, difficulty in selecting parameters, easily converging to local minima, and possibly lengthy or oscillatory learning process. In addition, the above Fourier neural network and algorithm have good properties of high-precision learning, noise-suppressing, and discontinuous-function approximation. So, left for interested readers as a topic of exercise, the pseudoinverse computation of a complex-valued matrix can be extended from this chapter and regarded as one kind of application of the complex-valued neural network.

17.3   Illustrative Examples

In this section, several representative and illustrative examples are shown to verify the proposed CVDTZD model (17.4) for generating new fractals by solving nonlinear equations in the complex

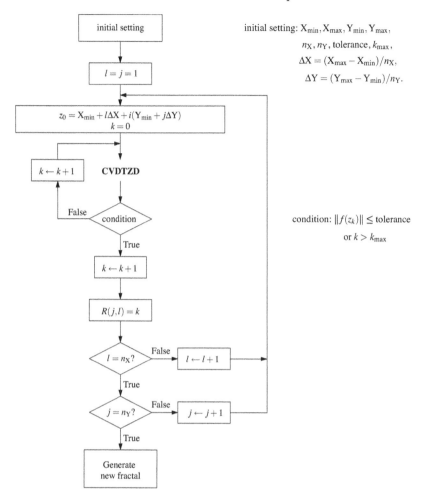

FIGURE 17.1 Flowchart of using CVDTZD method to generating new fractal via solving nonlinear equation in complex domain.

domain. We exploit CVDTZD model (17.4) with different values of step size h and different activation functions $\phi(\cdot)$ to generate the new fractals. Before that, the flowchart of the CVDTZD method for generating a fractal via solving a nonlinear equation in the complex domain is depicted in Figure 17.1.

17.3.1 By Solving $f(z) = z^3 + i = 0$

Let us consider the following complex-valued nonlinear equation:

$$f(z) = z^3 + i = 0. \tag{17.7}$$

We exploit CVDTZD model (17.4) activated by power-sum function (with $N \geqslant 2$)

$$\phi(e) = \sum_{\kappa=1}^{N} e^{2\kappa-1}$$

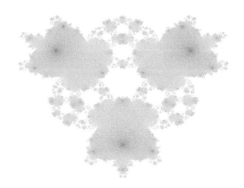

(a) Via CVDTZD model (17.4) activated by power-sum function

(b) Via CVDTZD model (17.4) activated by power-sigmoid function

FIGURE 17.2 New fractals generated by solving nonlinear Equation (17.7). *Reproduced from H. Wu, F. Li, Z. Li, et al., Zhang fractals yielded via solving nonlinear equations by discrete-time complex-valued ZD, Figure 2, Proceedings of IEEE International Conference on Automation and Logistics, pp. 1–6, 2012. ©IEEE 2012. With kind permission of IEEE.*

and power-sigmoid function (with $p = 3$ and $\xi = 4$)

$$\phi(e) = \begin{cases} e^p, & \text{if } |e| \geqslant 1 \\ \frac{1+\exp(-\xi)}{1-\exp(-\xi)} \frac{1-\exp(-\xi e)}{1+\exp(-\xi e)}, & \text{otherwise} \end{cases}$$

to generate new fractals, which are shown in Figure 17.2(a) and (b), respectively. Meanwhile, for comparison, the fractal generated by the generalized Newton iteration (17.5) in this example is

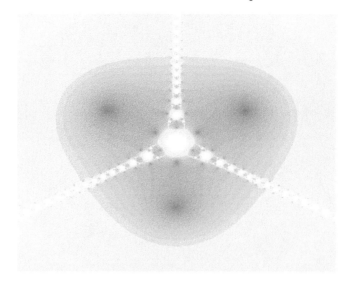

FIGURE 17.3 Newton fractal generated by solving nonlinear Equation (17.7). *Reproduced from H. Wu, F. Li, Z. Li, et al., Zhang fractals yielded via solving nonlinear equations by discrete-time complex-valued ZD, Figure 3, Proceedings of IEEE International Conference on Automation and Logistics, pp. 1–6, 2012. ©IEEE 2012. With kind permission of IEEE.*

shown in Figure 17.3. Evidently, we can observe that the presented two kinds of fractals, i.e., 1) the new fractals and 2) Newton fractal, are completely different, although they are generated under the same parameter conditions and by solving the same nonlinear equation.

17.3.2   By Solving $f(z) = \cos(z) = 0$

Consider the following complex-valued nonlinear equation:

$$f(z) = \cos(z) = 0. \tag{17.8}$$

The resultant new fractal generated by exploiting CVDTZD model (17.4) with $h = 0.2$ and using the power-sum activation function is shown in Figure 17.4(a). Additionally, Newton fractal generated under the same conditions is illustrated in Figure 17.4(b).

17.3.3   By Solving $f(z) = \varphi(z)\exp(\psi(z)) = 0$

Consider the following nonlinear equation:

$$f(z) = z\exp(z^{3+i}) = 0. \tag{17.9}$$

The beautiful new fractals generated by using CVDTZD model (17.4), with $h = 1$ and activated by the power-sum function and the power-sigmoid function, are shown in Figure 17.5(a) and (b), respectively.

Besides, we consider the following complex-valued nonlinear equation as another example, which belongs to the type of the aforementioned complex exponential function $f(z) = \varphi(z)\exp(\psi(z))$, as well:

$$f(z) = z^4\exp(z^2 + i) = 0. \tag{17.10}$$

(a) New fractal via CVDTZD model (17.4) activated by power-sum function

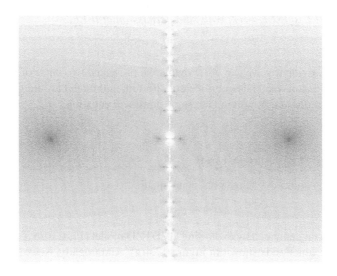

(b) Newton fractal

FIGURE 17.4 Fractals generated by solving nonlinear Equation (17.8). *Reproduced from H. Wu, F. Li, Z. Li, et al., Zhang fractals yielded via solving nonlinear equations by discrete-time complex-valued ZD, Figure 4, Proceedings of IEEE International Conference on Automation and Logistics, pp. 1–6, 2012. ©IEEE 2012. With kind permission of IEEE.*

We can obtain another beautiful new fractal by exploiting CVDTZD model (17.4) with $h = 0.2$ and using the power-sum activation function, which is displayed in Figure 17.6(a). Similar to the previous examples, the conventional Newton fractal generated under the same conditions is illustrated in Figure 17.6(b) for comparison.

(a) Via CVDTZD model (17.4) activated by power-sum function

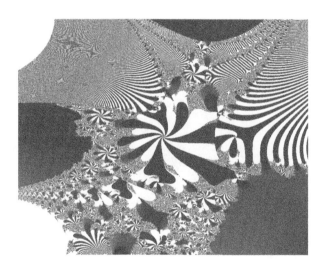

(b) Via CVDTZD model (17.4) activated by power-sigmoid function

FIGURE 17.5 New fractals generated by solving nonlinear Equation (17.9). *Reproduced from H. Wu, F. Li, Z. Li, et al., Zhang fractals yielded via solving nonlinear equations by discrete-time complex-valued ZD, Figure 5, Proceedings of IEEE International Conference on Automation and Logistics, pp. 1–6, 2012. ©IEEE 2012. With kind permission of IEEE.*

17.3.4 By Solving $f(z) = \exp(\psi(z)) = 0$

In the last example of the section, let us consider the complex exponential function $f(z) = \exp(z^5 + i)$, and its corresponding nonlinear equation is

$$f(z) = \exp(z^5 + i) = 0. \qquad (17.11)$$

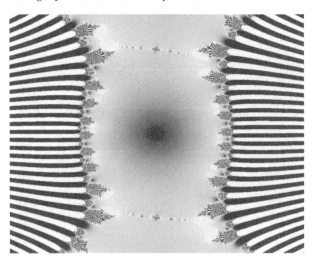

(a) New fractal generated via CVDTZD (17.4) activated by power-sum function

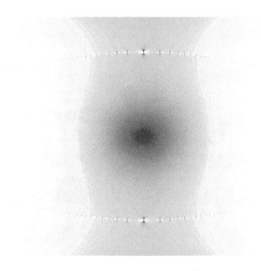

(b) Newton fractal

FIGURE 17.6 Fractals generated by solving nonlinear Equation (17.10). *Reproduced from H. Wu, F. Li, Z. Li, et al., Zhang fractals yielded via solving nonlinear equations by discrete-time complex-valued ZD, Figure 6, Proceedings of IEEE International Conference on Automation and Logistics, pp. 1–6, 2012. ©IEEE 2012. With kind permission of IEEE.*

The beautiful new fractals generated by using CVDTZD model (17.4), with $h = 0.2$ and activated by the power-sum function and the power-sigmoid function, are shown in Figure 17.7(a) and (b), respectively.

In summary, from the examples, we can conclude that the new fractals generated are completely different from Newton fractals (i.e., the fractals generated by the conventional or generalized

(a) Via CVDTZD model (17.4) activated by power-sum function

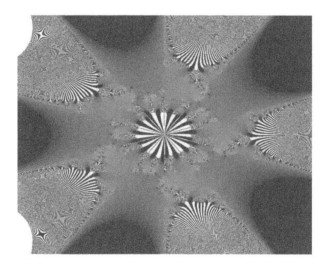

(b) Via CVDTZD model (17.4) activated by power-sigmoid function.

FIGURE 17.7 New fractals generated by solving nonlinear Equation (17.11). *Reproduced from H. Wu, F. Li, Z. Li, et al., Zhang fractals yielded via solving nonlinear equations by discrete-time complex-valued ZD, Figure 7, Proceedings of IEEE International Conference on Automation and Logistics, pp. 1–6, 2012. ©IEEE 2012. With kind permission of IEEE.*

Newton iteration). That is, the novel beautiful fractals yielded via the proposed CVDTZD model (17.4) incorporate such conventional Newton fractals as special cases. Additionally, by exploiting different activation functions $\phi(\cdot)$ and using different values of step size h, different new fractals can be generated correspondingly and abundantly.

17.4   Summary

In this chapter, some interesting and beautiful new fractals have been yielded by the novel CVDTZD for solving the nonlinear equations in the complex domain. Such a novel CVDTZD model has been designed based on an indefinite complex-valued error function and the ZD design formula. It has been found that the new fractals generated by the novel CVDTZD model are completely different from the fractals generated by (generalized) Newton iteration, i.e., Newton fractals. In addition, the well-known Newton fractals have been found to be special cases of new fractals generated via the novelly proposed CVDTZD model. Numerical-experiment results have further illustrated that the proposed CVDTZD model can be viewed as a new iterative algorithm to generate new fractals. Two future research directions may lie in the more detailed theoretical analysis of the CVDTZD model using presented or new activation functions as well as the deeper theoretical investigation of the new kind of fractals (which, similar to the previous ones pointed out in the book, are left to interested readers as topics of exercise).

Chapter 18

Fractals Yielded via Time-Varying Nonlinear Equation

Abstract

In this chapter, new fractals are yielded by using the complex-valued discrete-time zeroing dynamics (CVDTZD) to solve time-varying nonlinear equations in the complex domain (as extended from and compared with the solution of static nonlinear equations in the previous chapter). Such a CVDTZD model is designed by zeroing an indefinite complex-valued error function as well. The new fractals generated by the CVDTZD model in this chapter are quite different from the famous fractals generated by Newton iteration (i.e., Newton fractals) as well as the fractals generated in the previous chapter that relate to static nonlinear equations solving. The presented CVDTZD model with different types of activation functions usable can be seen as a new updating-type numerical algorithm to produce fractals. In addition, by comparing the area and degree of blue color in new fractals under the same conditions, the effectiveness of the CVDTZD model using different activation functions for solving time-varying nonlinear complex equations is reflected.

18.1 Introduction

As discussed in Chapter 17, the generation of fractals is an important and interesting issue in the fields of signal recovery [97] as well as computer geometry and graphics [76]. The fractals may show and reveal the chaotic phenomena [75] appearing in the natural or artificial architectures of microscopic self-computing or self-organization. Generation of many fractals is closely related to the method of Newton iteration for solving certain types of nonlinear equations in the complex domain, e.g., complex root finding, complex polynomials solving, complex trigonometric functions solving, and complex-exponential functions solving. Review the previous chapter and think. Then, by the authors' reasoning, time-varying nonlinear equations solving may output very different figures while generating new fractals.

In this chapter, by dicretizing the complex-valued continuous-time zeroing dynamics (CVCTZD) model for time-varying nonlinear equations solving in the complex domain, the corresponding CVDTZD model can be obtained [90, 109]. In addition, such a CVDTZD model is applied to generating new fractals, which are evidently different from the fractals generated by Newton iteration (i.e., Newton fractals).

18.2 Complex-Valued ZD Models

In this chapter, we consider the following time-varying nonlinear equation in the complex domain for generating new fractals [109]:

$$f(z(t),t) = g(z(t),t) - a(t) = 0 \in \mathbb{C}, t \in [0,+\infty), \qquad (18.1)$$

where $g(\cdot,\cdot) : \mathbb{C} \times [0,\infty) \to \mathbb{C}$ denotes a nonlinear complex mapping, $a(t) \in \mathbb{C}$ is a smoothly time-varying scalar in the complex domain, and $z(t) \in \mathbb{C}$ denotes the unknown time-varying complex-valued scalar to be solved for.

As the previous chapter shows, the well-known Newton iteration method in the complex domain has been applied to computing the root of complex-valued nonlinear equation $f(z) = g(z) - a = 0 \in \mathbb{C}$, which is evidently constant. The corresponding fractals (i.e., Newton fractals) can thus be generated by the method at the same time. Different from the conventional Newton iteration method, we further develop and exploit a CVDTZD model in this chapter to find the root of complex-valued time-varying nonlinear Equation (18.1), and thus obtain new fractals. Specifically, the time-varying nonlinear equation considered to be solved for yielding fractals in this chapter is $f(z(t),t) = z^\rho(t) - a(t)$, where $\rho \geqslant 3$ is a positive integer and $a(t)$ is a complex-valued time-varying scalar.

18.2.1 CVCTZD Model

By following Zhang *et al.*'s design method [102,104,107,117,118,122,129,134], a CVCTZD model is first constructed to solve time-varying nonlinear Equation (18.1). To monitor and control the solution process of time-varying nonlinear Equation (18.1), the following indefinite lower-unbounded complex-valued error function is defined:

$$e(t) = f(z(t),t) \in \mathbb{C}, \qquad (18.2)$$

where, evidently, $z(t) \in \mathbb{C}$ converges to a theoretical solution $z^*(t)$ of time-varying nonlinear Equation (18.1) as error function $e(t)$ converges to zero. To make error function $e(t)$ diminish to zero, the following zeroing dynamics design formula is employed:

$$\dot{e}(t) = -\gamma\phi(e(t)), \qquad (18.3)$$

where, as stated before, design parameter $\gamma > 0$ and $\phi(\cdot) : \mathbb{C} \to \mathbb{C}$ denotes the activation function. In this chapter, five types of activation functions (i.e., the linear, power-sigmoid, power-sum, hyperbolic sine, and bipolar sigmoid activation functions) are introduced and investigated. Other types of activation functions can then be readily generalized from these five basic types of activation functions.

Expanding ZD design formula (18.3), we have the following complex-valued differential equation for the so-called CVCTZD model:

$$\frac{\partial f(z(t),t)}{\partial z(t)} \frac{\mathrm{d}z(t)}{\mathrm{d}t} + \frac{\partial f(z(t),t)}{\partial t} = -\gamma\phi(f(z(t),t));$$

and, with $\partial f(z(t),t)/\partial z(t) \neq 0$ assumed, we further have

$$\dot{z}(t) = \frac{-\gamma\phi(f(z(t),t)) - f_t'(z(t),t)}{f_z'(z(t),t)}, \tag{18.4}$$

where $f_t'(z(t),t) = \partial f(z(t),t)/\partial t$ and $f_z'(z(t),t) = \partial f(z(t),t)/\partial z(t)$, with $z(t) \in \mathbb{C}$ denoting the neural state of the CVCTZD model corresponding to a solution of time-varying nonlinear Equation (18.1).

18.2.2 CVDTZD Model

In this subsection, we present the discrete-time form of the CVCTZD model to solve time-varying nonlinear Equation (18.1), which can be viewed as the essential step for the new fractal generation. By applying Euler forward-difference rule to the CVCTZD model, the corresponding CVDTZD model is obtained via the intermediate result:

$$\frac{z_{k+1} - z_k}{\tau} = \frac{-\gamma\phi(f(z_k,k\tau)) - f_t'(z_k,k\tau)}{f_z'(z_k,k\tau)}, \text{ with } f_z'(z_k,k\tau) \neq 0;$$

that is,

$$z_{k+1} = z_k - \frac{h\phi(f(z_k,k\tau)) + \tau f_t'(z_k,k\tau)}{f_z'(z_k,k\tau)}, \text{ with } f_z'(z_k,k\tau) \neq 0, \tag{18.5}$$

where $k = 0,1,2,\cdots$ denotes the update index, and design parameter $h = \gamma\tau \in \mathbb{R}^+$ denotes the step size. Note that different choices of step size h and activation function $\phi(\cdot)$ lead to different convergence performance of CVDTZD model (18.5).

Additionally, let us show the link of CVDTZD model (18.5) to Newton iteration. For simplicity, $\tau f_t'(z_k,k\tau)$ is omitted temporarily, and CVDTZD model (18.5) reduces to

$$z_{k+1} = z_k - h\frac{\phi(f(z_k,k\tau))}{f_z'(z_k,k\tau)}, \text{ with } f_z'(z_k,k\tau) \neq 0, \tag{18.6}$$

which is a simplified complex-valued DTZD (S-CVDTZD) model. It follows that, if we utilize linear activation function $\phi(e) = e$, S-CVDTZD model (18.6) becomes

$$z_{k+1} = z_k - h\frac{f(z_k,k\tau)}{f_z'(z_k,k\tau)}. \tag{18.7}$$

Furthermore, if $h = 1$ and $f(z(t),t) = f(z)$ (i.e., static), Equation (18.7) becomes

$$z_{k+1} = z_k - \frac{f(z_k)}{f'(z_k)}, \tag{18.8}$$

which is exactly Newton iteration for the nonlinear equation solving in the complex domain. This implies again that Newton iteration can be viewed as a special case of the general CVDTZD model (18.5). Newton iteration (18.8) is now also applied to generating different fractals by solving different types of nonlinear equations, and the fractals yielded are thus called Newton fractals. In the ensuing section, we show that the fractals generated by CVDTZD model (18.5) for solving time-varying nonlinear equations are quite different from those generated by Newton iteration.

As proposed and discussed slightly differently in Subsection 15.6.2, the corresponding residual error of the above CVDTZD model [which is derived from error function (18.2)] can be defined as

$$e_k = |f(z_k,k\tau)|,$$

where symbol $|\cdot|$ denotes the modulus of a complex number (and also the absolute value of a real

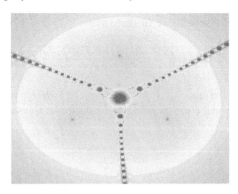

(a) Using linear activation function

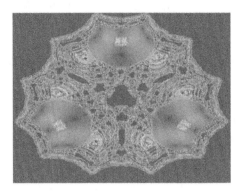

(b) Using power-sigmoid activation function

FIGURE 18.1 New fractals generated by solving time-varying complex-valued nonlinear Equation (18.10) via CVDTZD model (18.5), with $h = 0.2$ and using different activation functions. *Reproduced from Y. Zhang, L. Jin, Z. Zhang, et al., Zhang fractals yielded via solving time-varying nonlinear complex equations by discrete-time complex-valued ZD, Figure 1, LNAI 7530, pp. 596–603, 2012. ©Springer-Verlag Berlin Heidelberg 2012. With kind permission of Springer Science+Business Media.*

number). So, the minimum number of updates (MNU) for a model's residual error converging to its steady state can be defined and used to determine the color degree corresponding to an initial state in the new fractal generation. The MNU decision is based on the following practical criterion:

$$|e_{k+1} - e_k|/e_k \leqslant 10^{-3}, \tag{18.9}$$

where we have MNU $= k + 1$ when criterion (18.9) is satisfied for the first time during the solving process. For the purpose of better demonstration, the upper limit of MNU is set to 80; i.e., MNU $=$ 80 if it is larger than 80. Note that calculating MNU is the key step in producing a new fractal, and a similar criterion can be seen comparatively in Subsection 15.6.2, i.e., criterion (15.30).

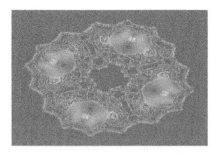

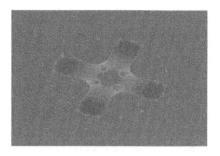

(a) Using power-sigmoid activation function

(b) Using power-sum activation function

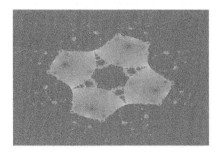

(c) Using hyperbolic sine activation function

(d) Using bipolar sigmoid activation function

FIGURE 18.2 New fractals generated by solving (18.11) via CVDTZD model (18.5), with $h = 0.2$ and using different activation functions, where $z_0 \in [-2, 2] + i[-2, 2]$. *Reproduced from Y. Zhang, L. Jin, Z. Zhang, et al., Zhang fractals yielded via solving time-varying nonlinear complex equations by discrete-time complex-valued ZD, Figure 3, LNAI 7530, pp. 596–603, 2012. ©Springer-Verlag Berlin Heidelberg 2012. With kind permission of Springer Science+Business Media.*

18.3 Illustrative Examples

In this section, some representative and illustrative examples are shown to verify CVDTZD model (18.5) for generating fractals, which solves time-varying nonlinear equations in the complex domain. We exploit the CVDTZD model with $h = 0.2$ and using different activation functions to generate the new fractals.

18.3.1 By Solving $z^3(t) - a(t) = 0$

Let us consider the following complex-valued time-varying nonlinear equation:

$$z^3(t) - \sin(t) - i\cos(t) = 0. \tag{18.10}$$

We exploit the CVDTZD model activated by the power-sigmoid function to generate new fractals, where the power-sigmoid function is

$$\phi(e) = \begin{cases} e^p, & \text{if } |e| \geq 1, \\ \dfrac{1+\exp(-\xi)}{1-\exp(-\xi)} \dfrac{1-\exp(-\xi e)}{1+\exp(-\xi e)}, & \text{otherwise,} \end{cases}$$

with parameters $\xi = 4$ and $p = 3$. For illustrating the effectiveness of using different activation functions for generating different fractals, Figure 18.1 shows two different new fractals generated by solving (18.10) via the CVDTZD model, with $h = 0.2$. In the figure, the horizontal and vertical axes correspond to the real and imaginary parts of initial state $z_0 \in [-2,2] + i[-2,2]$, respectively, while the color degree corresponds to the MNU required by CVDTZD model (18.5), starting with such an initial state z_0.

For comparison, let us consider $f(z(t),t) = z^3 - i$ [i.e., the constant case that can be viewed as a very simple problem of (18.10) with $t = 0$], where Newton iteration (18.8) is used. Newton fractal is thus generated as a special case of the new fractals. However, the graphical result of this Newton fractal is very similar to that in Figure 17.3 [which relates to $f(z(t),t) = z^3 + i$] and thus omitted here.

18.3.2   By Solving $z^4(t) - a(t) = 0$

Let us consider the following complex-valued time-varying nonlinear equation:

$$z^4(t) - \sin(t) - i\cos(t) = 0. \tag{18.11}$$

For the purpose of generating new more fractals and also for comparison, another three nonlinear activation functions (i.e., power-sum, hyperbolic sine, and bipolar sigmoid activation functions) are applied to the CVDTZD model, where
1) the power-sum activation function is

$$\phi(e) = \sum_{\kappa=1}^{N} e^{2\kappa-1}, \text{ with } N \geq 2;$$

2) the hyperbolic sine activation function is

$$\phi(e) = \frac{\exp(\xi e) - \exp(-\xi e)}{2}, \text{ with } \xi = 2; \text{ and}$$

3) the bipolar sigmoid activation function is

$$\phi(e) = \frac{1+\exp(-\xi)}{1-\exp(-\xi)} \frac{1-\exp(-\xi e)}{1+\exp(-\xi e)}, \text{ with } \xi = 4.$$

Seeing Figure 18.2 and comparing the four new fractals that are generated by solving time-varying nonlinear Equation (18.11) with different activation functions, we get a general conclusion that the power-sigmoid activation function, power-sum activation function, and hyperbolic sine activation function have relatively better effects on solving time-varying nonlinear Equation (18.11) via CVDTZD (18.5), as compared with the bipolar sigmoid activation function. Note that the better efficacy is reflected by the darker degree and area of blue color in the fractals.

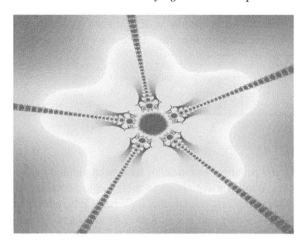

(a) $z_0 \in [-2,2] + i[-2,2]$

(b) $z_0 \in [-0.6, -0.1] + i[-0.2, 0.1]$

FIGURE 18.3 New fractal generated by solving (18.12) via CVDTZD model (18.5) with $h = 0.2$. *Reproduced from Y. Zhang, L. Jin, Z. Zhang, et al., Zhang fractals yielded via solving time-varying nonlinear complex equations by discrete-time complex-valued ZD, Figure 1, LNAI 7530, pp. 596–603, 2012. ©Springer-Verlag Berlin Heidelberg 2012. With kind permission of Springer Science+Business Media.*

18.3.3 By Solving $z^5(t) - a(t) = 0$

Now consider the following complex-valued time-varying nonlinear equation:

$$z^5(t) - \sin(t) - it = 0. \tag{18.12}$$

For further illustrating the effectiveness and the wide applications of the proposed model, we use CVDTZD model (18.5) with a linear activation function and $h = 0.2$ to solve Equation (18.12). The corresponding result is shown in Figure 18.3.

18.4   Summary

In this chapter, interesting and beautiful new fractals have been yielded by the presented CVDTZD for solving the time-varying nonlinear equations in the complex domain. In addition, the darker degree and area of blue color in the new fractals that have been generated by using different activation functions can be used to compare and reflect the solution effect of the time-varying nonlinear complex equations via the CVDTZD model. The provided graphical results have illustrated that the CVDTZD model can be viewed as a new numerical algorithm to generate many different new fractals.

Glossary

CZD: Conventional zeroing dynamics.

CTZD: Continuous-time zeroing dynamics.

CVZD: Complex-valued zeroing dynamics.

CVCTZD: Complex-valued continuous-time zeroing dynamics.

CVDTZD: Complex-valued discrete-time zeroing dynamics.

DTZD: Discrete-time zeroing dynamics.

GD: Gradient dynamics.

MSSRE: Maximal steady-state residual error.

MZD: Modified zeroing dynamics.

QM: Quadratic minimization.

QP: Quadratic programming.

RMP: Repetitive motion planning.

S-CTZD: Simplified continuous-time zeroing dynamics.

S-DTZD: Simplified discrete-time zeroing dynamics.

ZD: Zeroing dynamics.

Bibliography

[1] S. Abbasbandy. Improving Newton-Raphson method for nonlinear equations by modified adomian decomposition method. *Applied Mathematics and Computation*, 145:887–893(2–3), 2003.

[2] E.L. Allgower and K. Georg. *Numerical Continuation Methods: An Introduction*. Springer-Verlag, New York, USA, 1990.

[3] M. Basto, V. Semiao, and F. Calheiros. A new iterative method to compute nonlinear equations. *Applied Mathematics and Computation*, 173:468–483(1), 2006.

[4] W. Bian and X. Xue. Subgradient-based neural networks for nonsmooth nonconvex optimization problems. *IEEE Transactions on Neural Networks*, 20:1024–1038, 2009.

[5] D.A. Bini, G. Codevico, and M. Van Barel. Solving toeplitz least squares problems by means of Newton's iteration. *Numerical Algorithms*, 33:93–103(1–4), 2003.

[6] S. Boyd and L. Vandenberghe. *Convex Optimization*. Cambridge University Press, New York, USA, 2004.

[7] C.G. Broyden. A class of methods for solving nonlinear simultaneous equations. *Mathematics of Computation*, 19:577–593(92), 1965.

[8] Mead. C. and M. Ismail. *Analog VLSI Implementation of Neural Systems*. Springer, New York, USA, 1989.

[9] P.W. Carlson. Two artistic orbit trap rendering methods for Newton m-set fractals. *Computers and Graphics*, 23:925–931, 1999.

[10] R.J. Charles, O. Kazuyoshi, and R. Robert. Uniqueness of matrix square roots and an application. *Linear Algebra and Its Applications*, 323:51–60(1), 2001.

[11] C. Chen and O.L. Mangasarian. Smoothing methods for convex inequalities and linear complementarity problems. *Mathematical Programming*, 71:51–69, 1995.

[12] L. Chen, E.V. Krishnarnurthy, and I. Macleod. Generalised matrix inversion and rank computation by successive matrix powering. *Parallel Computing*, 20:297–311(3), 1994.

[13] N. Chen, X. Zhua, and K. Chung. m and j sets from Newton's transformation of the transcendental mapping $f(z) = e^{z^w + c}$ with VCPS. *Computers and Graphics*, 26:371–383, 2002.

[14] W.S. Cheung. Some new nonlinear inequalities and applications to boundary value problems. *Nonlinear Analysis: Theory, Methods and Applications*, 64:2112–2128, 2006.

[15] L.O. Chua and G. Lin. Nonlinear programming without computation. *IEEE Transactions on Circuits and Systems II: Express Briefs*, 31:182–188, 1984.

[16] C. Chun. Construction of Newton-like iteration methods for solving nonlinear equations. *Numerische Mathematik*, 104:297–315(3), 2006.

[17] A. Cichocki and A. Bargiela. Neural network for solving linear inequality systems. *Parallel Computing*, 22:1455–1475, 1997.

[18] T.F. Coleman, J. Liu, and W. Yuan. A quasi-Newton quadratic penalty method for minimization subject to nonlinear equality constraints. *Computational Optimization and Applications*, 15:103–123(2), 2000.

[19] J.W. Daniel. Newton's method for nonlinear inequalities. *Numerische Mathematik*, 21:381–387, 1973.

[20] E. Denman and N. Beavers. The matrix sign function and computations in systems. *Applied Mathematics and Computation*, 2:63–94, 1976.

[21] P. Deuflhard. *Newton Methods for Nonlinear Problems: Affine Invariance and Adaptive Algorithms*. Springer, New York, USA, 2011.

[22] D. DTakahashi. Implementation of multiple-precision parallel division and square root on distributed-memory parallel computers. In *Proceedings of Parallel Processing*, pages 229–235, 2000.

[23] A. El-Amawy. A systolic architecture for fast dense matrix inversion. *IEEE Transactions on Computers*, 38:449–455(3), 1989.

[24] M.D. Ercegovac. On digit-by-digit methods for computing certain functions. In *Proceedings of Asilomar Conference on Signals, Systems, and Computers*, pages 338–342, 2007.

[25] J. Feng, G. Che, and Y. Nie. *Principles of Numerical Analysis*. Science Press, Beijing, China, 2001.

[26] P.W. Fieguth, D. Menemenlis, and I. Fukumori. Mapping and pseudoinverse algorithms for ocean data assimilation. *IEEE Transactions on Geoscience and Remote Sensing*, 4:43–51, 2003.

[27] R. Fletcher. *Practical Methods of Optimization*. John Wiley & Sons, New Jersey, USA, 1987.

[28] M. Forti, P. Nistri, and M. Quincampoix. Generalized neural network for nonsmooth nonlinear programming problems. *IEEE Transactions on Circuits and Systems II: Express Briefs*, 51:1741–1754, 2004.

[29] B. Guo, D. Wang, Y. Shen, and Z. Li. A hopfield neural network approach for power optimization of real-time operating systems. *Neural Computing and Applications*, 17:11–17, 2008.

[30] F.M. Ham and I. Kostanic. *Principles of Neurocomputing for Science and Engineering*. McGraw-Hill, New York, USA, 2001.

[31] N. J. Higham. *Functions of Matrices: Theory and Computation*. Society for Industrial and Applied Mathematics, USA, 2008.

[32] N.J. Higham. Stable iterations for the matrix square root. *Numerical Algorithm*, 15:227–242(2), 1997.

[33] N.S. Hoang and A.G. Ramm. A nonlinear inequality and applications. *Nonlinear Analysis: Theory, Methods and Applications*, 71:2744–2752, 2009.

[34] J. Hu, S. Qian, and Y. Ding. Improved pseudoinverse algorithm and its application in controlling acoustic field generated by phased array. *Journal of System Simulation*, 22:1111–1116, 2010.

[35] J.L. Hu, Z. Wu, H. McCann, L.E. Davis, and C.G. Xie. Bfgs quasi-Newton method for solving electromagnetic inverse problems. *IET Microwaves, Antennas and Propagation*, 153:199–204(2), 2006.

[36] L.D. Jasio. *Programming 16-bit PIC Microcontrollers in C: Learning to Fly the PIC 24*. Elsevier, Boston, USA, 2011.

[37] R.E. Kalaba, E. Zagustin, W. Holbrom, and R. Huss. A modification of Davidenko's method for nonlinear systems. *Computational and Applied Mathematics*, 3:315–319, 1997.

[38] Y. Kasahara. The jones representation of genus 2 at the 4th root of unity and the Torelli group. *Topology and Its Applications*, 124:129–138, 2002.

[39] Z. Ke, Y. Yang, and Y. Zhang. Discrete-time ZNN algorithms for time-varying quadratic programming subject to time-varying equality constraint. In *Proceedings of International Symposium on Neutral Networks*, pages 47–54, 2012.

[40] F. Kong, Z. Cai, J. Yu, and D.X. Li. Improved generalized atkin algorithm for computing square roots in finite fields. *Information Processing Letters*, 98:1–5(1), 2006.

[41] P. Kovesi. Phase congruency detects corners and edges. In *Proceedings of 7th Digital Image Computing: Techniques and Applications*, pages 10–12, 2003.

[42] G. Labonte. On solving systems of linear inequalities with artificial neural network. *IEEE Transactions on Neural Networks*, 8:590–600(3), 1997.

[43] S.-J. Lee and B.-S. Lee. Existence results for vector nonlinear inequalities. *Communications of the Korean Mathematical Society*, 18:737–743(4), 2003.

[44] F. Leibfritz and E.W. Sachs. Inexact SQP interior point methods and large scale optimal control problems. *SIAM Journal on Control and Optimization*, 38:272–293, 1999.

[45] W.E. Leithead and Y. Zhang. $o(n^2)$-operation approximation of covariance matrix inverse in gaussian process regression based on quasi-Newton BFGS method. *Communications in Statistics-Simulation and Computation*, 36:367–380(2), 2007.

[46] C. Li and X. Liao. Robust stability and robust periodicity of delayed recurrent neural networks with noise disturbance. *IEEE Transactions on Circuits and Systems I: Regular Papers*, 53:2265–2273(10), 2006.

[47] D. Li, M. Fukushima, L. Qi, and N. Yamashita. Regularized Newton methods for convex minimization problems with singular solutions. *Computational Optimization and Applications*, 28:131–147(2), 2004.

[48] W. Li. Error bounds for piecewise convex quadratic programs and applications. *SIAM Journal on Control and Optimization*, 33:1510–1529, 1995.

[49] W. Li and J. Swetits. A new algorithm for solving strictly convex quadratic programs. *SIAM Journal on Control and Optimization*, 7:595–619, 1997.

[50] Y. Li and D. Li. Truncated regularized Newton method for convex minimizations. *Computational Optimization and Applications*, 43:119–131(1), 2007.

[51] C. Lin. *Numerical Computation Methods*. Science Press, Beijing, China, 2005.

[52] C. Lin, C. Lai, and T. Huang. A neural network for linear matrix inequality problems. *IEEE Transactions on Neural Networks*, 11:1078–1092(5), 2000.

[53] J. Long, X. Hu, and L. Zhang. Newton's method with exact line search for the square root of a matrix. *Journal of Physics Conference Series*, 96:1–5, 2008.

[54] R.K. Manherz, B.W. Jordan, and S.L. Hakimi. Analog methods for computation of the generalized inverse. *IEEE Transactions on Automatic Control*, 13:582–585(3), 1968.

[55] R.K. Manherz, B.W. Jordan, and S.L. Hakimi. Exploiting Hessian matrix and trust-region algorithm in hyperparameters estimation of gaussian process. *Applied Mathematics and Computation*, 171:1264–1281(2), 2005.

[56] T. Martyn. Realistic rendering 3d ifs fractals in real-time with graphics accelerators. *Computers and Graphics*, 34:167–175, 2010.

[57] J.H. Mathews and K.D. Fink. *Numerical Methods Using MATLAB*. Publishing House of Electronics Industry, Beijing, China, 2005.

[58] D.Q. Mayne, E. Polak, and A.J. Heunis. Solving nonlinear inequalities in a finite number of iterations. *Journal of Optimization Theory and Applications*, 33:207–221, 1981.

[59] B. Meini. *The matrix square root from a new functional perspective: Theoretical results and computational issues*. Technical Report, Dipartimento di Matematica, Università di Pisa, Pisa, 2003.

[60] A.H. Mohammed, A.H. Ali, and R. Syed. Fixed point iterations for computing square roots and the matrix sign function of complex matrices. In *Proceedings of IEEE Conference on Decision and Control*, pages 4253–4258, 2000.

[61] H. Moriyama, N. Yamashita, and M. Fukushima. The incremental Gauss-Newton algorithm with adaptive step-size rule. *Computational Optimization and Applications*, 26:107–141(2), 2003.

[62] H. Myung and J. Kim. Time-varying two-phase optimization and its application to neural-network learning. *IEEE Transactions on Neural Networks*, 8:1293–1300, 1997.

[63] D. Panario and D. Thomson. Efficient pth root computations in finite fields of characteristic p. *Designs, Codes and Cryptography*, 50:351–358, 2009.

[64] J. Park, Y. Choi, W.K. Chung, and Y. Youm. Multiple tasks kinematics using weighted pseudo-inverse for kinematically redundant manipulators. In *Proceedings of IEEE Conference on Robotics and Automation*, pages 4041–4047, 2001.

[65] H.O. Peitgen, D. Saupe, and M.F. Barnsley. *The Science of Fractal Images*. Springer-Verlag, New York, USA, 1988.

[66] H. Peng. Algorithms for extracting square roots and cube roots. In *Proceedings of the Fifth IEEE International Symposium on Computer Arithmetic*, pages 121–126, 1981.

[67] J.-A. Piñeiro, J.D. Bruguera, P. Lamberti, and F. Montuschi. A radix-2 digit-by-digit architecture for cube root. *IEEE Transactions on Computers*, 57:562–566(4), 2008.

[68] J. Qian and C. Wang. How much precision is needed to compare two sums of square roots of integers. *Information Processing Letters*, 100:194–198(5), 2006.

[69] S.M. Robinson. Stability theory for systems of inequalities, part II: differentiable nonlinear systems. *SIAM Journal on Numerical Analysis*, 13:497–513(4), 1976.

[70] M. Sahba. On the solution of nonlinear inequalities in a finite number of iterations. *Numerische Mathematik*, 46:229–236, 1985.

[71] T. Sarkar, K. Siarkiewicz, and R. Stratton. Survey of numerical methods for solution of large systems of linear equations for electromagnetic field problems. *IEEE Transactions on Antennas and Propagation*, 29:847–856(6), 1981.

[72] J.R. Sharma. A composite third order Newton-Steffensen method for solving nonlinear equations. *Applied Mathematics and Computation*, 169:242–246, 2005.

[73] Y. Shi. Globally convergent algorithms for unconstrained optimization. *Computational Optimization and Applications*, 16:295–308(3), 2000.

[74] H. Shin, J. Lee, and L. Kim. A minimized hardware architecture of fast phong shader using Taylor series approximation in 3D graphics. In *Proceedings of IEEE International Conference on Computer Design: VLSI in Computers and Processors*, pages 286–291, 1998.

[75] G. Singh. Beauty in chaos. *IEEE Computer Graphics and Applications*, 27:4–5, 2007.

[76] G. Singh. Fun with fractal art. *IEEE Computer Graphics and Applications*, 29:4–5, 2009.

[77] V. Singh. A stability inequality for nonlinear discrete-time systems with slope-restricted nonlinearity. *IEEE Transactions on Circuits and Systems*, 31:1058–1060(12), 1984.

[78] J.-J. E. Slotine and W. Li. *Applied Nonlinear Control*. Prentice-Hall, Englewood Cliffs, New Jersey, USA, 2005.

[79] M. Staroswiecki. Fault tolerant control: The pseudo-inverse method revisited. In *Proceedings of 16th IFAC World Congress, Prague, Czech Republic*, 2005.

[80] R.J. Steriti and M.A. Fiddy. Regularized image reconstruction using SVD and a neural network method for matrix inversion. *IEEE Transactions on Signal Processing*, 41:3074–3077(10), 1993.

[81] R.H. Sturges Jr. Analog matrix inversion (robot kinematics). *IEEE Journal of Robotics and Automation*, 4:157–162(2), 1989.

[82] K. Turkowski. *Computing the Cube Root*. Technical Report, Apple Computer, 1998.

[83] N. Ujevic. A method for solving nonlinear equations. *Applied Mathematics and Computation*, 174:1416–1426, 2006.

[84] P. Underwood. Dynamic relaxation. In *Computational Methods for Transient Analysis*, pages 245–265. Elsevier, Boston, USA, 1983.

[85] S. Van Huffel and J. Vandewalle. Analysis and properties of the generalized total least squares problem $ax \approx b$ when some or all columns in a are subject to errors. *SIAM Journal on Matrix Analysis and Applications*, 10:294–315, 1989.

[86] H. Wang, J. Li, and H. Liu. Practical limitations of an algorithm for the singular value decomposition as applied to redundant manipulators. In *Proceedings of IEEE Conference on Robotics, Automation and Mechatronics*, pages 1–6, 2006.

[87] X. Wang, Y. Li, Y. Sun, J. Song, and F. Ge. Julia sets of Newton's method for a class of complex-exponential function $f(z) = p(z)e^{Q(z)}$. *Nonlinear Dynamics*, 62:955–966, 2010.

[88] Y.Q. Wang and H.B. Gooi. New ordering methods for space matrix inversion via diagonalization. *IEEE Transactions on Power Systems*, 12:1298–1305(3), 1997.

[89] Y. Wei, J. Cai, and M.K. Ng. Computing Moore–Penrose inverses of toeplitz matrices by Newton's iteration. *Mathematical and Computer Modelling*, 40:181–191, 2004.

[90] H. Wu, F. Li, Z. Li, and Y. Zhang. Zhang fractals yielded via solving nonlinear equations by discrete-time complex-valued zd. In *Proceedings of IEEE International Conference on Automation and Logistics*, pages 1–6, 2012.

[91] J.H. Xia and A.S. Kumta. Feed-forward neural network trained by BFGS algorithm for modeling plasma etching of silicon carbide. *IEEE Transactions on Plasma Science*, 38:142–148(2), 2010.

[92] Y. Xia and J. Wang. A recurrent neural network for nonlinear convex optimization subject to nonlinear inequality constraints. *IEEE Transactions on Circuits and Systems I: Regular Papers*, 51:1385–1394(7), 2004.

[93] Y. Xia, J. Wang, and D.L. Hung. Recurrent neural networks for solving linear inequalities and equations. *IEEE Transactions on Circuits and Systems I: Regular Papers*, 46:452–462(4), 1999.

[94] L. Xiao and Y. Zhang. Zhang neural network versus gradient neural network for solving time-varying linear inequalities. *IEEE Transactions on Neural Networks*, 22:1676–1684(10), 2011.

[95] L. Xiao and Y. Zhang. Two new types of Zhang neural networks solving systems of time-varying nonlinear inequalities. *IEEE Transactions on Circuits and Systems I: Regular Papers*, 59:2363–2373(10), 2012.

[96] K.S. Yeung and F. Kumbi. Symbolic matrix inversion with application to electronic circuits. *IEEE Transactions on Circuits and Systems I: Regular Papers*, 35:235–238(2), 1988.

[97] M. Zhai. Signal recovery in power-line communications systems based on the fractals. *IEEE Transactions on Power Delivery*, 26:1864–1872, 2011.

[98] Y. Zhang. On the LVI-based primal-dual neural network for solving online linear and quadratic programming problems. In *Proceedings of American Control Conference*, pages 1351–1356, 2005.

[99] Y. Zhang. Revisit the analog computer and gradient-based neural system for matrix inversion. In *Proceedings of IEEE International Symposium on Intelligent Control*, pages 1411–1416, 2005.

[100] Y. Zhang. A set of nonlinear equations and inequalities arising in robotics and its online solution via a primal neural network. *Neurocomputing*, 70:513–524, 2006.

[101] Y. Zhang. Towards piecewise-linear primal neural networks for optimization and redundant robotics. In *Proceedings of IEEE International Conference on Networking, Sensing and Control*, pages 374–379, 2006.

[102] Y. Zhang, K. Chen, and B. Cai. Zhang neural network without using time-derivative information for constant and time-varying matrix inversion. In *Proceedings of International Joint Conference on Neural Networks*, pages 142–146, 2008.

[103] Y. Zhang, K. Chen, X. Li, C. Yi, and H. Zhu. Simulink modeling and comparison of Zhang neural networks and gradient neural networks for time-varying Lyapunov equation solving. In *Proceedings of the 4th International Conference on Natural Computation*, pages 521–525, 2008.

[104] Y. Zhang and S. S. Ge. Design and analysis of a general recurrent neural network model for time-varying matrix inversion. *IEEE Transactions on Neural Network*, 16:1477–1490(6), 2005.

[105] Y. Zhang, D. Guo, C. Yi, L. Li, and Z. Ke. More than Newton iterations generalized from Zhang neural network for constant matrix inversion aided with line-search algorithm. In *Proceedings of 8th IEEE International Conference on Control and Automation*, pages 399–404, 2010.

[106] Y. Zhang and Wang J. Obstacle avoidance of kinematically redundant manipulators using a dual neural network. *IEEE Transactions on Systems, Man, and Cybernetics, Part B: Cybernetics*, 34:752–759, 2004.

[107] Y. Zhang, D. Jiang, and J. Wang. A recurrent neural network for solving sylvester equation with time-varying coefficients. *IEEE Transactions on Neural Network*, 13:1053–1063(5), 2002.

[108] Y. Zhang, L. Jin, and Ke Z. Superior performance of using hyperbolic sine activation functions in ZNN illustrated via time-varying matrix square roots finding. *Computer Science and Information Systems*, 9:1603–1625(4), 2012.

[109] Y. Zhang, L. Jin, Z. Zhang, L. Xiao, and S. Fu. Zhang fractals yielded via solving time-varying nonlinear complex equations by discrete-time complex-valued zd. In *Proceedings of Artificial Intelligence and Computational Intelligence*, pages 596–603, 2012.

[110] Y. Zhang and Z. Ke. On hyperbolic sine activation functions used in ZNN for time-varying matrix square roots finding. In *Proceedings of International Conference on Systems and Informatics*, pages 740–744, 2012.

[111] Y. Zhang, Z. Ke, D. Guo, and F. Li. Solving for time-varying and static cube roots in real and complex domains via discrete-time ZD models. *Neural Computing and Applications*, 23:255–268, 2013.

[112] Y. Zhang, Z. Ke, K. Li, and Z. Li. Zhang dynamics with modified error-functions for online solution of nonlinear equations so as to avoid local minima. In *Proceedings of Chinese Control and Decision Conference*, pages 3864–3869, 2011.

[113] Y. Zhang, Z. Ke, Z. Li, and D. Guo. Comparison on continuous-time Zhang dynamics and Newton-Raphson iteration for online solution of nonlinear equations. In *Proceedings of the 8th International Symposium on Neural Networks*, pages 393–402, 2011.

[114] Y. Zhang, Z. Ke, P. Xu, and C. Yi. Time-varying square roots finding via Zhang dynamics versus gradient dynamics and the former's link and new explanation to Newton-Raphson iteration. *Information Processing Letters*, 110:1103–1109, 2010.

[115] Y. Zhang, W.E. Leithead, and D.J. Leith. Time-series Gaussian process regression based on Toeplitz computation of $O(N^2)$ operations and $O(N)$-level storage. In *Proceedings of the 44th IEEE Conference on Decision and Control*, pages 3711–3716, 2000.

[116] Y. Zhang and F. Li. Zhang dynamics solving scalar-valued time-varying linear inequalities using different activation functions. In *Proceedings of 5th IEEE International Conference on Cybernetics and Intelligent Systems*, pages 340–345, 2011.

[117] Y. Zhang and Z. Li. Zhang neural network for online solution of time-varying convex quadratic program subject to time-varying linear-equality constraints. *Physics Letters A*, 373:1639–1643(18–19), 2009.

[118] Y. Zhang, W. Ma, and B. Cai. From Zhang neural network to Newton iteration for matrix inversion. *IEEE Transactions on Circuits and Systems I: Regular Papers*, 56:1405–1415(7), 2009.

[119] Y. Zhang, W. Ma, K. Li, and C. Yi. A new iterative method to compute nonlinear equations. *China Academic Journal Electronic Publishing House*, 13:115–117, 2008.

[120] Y. Zhang, B. Mu, and H. Zheng. Discrete-time Zhang neural network and numerical algorithm for time-varying quadratic minimization. In *Proceedings of International Conference on Cybernetics and Intelligent Systems*, pages 346–351, 2011.

[121] Y. Zhang, C. Peng, W. Li, Y. Shi, and Y. Ling. Broyden-method aided discrete ZNN solving the systems of time-varying nonlinear equations. In *Proceedings of International Conference on Control Engineering and Communication Technology*, pages 492–495, 2012.

[122] Y. Zhang and H. Peng. Zhang neural network for linear time-varying equation solving and its robotic application. In *Proceedings of the 6th International Conference on Machine Learning and Cybernetics*, pages 3543–3548, 2007.

[123] Y. Zhang, Y. Shi, L. Xiao, and B. Mu. Convergence and stability results of Zhang neural network solving systems of time-varying nonlinear equations. In *Proceedings of IEEE International Conference Natural Computation*, pages 143–147, 2012.

[124] Y. Zhang, N. Tan, B. Cai, and Z. Chen. MATLAB Simulink modeling of Zhang neural network solving for time-varying pseudoinverse in comparison with gradient neural network. In *Proceedings of the 2nd International Symposium on Intelligent Information Technology Application*, pages 39–43, 2008.

[125] Y. Zhang, Y. Wang, L. Jin, J. Chen, and Y. Yang. Simulations and experiments of ZNN for online quadratic programming applied to manipulator inverse kinematics. In *Proceedings of International Conference on Information Science and Technology*, pages 265–270, 2013.

[126] Y. Zhang, L. Xiao, G. Ruan, and Z. Li. Continuous and discrete time Zhang dynamics for time-varying 4th root finding. *Numerical Algorithm*, 57:35–51, 2011.

[127] Y. Zhang, P. Xu, and N. Tan. Further studies on Zhang neural-dynamics and gradient dynamics for online nonlinear equations solving. In *Proceedings of the IEEE International Conference on Automation and Logistics*, pages 566–571, 2009.

[128] Y. Zhang, P. Xu, and N. Tan. Solution of nonlinear equations by continuous- and discrete-time Zhang dynamics and more importantly their links to Newton iteration. In *Proceedings of IEEE International Conference on Information*, pages 1–5, 2009.

[129] Y. Zhang and Y. Yang. Simulation and comparison of Zhang neural network and gradient neural network solving for time-varying matrix square roots. In *Proceedings of the 2nd International Symposium on Intelligent Information Technology Application*, pages 966–970, 2008.

[130] Y. Zhang, Y. Yang, B. Cai, and D. Guo. Zhang neural network and its application to Newton iteration for matrix square root estimation. *Neural Computing and Applications*, 21:453–460(3), 2012.

[131] Y. Zhang, Y. Yang, N. Tan, and B. Cai. Zhang neural network solving for time-varying full-rank matrix Moore–Penrose inverse. *Computing*, 92:97–121(2), 2011.

[132] Y. Zhang and C. Yi. *Zhang Neural Networks and Neural-Dynamic Method.* Nova Science Publishers, New York, USA, 2011.

[133] Y. Zhang, C. Yi, D. Guo, and J. Zheng. Comparison on Zhang neural dynamics and gradient-based neural dynamics for online solution of nonlinear time-varying equation. *Neural Computing and Applications*, 20:1–7, 2011.

[134] Y. Zhang, C. Yi, and W. Ma. Comparison on gradient-based neural dynamics and Zhang neural dynamics for online solution of nonlinear equations. *Lecture Notes in Computer Science*, 5370:269–279, 2008.

[135] Y. Zhang, Q. Zeng, X. Xiao, X. Jiang, and A. Zou. Complex-exponential Fourier neural network and its hidden-neuron growing algorithm. *Journal of Computer Applications*, 28:2503–2506(10), 2008.

[136] Y. Zhang and Z. Zhang. *Repetitive Motion Planning and Control of Redundant Robot Manipulators.* Springer-Verlag, New York, USA, 2013.

[137] J. Zhou, Y.M. Zhu, X.R. Li, and Z.S. You. Variants of the Greville formula with applications to exact recursive least squares. *SIAM Journal on Matrix Analysis and Applications*, 24:150–164(1), 2002.

Index

Index

Printed and bound by CPI Group (UK) Ltd, Croydon, CR0 4YY

23/10/2024

01777691-0008